Solutions Manual

Analysis and Design of Digital Systems with VHDL

Allen Dewey
IBM Corporation

PWS Publishing Company

I(T)P

An International Thomson Publishing Company

Boston • Albany • Bonn • Cincinnati • Detroit • London • Madrid • Melbourne
Mexico City • New York • Pacific Grove • Paris • San Francisco • Singapore • Tokyo • Toronto • Washington

PWS PUBLISHING COMPANY
20 Park Plaza, Boston, MA 02116-4324

For more information, contact:

PWS Publishing Company
20 Park Plaza
Boston, MA 02116

International Thomson Publishing Europe
Berkshire House I68-I73
High Holborn
London WC1V 7AA
England

Thomas Nelson Australia
102 Dodds Street
South Melbourne, 3205
Victoria, Australia

Nelson Canada
1120 Birchmount Road
Scarborough, Ontario
Canada M1K 5G4

International Thomson Editores
Campos Eliseos 385, Piso 7
Col. Polanco
11560 Mexico D.F., Mexico

International Thomson Publishing GmbH
Königswinterer Strasse 418
53227 Bonn, Germany

International Thomson Publishing Asia
221 Henderson Road
#05-10 Henderson Building
Singapore 0315

International Thomson Publishing Japan
Hirakawacho Kyowa Building, 31
2-2-1 Hirakawacho
Chiyoda-ku, Tokyo 102
Japan

Sponsoring Editor: Bill Barter
Marketing Manager: Nathan Wilbur
Production Editor: Pamela Rockwell
Cover Designer: Julia Gecha
Manufacturing Coordinator: Andrew Christensen
Text /Cover Printer and Binder: Patterson Printing Company

Printed and bound in the United States of America.   96 97 98 99 00 — 10 9 8 7 6 5 4 3 2 1

ISBN: 0-534-95119-8

Chapter 1. Introduction

Chapter 2. Representing Information

2-1. Solution.

$$0_{10} = 0 \times 2^0$$

$$1_{10} = 1 \times 2^0$$

$$2_{10} = (1 \times 2^1) + (0 \times 2^0)$$

$$3_{10} = (1 \times 2^1) + (1 \times 2^0)$$

$$4_{10} = (1 \times 2^2) + (0 \times 2^1) + (0 \times 2^0)$$

$$5_{10} = (1 \times 2^2) + (0 \times 2^1) + (1 \times 2^0)$$

$$6_{10} = (1 \times 2^2) + (1 \times 2^1) + (0 \times 2^0)$$

$$7_{10} = (1 \times 2^2) + (1 \times 2^1) + (1 \times 2^0)$$

$$8_{10} = (1 \times 2^3) + (0 \times 2^2) + (0 \times 2^1) + (0 \times 2^0)$$

$$9_{10} = (1 \times 2^3) + (0 \times 2^2) + (0 \times 2^1) + (1 \times 2^0)$$

$$10_{10} = (1 \times 2^3) + (0 \times 2^2) + (1 \times 2^1) + (0 \times 2^0)$$

2-2. Solution.

a. 0 to 15 contains 16 numbers which requires 4 bits because 4 bits provide$16 = 2^4$ unique encodings.

b.

Decimal Number	Binary Number
0	0000
1	0001
2	0010
3	0011
4	0100
5	0101
6	0110
7	0111
8	1000
9	1001
10	1010
11	1011
12	1100
13	1101
14	1110
15	1111

c. The first, rightmost column alternates 0's and 1's, starting with a 0. The next column to the left alternates two 0's and two 1's, starting with two 0's. The next column to the left alternates four 0's and four 1's, starting with four 0's. Finally, the last, leftmost column contains eight 0's followed by eight 1's.

2-3. Solution.

$2^6 = 64$ and $2^7 = 128$, so the minimum required number of bits is 7.

2-4. Solution.

a.

101.101_2

$$
\begin{array}{r|l l}
2 & 5 & 1 \\
2 & 2 & 0 \\
2 & 1 & 1 \\
 & 0 &
\end{array}
$$

$(0.625) \times (2) = \underline{1}.25$

$(0.25) \times (2) = \underline{0}.50$

$(0.50) \times (2) = \underline{1}.00$

$$5.625_{10} = (1 \times 2^2) + (0 \times 2^1) + (1 \times 2^0) + (1 \times 2^{-1}) + (0 \times 2^{-2}) + (1 \times 2^{-3})$$

b.

0.01101_2

$(0.40625) \times (2) = \underline{0}.8125$

$(0.8125) \times (2) = \underline{1}.625$

$(0.625) \times (2) = \underline{1}.25$

$(0.25) \times (2) = \underline{0}.50$

$(0.50) \times (2) = \underline{1}.00$

$$0.40625_{10} = (0 \times 2^{-1}) + (1 \times 2^{-2}) + (1 \times 2^{-3}) + (0 \times 2^{-4}) + (1 \times 2^{-5})$$

c.

$0.\overline{1001}_2$

$(0.60) \times (2) = \underline{1}.20$

$(0.20) \times (2) = \underline{0}.40$

$(0.40) \times (2) = \underline{0}.80$

$(0.80) \times (2) = \underline{1}.60$

$(0.60) \times (2) = \underline{1}.20$

$(0.20) \times (2) = \underline{0}.40$

$(0.40) \times (2) = \underline{0}.80$

$\vdots$

$$0.6_{10} = (1 \times 2^{-1}) + (0 \times 2^{-2}) + (0 \times 2^{-3}) + (1 \times 2^{-4}) + (1 \times 2^{-5}) \cdots$$

d.

Division	Remainder
2 ⌊ 507	1
2 ⌊ 253	1
2 ⌊ 126	0
2 ⌊ 63	1
2 ⌊ 31	1
2 ⌊ 15	1
2 ⌊ 7	1
2 ⌊ 3	1
2 ⌊ 1	1
0	

111111011_2

$$507_{10} = (1 \times 2^8) + (1 \times 2^7) + (1 \times 2^6) + (1 \times 2^5) + (1 \times 2^4) + (1 \times 2^3) + (0 \times 2^2) + (1 \times 2^1) + (1 \times 2^0)$$

e.

11111111_2

Division	Remainder
2 ⌊ 255	1
2 ⌊ 127	1
2 ⌊ 63	1
2 ⌊ 31	1
2 ⌊ 15	1
2 ⌊ 7	1
2 ⌊ 3	1
2 ⌊ 1	1
0	

$$255_{10} = (1 \times 2^7) + (1 \times 2^6) + (1 \times 2^5) + (1 \times 2^4) + (1 \times 2^3) + (1 \times 2^2) + (1 \times 2^1) + (1 \times 2^0)$$

f.

111111.011_2

$$
\begin{array}{r|l l}
2 & 63 & 1 \\
2 & 31 & 1 \\
2 & 15 & 1 \\
2 & 7 & 1 \\
2 & 3 & 1 \\
2 & 1 & 1 \\
 & 0 &
\end{array}
$$

$$(0.375) \times (2) = \underline{0}.75$$

$$(0.75) \times (2) = \underline{1}.50$$

$$(0.50) \times (2) = \underline{1}.00$$

$$63.375_{10} = (1 \times 2^5) + (1 \times 2^4) + (1 \times 2^3) + (1 \times 2^2) + (1 \times 2^1) + (1 \times 2^0) + (0 \times 2^{-1}) + (1 \times 2^{-2}) + (1 \times 2^{-3})$$

2-5. Solution.

a.

5.5_8

$$
\begin{array}{r|l l}
8 & 5 & 5 \\
 & 0 &
\end{array}
$$

$$(0.625) \times (8) = \underline{5}.00$$

$$\underbrace{101}.\underbrace{101}_2$$

$$5.5_8$$

$5.A_{16}$

$$
\begin{array}{r|l l}
16 & 5 & 5 \\
 & 0 &
\end{array}
$$

$$(0.625) \times (16) = \underline{10}.00$$

$$\underbrace{0101}.\underbrace{1010}_{16}$$

$$5.A_{16}$$

b.

0.32_8

$$(0.40625) \times (8) = \underline{3}.25$$

$$(0.25) \times (8) = \underline{2}.00$$

$$\underbrace{0.011010_2}_{0.32_8}$$

0.68_{16}

$$(0.40625) \times (16) = \underline{6}.50$$

$$(0.50) \times (16) = \underline{8}.00$$

$$\underbrace{0.01101000_2}_{0.68_{16}}$$

c.

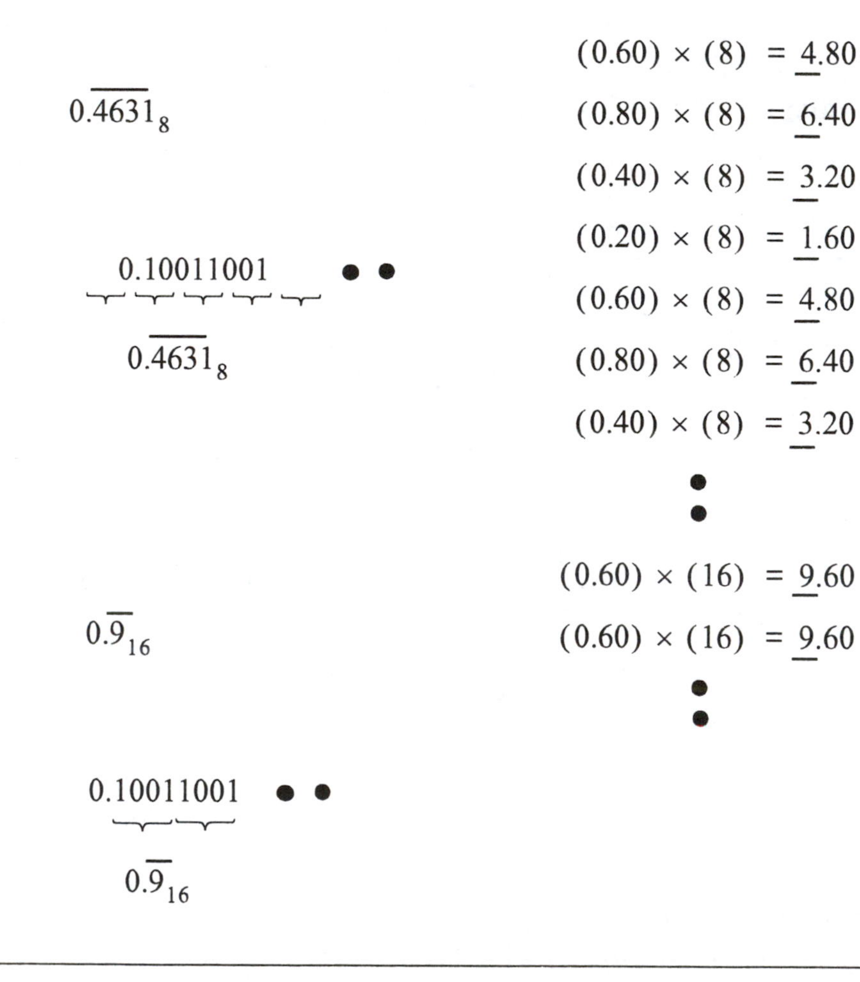

$0.\overline{4631}_8$

$$(0.60) \times (8) = \underline{4}.80$$

$$(0.80) \times (8) = \underline{6}.40$$

$$(0.40) \times (8) = \underline{3}.20$$

$$(0.20) \times (8) = \underline{1}.60$$

$$(0.60) \times (8) = \underline{4}.80$$

$$(0.80) \times (8) = \underline{6}.40$$

$$(0.40) \times (8) = \underline{3}.20$$

$\vdots$

$$\underbrace{0.10011001 \cdots}_{0.\overline{4631}_8}$$

$0.\overline{9}_{16}$

$$(0.60) \times (16) = \underline{9}.60$$

$$(0.60) \times (16) = \underline{9}.60$$

$\vdots$

$$\underbrace{0.10011001 \cdots}_{0.\overline{9}_{16}}$$

d.

773_8

$\underbrace{111}\underbrace{111}\underbrace{011}_2$

773_8

8	507		3
8	63		7
8	7		7
	0		

$1FB_{16}$

$\underbrace{0001}\underbrace{1111}\underbrace{1011}_2$

$1FB_{16}$

16	507		11
16	31		15
16	1		1
	0		

e.

377_8

$\underbrace{011}\underbrace{111}\underbrace{111}_2$

377_8

8	255		7
8	31		7
8	3		3
	0		

FF_{16}

$\underbrace{1111}\underbrace{1111}_2$

FF_{16}

16	255		15
16	15		15
	0		

f.

77.3_8

$$
\begin{array}{r|l}
8 & 63 \\ \hline
8 & 7 \\ \hline
 & 0
\end{array}
\qquad
\begin{array}{c}
7 \\ 7
\end{array}
$$

$(0.375) \times (8) = \underline{3}.00$

111111.011_2

77.3_8

$3F.6_{16}$

$$
\begin{array}{r|l}
16 & 63 \\ \hline
16 & 3 \\ \hline
 & 0
\end{array}
\qquad
\begin{array}{c}
15 \\ 3
\end{array}
$$

$(0.375) \times (16) = \underline{6}.00$

00111111.0110_2

$3F.6_{16}$

2-6. Solution.

a.

$$
\begin{array}{r}
1\ 11\quad \\
10110_2 \\
+\ 10011_2 \\ \hline
101001_2
\end{array}
$$

b.

$$
\begin{array}{r}
1111\quad\quad \\
110.110_2 \\
+\ \ 11.011_2 \\ \hline
1010.001_2
\end{array}
$$

c.

$$
\begin{array}{r}
1\quad\quad\quad\quad \\
0.10101_2 \\
+\ 0.1001_2 \\ \hline
1.00111_2
\end{array}
$$

d.

$$\begin{array}{r} \scriptstyle 0\,2 \\ 1\not{1}01_2 \\ -\quad 11_2 \\ \hline 1010_2 \end{array}$$

e.

$$\begin{array}{r} \scriptstyle 1\,1\,1 \\ \scriptstyle \not{2}\not{2}\,\not{2}\,2 \\ \not{1}00.00_2 \\ -\quad 1.01_2 \\ \hline 10.11_2 \end{array}$$

f.

$$\begin{array}{r} \scriptstyle 0\,2 \\ 0.10\not{1}01_2 \\ -\ 0.1001_2 \\ \hline 0.00011_2 \end{array}$$

2-7. Solution.

$$\begin{array}{rr} & \$\ 110110.10_2 \\ - & \$\ \quad 1010.01_2 \\ \hline & \$\ 101100.01_2 \end{array} \quad = \quad \$\ 44.25_{10}$$

2-8. Solution.

a.

$$\begin{array}{r} 110_2 \\ \times\quad 10_2 \\ \hline 000_2 \\ +\ 110_2 \\ \hline 1100_2 \end{array}$$

b.

$$
\begin{array}{r}
0110_2 \\
\times \;\; 1010_2 \\
\hline
000_2 \\
110_2 \;\; \\
000_2 \;\;\;\; \\
+ \;\; 110_2 \;\;\;\;\;\; \\
\hline
111100_2
\end{array}
$$

c.

$$
\begin{array}{r}
0.1_2 \\
\times \;\; 0.1_2 \\
\hline
0.01_2
\end{array}
$$

d.

$$
\begin{array}{r}
101_2 \\
\times \;\; 101_2 \\
\hline
101_2 \\
000_2 \;\; \\
+ \;\; 101_2 \;\;\;\; \\
\hline
11001_2
\end{array}
$$

2-9. Solution.

Decimal Number	Binary Representation		
	Two's Complement	One's Complement	Signed-Magnitude
15	01111	01111	01111
14	01110	01110	01110
13	01101	01101	01101
12	01100	01100	01100
11	01011	01011	01011
10	01010	01010	01010
9	01001	01001	01001
8	01000	01000	01000
7	00111	00111	00111
6	00110	00110	00110
5	00101	00101	00101
4	00100	00100	00100
3	00011	00011	00011
2	00010	00010	00010
1	00001	00001	00001
0	00000	00000 11111	00000 10000
-1	11111	11110	10001
-2	11110	11101	10010
-3	11101	11100	10011
-4	11100	11011	10100
-5	11011	11010	10101
-6	11010	11001	10110
-7	11001	11000	10111
-8	11000	10111	11000
-9	10111	10110	11001
-10	10110	10101	11010
-11	10101	10100	11011

Decimal Number	Binary Representation		
	Two's Complement	One's Complement	Signed-Magnitude
-12	10100	10011	11100
-13	10011	10010	11101
-14	10010	10001	11110
-15	10001	10000	11111
-16	10000	X	X

2-10. Solution.

a.

$$\begin{array}{r} 01011_2 \\ +\ 11001_2 \\ \hline 00100_2 \end{array}$$

b.

$$\begin{array}{r} 01010_2 \\ +\ 01010_2 \\ \hline 10100_2 \end{array}$$ Overflow

c.

$$\begin{array}{r} 11000_2 \\ +\ 11011_2 \\ \hline 10011_2 \end{array}$$

d.

$$\begin{array}{r} 00101_2 \\ +\ 11011_2 \\ \hline 00000_2 \end{array}$$

e.

$$\begin{array}{r} 10111_2 \\ +\ 10111_2 \\ \hline 01110_2 \end{array} \quad \text{Underflow}$$

f.

$$\begin{array}{r} 11101_2 \\ +\ 01011_2 \\ \hline 01000_2 \end{array}$$

2-11. Solution.

a.

$$0.125_{10} = 0.001_2 \rightarrow 0.110_2 + 0.001_2 = 0.111_2 = -0.125_{10}$$

Flip Bits   Add 1 to LSB

b.

$$0.25_{10} = 0.010_2 \rightarrow 0.101_2 + 0.001_2 = 0.110_2 = -0.25_{10}$$

Flip Bits   Add 1 to LSB

c.

$$0.5_{10} = 0.100_2 \rightarrow 0.011_2 + 0.001_2 = 0.100_2 = -0.5_{10}$$

Flip Bits   Add 1 to LSB

d.

$$0.75_{10} = 0.110_2 \rightarrow 0.001_2 + 0.001_2 = 0.010_2 = -0.75_{10}$$

Flip Bits   Add 1 to LSB

2-12. Solution.

a.

$$\underbrace{1011}.\underbrace{1100}_2$$

$$B.C_{16}$$

b.

$$35.2_8$$

$$\overbrace{11}\overbrace{101}.\overbrace{01}_2$$

c.

$$\underbrace{1111}\underbrace{1111}.\underbrace{1000}_2$$

$$FF.8_{16}$$

d.

$$\underbrace{11}\underbrace{111}\underbrace{111}.\underbrace{100}_2$$

$$377.4_8$$

e.

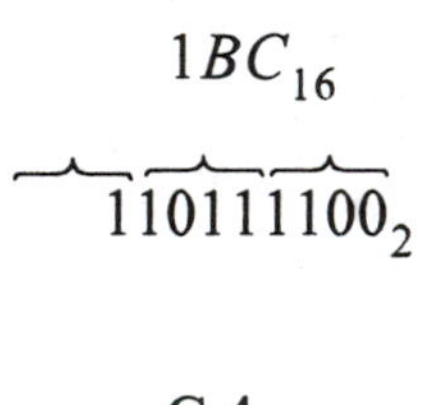

f.

$$C.4_{16}$$

$$\overbrace{1100}.\overbrace{01}_2$$

2-13. Solution.

$$BBB_{16} = (11 \times 16^2) + (11 \times 16^1) + (11 \times 16^0) = 3003_{10}$$

2-14. Solution.

a.

octal code	file permissions		
	owner	group	public
644_8	110 rw-	100 r--	100 r--

b.

octal code	file permissions		
	owner	group	public
444_8	100 r--	100 r--	100 r--

c.

octal code	file permissions		
	owner	group	public
711_8	111 rwx	001 --x	001 --x

d.

octal code	file permissions		
	owner	group	public
755_8	111 rwx	101 r-x	101 r-x

2-15. Solution.

a.

$$\begin{array}{r} {}^{1} \\ 72_8 \\ +\ 41_8 \\ \hline 133_8 \end{array}$$

b.

$$\begin{array}{r} 72_8 \\ -\ 41_8 \\ \hline 31_8 \end{array}$$

c.

$$\begin{array}{r} 7 \\ 0\not{8}\,8 \\ \not{1}00_8 \\ -\ \ 6_8 \\ \hline 72_8 \end{array}$$

d.

$$\begin{array}{r} {}^{1} \\ 7E_{16} \\ +\ 4C_{16} \\ \hline CA_{16} \end{array}$$

e.

$$\begin{array}{r} 7E_{16} \\ -\ 4C_{16} \\ \hline 32_{16} \end{array}$$

f.

$$\begin{array}{r} 15 \\ 0\not{1}\not{6}16 \\ \not{1}00_{16} \\ -\ \ 6_{16} \\ \hline FA_{16} \end{array}$$

2-16. Solution.

a.

Decimal Value	Base 18 Numeral
0	0
1	1
2	2
3	3
4	4
5	5
6	6
7	7
8	8
9	9
10	A
11	B
12	C
13	D
14	E
15	F
16	G
17	H

b. A base 18 number having an N-bit integer part and an M-bit fraction part,

$$e_{N-1}e_{N-2}\ldots e_1 e_0 \,.\, e_{-1}e_{-2}\ldots e_{-(M-1)}e_{-M}$$

is converted to decimal by

$$\sum_{i=-M}^{N-1} e_i \times 18^i$$

c. A decimal integer is converted to a base-18 integer by a series of divide-by-18 operations. The remainders from the divisions form the base-18 number, starting with the least significant digit and ending with the most significant digit. A decimal fraction is converted to a base-18 fraction by a series of multiply-by-18 operations. The integer part of the products form the base-18 fraction, starting with the most significant digit and ending with the least significant digit.

d. Addition: Two base-18 numbers are added column-by-column, moving right to left, starting with the least significant digits. Anytime a column sum equals or exceeds the base 18, 18 is subtracted from the column sum and a 1 is carried forward to the next column.

Subtraction: Two base-18 numbers are subtracted column-by-column, moving right to left, starting with the least significant digits. Anytime the subtrahend digit is more than the minuend digit, a borrow is made from the next column to the left. The minuend digit contributing the borrow is decremented by 1 and the minuend digit receiving the digit is incremented by the base 18.

Base 18 subtraction can also be defined by introducing the 18's complement notation. The 18's complement of an N-digit integer E is defined as

$$18\text{'s complement} = \begin{cases} 18^N - E, & (E \neq 0) \\ 0, & (E = 0) \end{cases}$$

2-17. Solution.

a. $0_{10} - 99_{10}$

b.

Decimal Number	BCD Counter
0	0000 0000
1	0000 0001

Decimal Number	BCD Counter
2	0000 0010
3	0000 0011
4	0000 0100
5	0000 0101
6	0000 0110
7	0000 0111
8	0000 1000
9	0000 1001
10	0001 0000
11	0001 0001
12	0001 0010
13	0001 0011
14	0001 0100
15	0001 0101
16	0001 0110
17	0001 0111
18	0001 1000
19	0001 1001
20	0010 0000

c. Illegal two-digit BCD codes.

$0A_{16}$ - $0F_{16}$, $1A_{16}$ - $1F_{16}$, $2A_{16}$ - $2F_{16}$, $3A_{16}$ - $3F_{16}$, $4A_{16}$ - $4F_{16}$, $5A_{16}$ - $5F_{16}$, $6A_{16}$ - $6F_{16}$, $7A_{16}$ - $7F_{16}$, $8A_{16}$ - $8F_{16}$, $9A_{16}$ - $9F_{16}$, AA_{16} - AF_{16}, BA_{16} - BF_{16}, CA_{16} - CF_{16}, DA_{16} - DF_{16}, EA_{16} - EF_{16}, FA_{16} - FF_{16}

2-18. Solution.

a.

Excess-3 Code	Decimal Digit
0000	Not Used
0001	
0010	

Excess-3 Code	Decimal Digit
0011	0
0100	1
0101	2
0110	3
0111	4
1000	5
1001	6
1010	7
1011	8
1100	9
1101	Not Used
1110	
1111	

b.

$$\begin{array}{r} 1010_2 \\ +\ 1000_2 \\ \hline 10010_2 \end{array}$$

2-19. Solution.

a.

Binary Number	Gray Code
00000	00000
00001	00001
00010	00011
00011	00010
00100	00110
00101	00111
00110	00101
00111	00100
01000	01100

Binary Number	Gray Code
01001	01101
01010	01111
01011	01110
01100	01010
01101	01011
01110	01001
01111	01000
10000	11000
10001	11001
10010	11011
10011	11010
10100	11110
10101	11111
10110	11101
10111	11100
11000	10100
11001	10101
11010	10111
11011	10110
11100	10010
11101	10011
11110	10001
11111	10000

b.

$$\underset{}{81_{10}} = \underset{\text{Binary}}{1010001_2} = \underset{\text{Gray}}{1111001_2}$$

c.

$$200_{10} = \underset{\text{Binary}}{11001000_2} = \underset{\text{Gray}}{10101100_2}$$

d.

$$44_{10} = 101100_2 = 111010_2$$

Binary Gray

2-20. Solution.

Text	ASCII Code (Hexadecimal)	Text	ASCII Code (Hexadecimal)
L	4C	space	20
e	65	i	69
a	61	m	6D
r	72	p	70
n	6E	l	6C
i	69	y	79
n	6E	space	20
g	67	u	75
space	20	n	6E
d	64	d	64
o	6F	e	65
e	65	r	72
s	73	s	73
space	20	t	74
n	6E	a	61
o	6F	n	6E
t	74	d	64
space	20	i	69
a	61	n	6E
l	6C	g	67
w	77	.	2E
a	61		
y	79		
s	73		

Chapter 3. Combinational Systems - Definition & Analysis

3-1. Solution.

a. Combinational Digital System.

b. A touch-tone keypad of a telephone is a combinational digital system because each input, a selected button, always yields the same output, an associated tone, regardless of any buttons that may have been previously selected.

3-2. Solution.

a. Sequential Digital System.

b. A telephone network is a sequential digital system because the order in which the inputs are received, i.e., buttons selected, determines the number being called and the particular destination phone activated.

3-3. Solution.

a. Use the definition of the **or** operator to prove P2a: $X + 0 = X$.

X	$X + 0 = X$
0	0
1	1

Use the definition of the **and** operator to prove P2b: $X \bullet 1 = X$

X	$X \bullet 1 = X$
0	0
1	1

b. Use the definition of the **or** operator to prove P3a: $X + Y = Y + X$

X	Y	$X+Y$	$Y+X$
0	0	0	0
0	1	1	1
1	0	1	1
1	1	1	1

Use the definition of the **and** operator to prove P3b: $X \bullet Y = Y \bullet X$

X	Y	$X \bullet Y$	$Y \bullet X$
0	0	0	0
0	1	0	0
1	0	0	0
1	1	1	1

c. Use the definition of the **or** and **and** operators to prove P4a:
$X + (Y \bullet Z) = (X + Y) \bullet (X + Z)$

X	Y	Z	$(Y \bullet Z)$	$X + (Y \bullet Z)$	$(X+Y)$	$(X+Z)$	$(X+Y) \bullet (X+Z)$
0	0	0	0	0	0	0	0
0	0	1	0	0	0	1	0
0	1	0	0	0	1	0	0
0	1	1	1	1	1	1	1
1	0	0	0	1	1	1	1
1	0	1	0	1	1	1	1
1	1	0	0	1	1	1	1
1	1	1	1	1	1	1	1

EQUAL

Use the definition of the **or** and **and** operators to prove P4b:
$X \bullet (Y+Z) = (X \bullet Y) + (X \bullet Z)$

X	Y	Z	$(Y+Z)$	$X \bullet (Y+Z)$	$(X \bullet Y)$	$(X \bullet Z)$	$(X \bullet Y) + (X \bullet Z)$
0	0	0	0	0	0	0	0
0	0	1	1	0	0	0	0
0	1	0	1	0	0	0	0
0	1	1	1	0	0	0	0
1	0	0	0	0	0	0	0
1	0	1	1	1	0	1	1
1	1	0	1	1	1	0	1
1	1	1	1	1	1	1	1

EQUAL

d. Use the definition of the **or** operator to prove P5a: $X+\overline{X} = 1$.

X	$X+\overline{X} = 1$
0	1
1	1

Use the definition of the **and** operator to prove P5b: $X \bullet \overline{X} = 0$

X	$X \bullet \overline{X} = 0$
0	0
1	0

3-4. Solution.

a. X — [AND gate] — Z

X	$Z = X \bullet X = X$
0	0
1	1

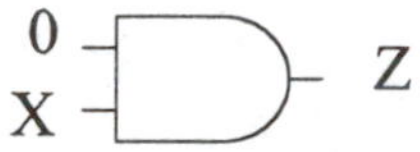

0, X → AND → Z

X	$Z = X \bullet 0 = 0$
0	0
1	0

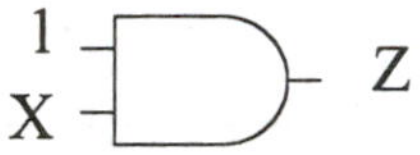

1, X → AND → Z

X	$Z = X \bullet 1 = X$
0	0
1	1

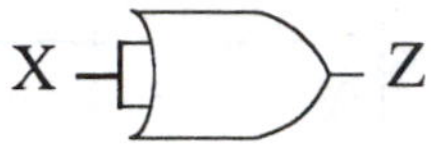

b. X → OR → Z

X	$Z = X + X = X$
0	0
1	1

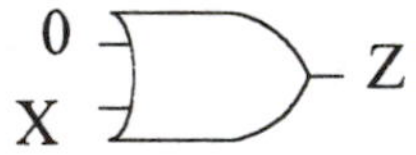

0, X → OR → Z

X	$Z = X + 0 = X$
0	0
1	1

1, X → OR → Z

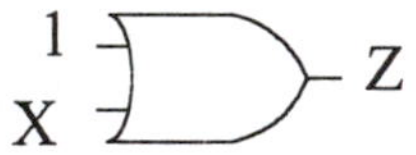

X	$Z = X + 1 = 1$
0	1
1	1

c. X → NAND → Z

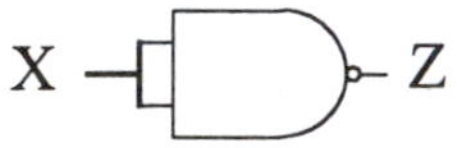

X	$Z = \overline{X \bullet X} = \overline{X}$
0	1
1	0

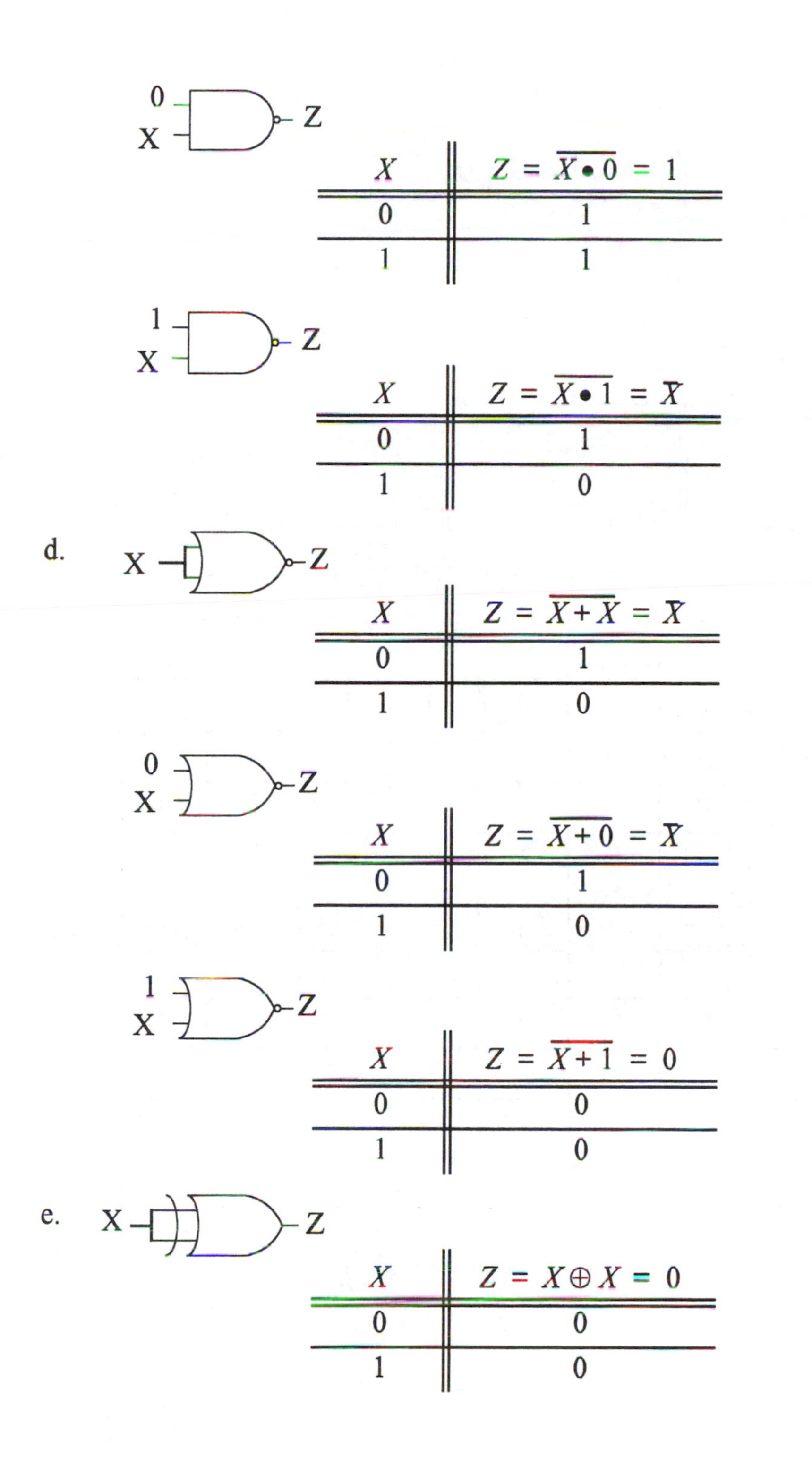

X	$Z = \overline{X \bullet 0} = 1$
0	1
1	1

X	$Z = \overline{X \bullet 1} = \overline{X}$
0	1
1	0

d.

X	$Z = \overline{X + X} = \overline{X}$
0	1
1	0

X	$Z = \overline{X + 0} = \overline{X}$
0	1
1	0

X	$Z = \overline{X + 1} = 0$
0	0
1	0

e.

X	$Z = X \oplus X = 0$
0	0
1	0

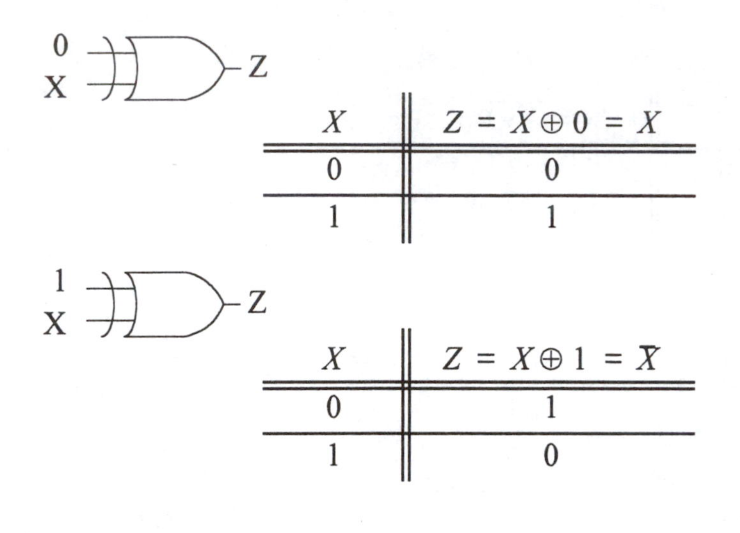

X	$Z = X \oplus 0 = X$
0	0
1	1

X	$Z = X \oplus 1 = \overline{X}$
0	1
1	0

3-5. Solution.

X	Y	F_1	F_2	F_3	F_4	F_5	F_6	F_7	F_8
0	0	0	0	0	0	0	0	0	0
0	1	0	0	0	0	1	1	1	1
1	0	0	0	1	1	0	0	1	1
1	1	0	1	0	1	0	1	0	1

X	Y	F_9	F_{10}	F_{11}	F_{12}	F_{13}	F_{14}	F_{15}	F_{16}
0	0	1	1	1	1	1	1	1	1
0	1	0	0	0	0	1	1	1	1
1	0	0	0	1	1	0	0	1	1
1	1	0	1	0	1	0	1	0	1

3-6. Solution.

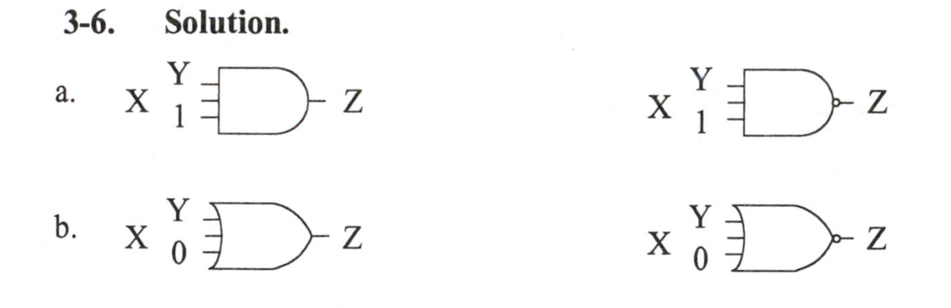

c.

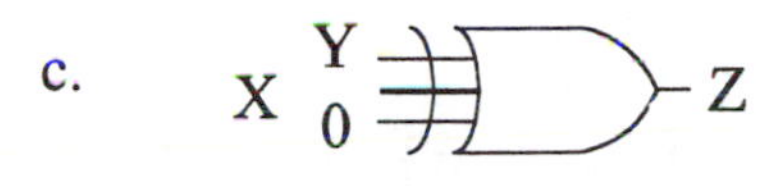

Y
X
0
Z

3-7. Solution.

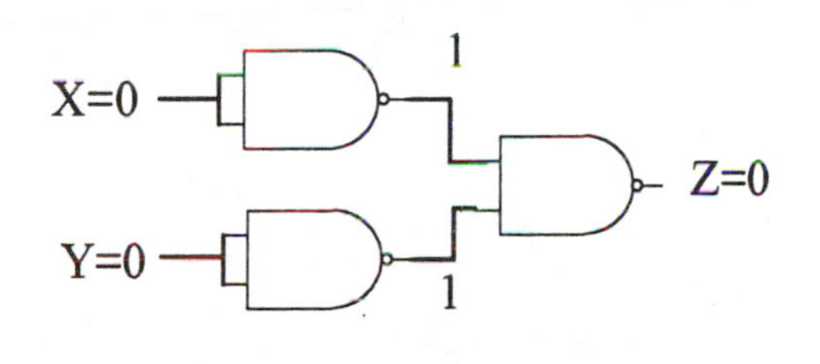

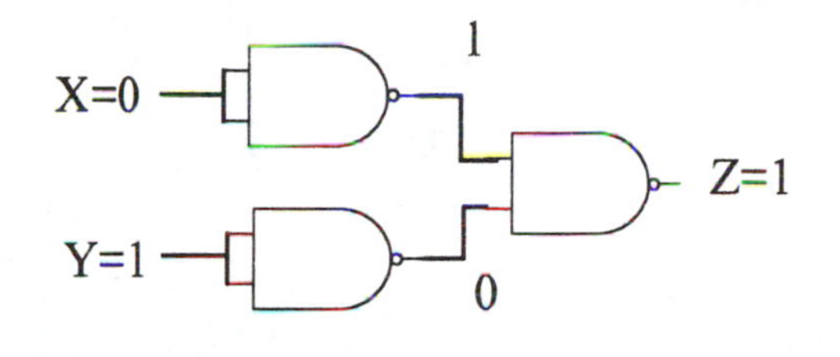

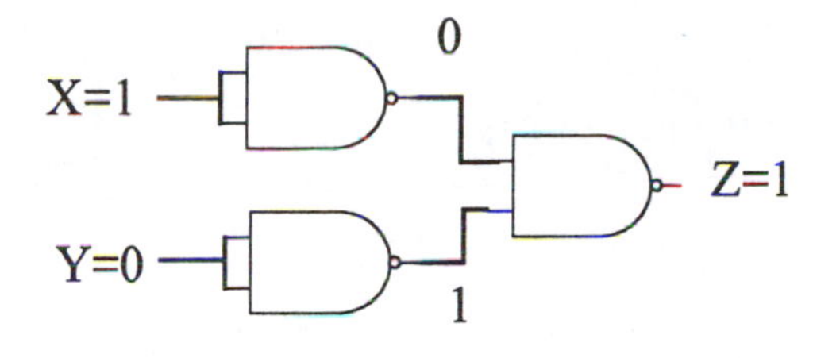

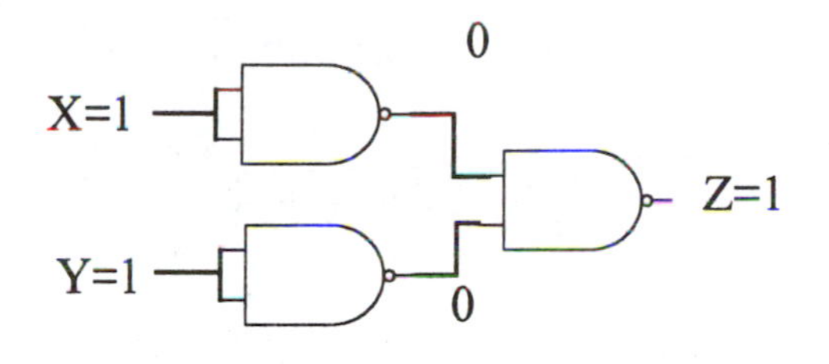

3-8. Solution.

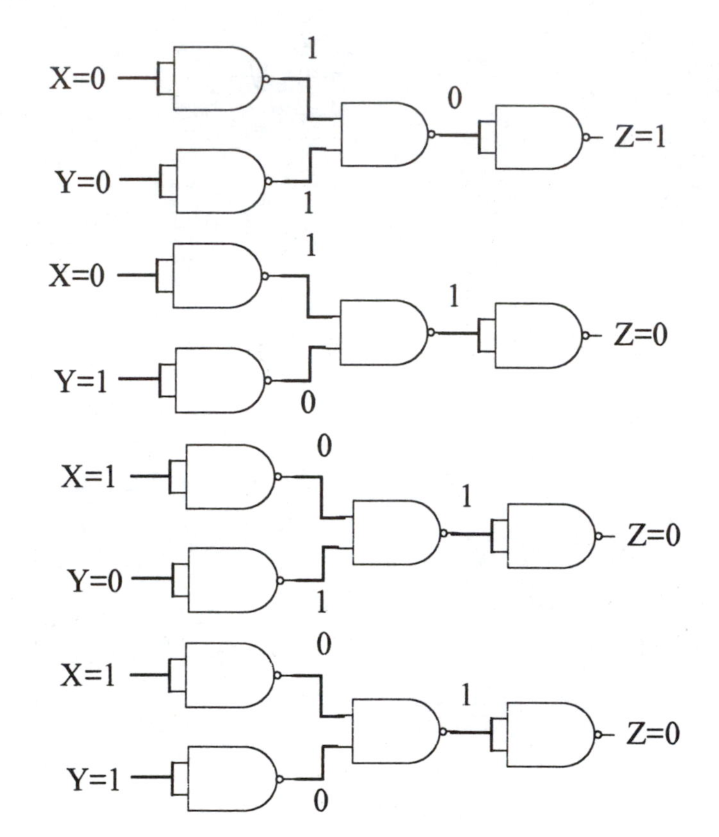

3-9. Solution.

a.

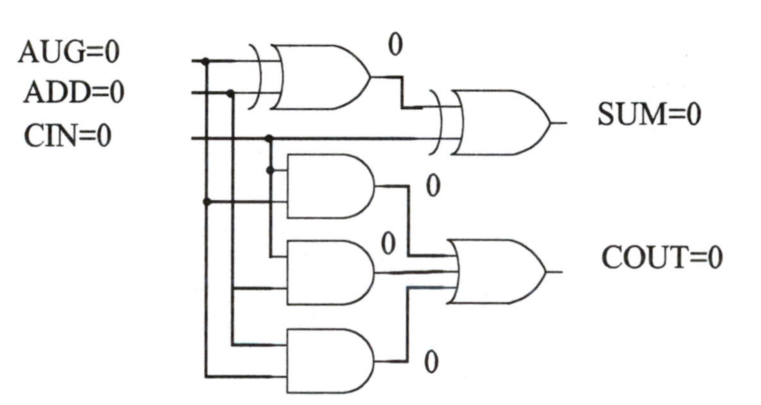

b.

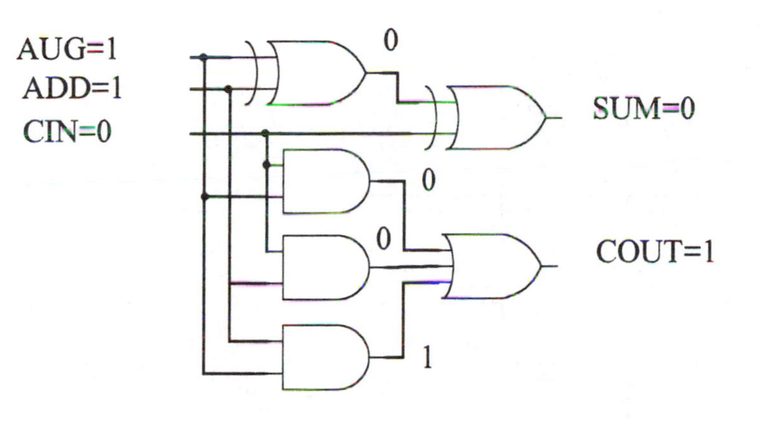
AUG=1
ADD=1
CIN=0
0
SUM=0
0
0
COUT=1
1

c.

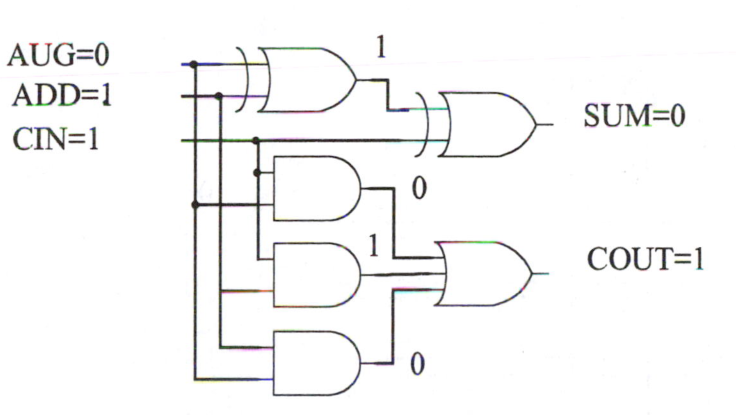
AUG=0
ADD=1
CIN=1
1
SUM=0
0
1
COUT=1
0

d.

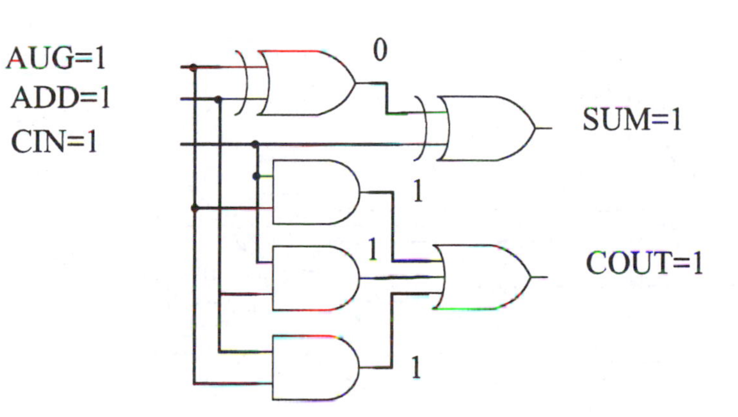
AUG=1
ADD=1
CIN=1
0
SUM=1
1
1
COUT=1
1

3-10. Solution.

a.

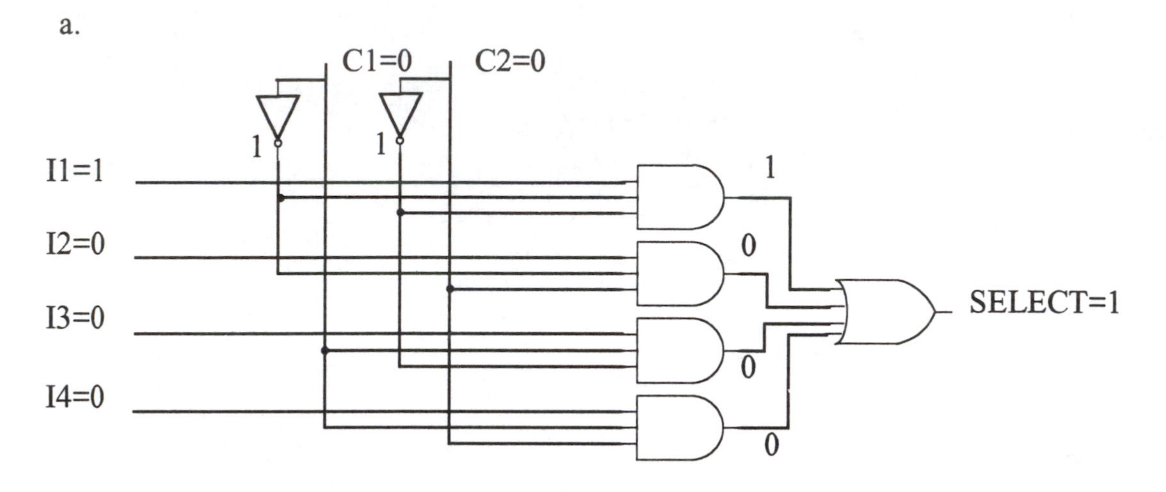

b.

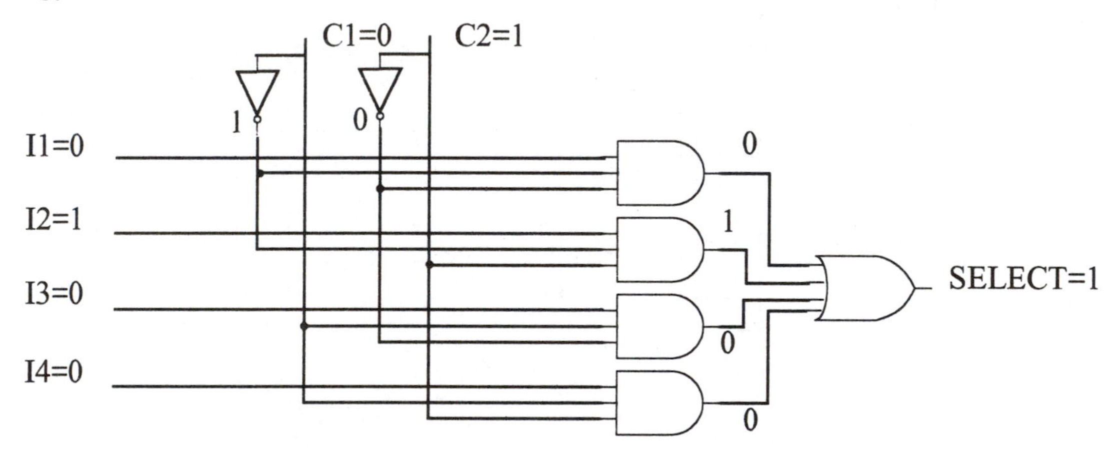

c.

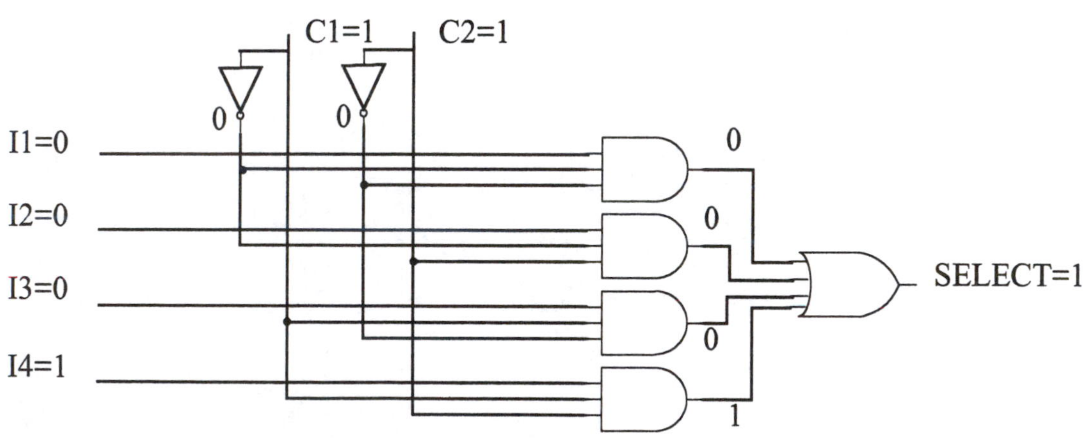

3-11. Solution.

a.

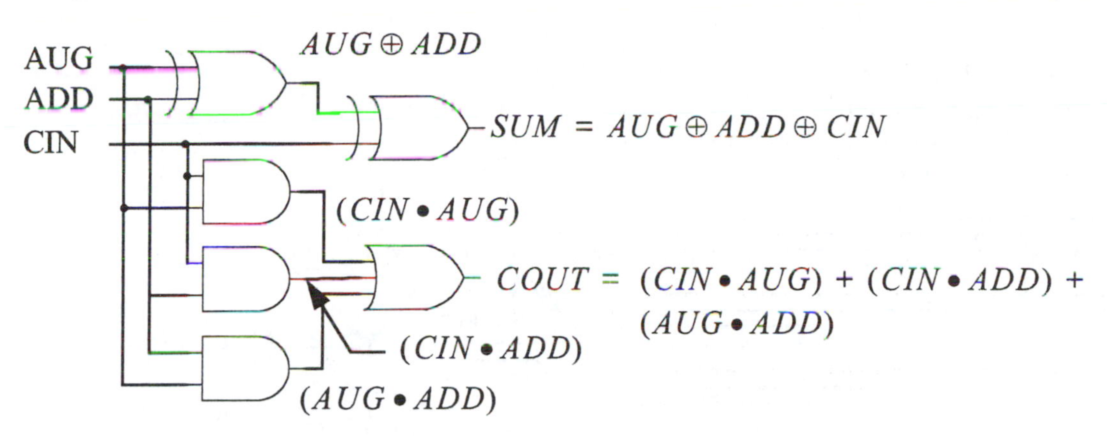

b.

AUG	ADD	CIN	SUM	COUT
0	0	0	0	0
0	0	1	1	0
0	1	0	1	0
0	1	1	0	1
1	0	0	1	0
1	0	1	0	1
1	1	0	0	1
1	1	1	1	1

c. Two-bit, binary full adder

3-12. Solution.

a.

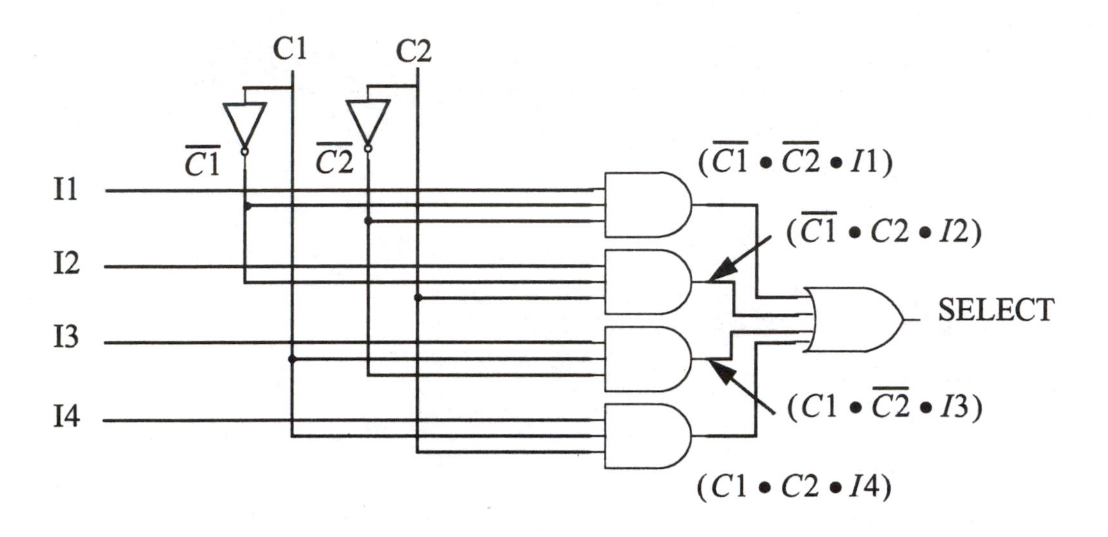

$$SELECT = (\overline{C1} \bullet \overline{C2} \bullet I1) + (\overline{C1} \bullet C2 \bullet I2) + (C1 \bullet \overline{C2} \bullet I3) + (C1 \bullet C2 \bullet I4)$$

b.

C1	C2	SELECT
0	0	I1
0	1	I2
1	0	I3
1	1	I4

c. 4-input multiplexer

3-13. Solution.

a. $\overline{0} + 1 \bullet 1 = 1$

b. $\overline{1 + 1 + 0} = 0$

c. $1 \bullet \overline{0} + \overline{(1 + 1)} = 1$

d. $\overline{(0 \bullet 1) \bullet \overline{(1 + 0)}} = 1$

3-14. Solution.

a.

A	B	C	Z	Minterms
0	0	0	1	$(\bar{A} \bullet \bar{B} \bullet \bar{C})$
0	0	1	0	
0	1	0	1	$(\bar{A} \bullet B \bullet \bar{C})$
0	1	1	1	$(\bar{A} \bullet B \bullet C)$
1	0	0	0	
1	0	1	0	
1	1	0	0	
1	1	1	1	$(A \bullet B \bullet C)$

b.

A	B	C	Z	Maxterms
0	0	0	1	
0	0	1	0	$(A + B + \bar{C})$
0	1	0	1	
0	1	1	1	
1	0	0	0	$(\bar{A} + B + C)$
1	0	1	0	$(\bar{A} + B + \bar{C})$
1	1	0	0	$(\bar{A} + \bar{B} + C)$
1	1	1	1	

c. $Z = (\bar{A} \bullet \bar{B} \bullet \bar{C}) + (\bar{A} \bullet B \bullet \bar{C}) + (\bar{A} \bullet B \bullet C) + (A \bullet B \bullet C)$

d. $Z = \sum m\,(0, 2, 3, 7)$

e. $Z = (A + B + \bar{C}) \bullet (\bar{A} + B + C) \bullet (\bar{A} + B + \bar{C}) \bullet (\bar{A} + \bar{B} + C)$

f. $\prod M\,(1, 4, 5, 6)$

3-15. Solution.

Yes

3-16. Solution.

Standard Product-of-Sums (F) = $\overline{\text{Standard Sum-of-Products } (\overline{F})}$

Standard Sum-of-Products (F) = $\overline{\text{Standard Product-of-Sums } (\overline{F})}$

3-17. Solution.

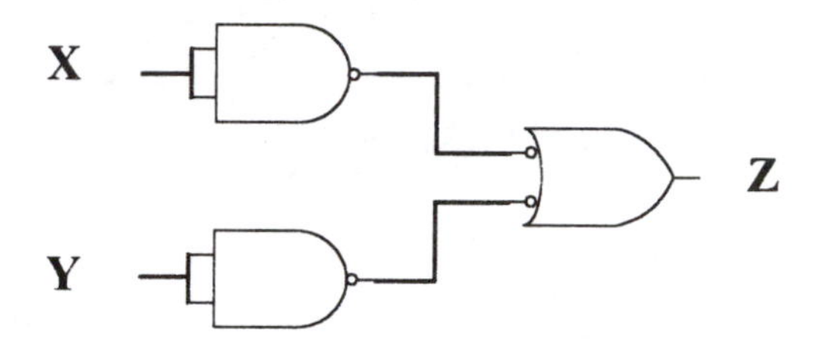

3-18. Solution.

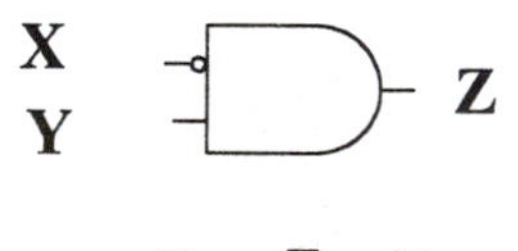

$Z = \overline{X} \bullet Y$

X	Y	Z
0	0	0
0	1	1
1	0	0
1	1	0

X
Y
Z

$$Z = \overline{X + \overline{Y}}$$

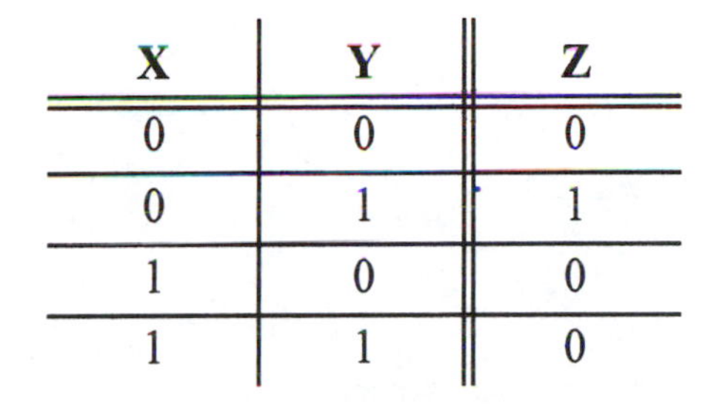

X	Y	Z
0	0	0
0	1	1
1	0	0
1	1	0

3-19. Solution.

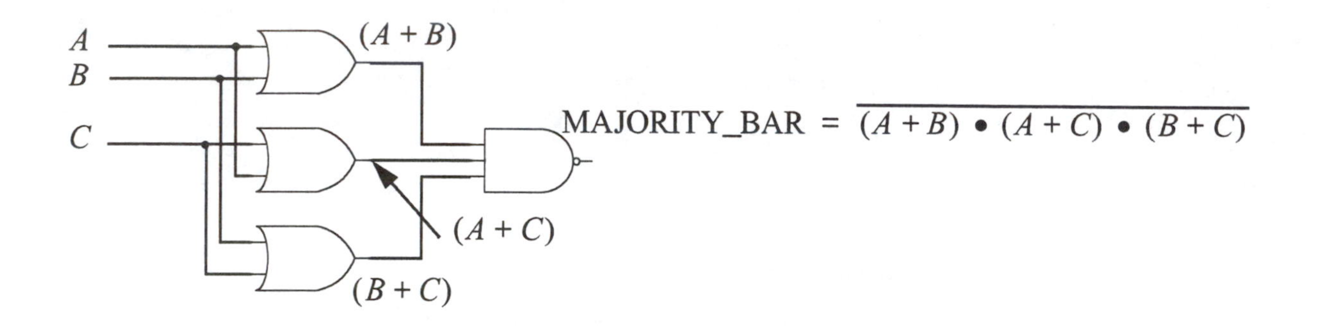

A	B	C	$(A+B)$	$(A+C)$	$(B+C)$	$\overline{(A+B) \bullet (A+C) \bullet (B+C)}$
0	0	0	0	0	0	1
0	0	1	0	1	1	1
0	1	0	1	0	1	1
0	1	1	1	1	1	0
1	0	0	1	1	0	1
1	0	1	1	1	1	0
1	1	0	1	1	1	0
1	1	1	1	1	1	0

The output MAJORITY_BAR is asserted active-0 whenever two or more, i.e., the majority, of inputs are asserted active-1.

Chapter 4. Combinational Design-Synthesis

4-1. Solution.

a.

A	B	C	ERROR
0	0	0	0
0	0	1	1
0	1	0	1
0	1	1	0
1	0	0	0
1	0	1	1
1	1	0	1
1	1	1	0

$$ERROR = (\overline{A} \bullet \overline{B} \bullet C) + (\overline{A} \bullet B \bullet \overline{C}) + (A \bullet \overline{B} \bullet C) + (A \bullet B \bullet \overline{C})$$

$$ERROR = (A + B + C) \bullet (A + \overline{B} + \overline{C}) \bullet (\overline{A} + B + C) \bullet (\overline{A} + \overline{B} + \overline{C})$$

b.

A	B	C	ERROR_BAR
0	0	0	1
0	0	1	0
0	1	0	0
0	1	1	1
1	0	0	1
1	0	1	0
1	1	0	0
1	1	1	1

$$\text{ERROR\_BAR} = (\bar{A} \bullet \bar{B} \bullet \bar{C}) + (\bar{A} \bullet B \bullet C) + (A \bullet \bar{B} \bullet \bar{C}) + (A \bullet B \bullet C)$$

$$\text{ERROR\_BAR} = (A + B + \bar{C}) \bullet (A + \bar{B} + C) \bullet (\bar{A} + B + \bar{C}) \bullet (\bar{A} + \bar{B} + C)$$

4-2. Solution.

a.

B3	B2	B1	B0	O1	O2	O3	O4	O5
0	0	0	0	1	1	0	1	1
0	0	0	1	1	1	1	0	1
0	0	1	0	1	1	0	1	1
0	0	1	1	1	1	1	0	1
0	1	0	0	1	1	0	1	1
0	1	0	1	1	1	1	0	1
0	1	1	0	1	1	0	1	1
0	1	1	1	1	1	1	0	1
1	0	0	0	1	1	0	1	1
1	0	0	1	1	1	1	0	1
1	0	1	0	1	1	0	1	1
1	0	1	1	1	1	1	0	1
1	1	0	0	1	1	0	1	1
1	1	0	1	1	1	1	0	1
1	1	1	0	1	1	0	1	1
1	1	1	1	1	1	1	0	1

$$O1 = O2 = O5 = 1$$

$$O3 = \sum m(1, 3, 5, 7, 9, 11, 13, 15) = \sum_{i=0}^{7} m_{2i+1}$$

$$O4 = \sum m(0, 2, 4, 6, 8, 10, 12, 14) = \sum_{i=0}^{7} m_{2i}$$

$$O1 = O2 = O5 = 1$$

$$O3 = \prod M(0, 2, 4, 6, 8, 10, 12, 14) = \prod_{i=0}^{7} M_{2i}$$

$$O4 = \prod M(1, 3, 5, 7, 9, 11, 13, 15) = \prod_{i=0}^{7} M_{2i+1}$$

b.

B3	B2	B1	B0	O1	O2	O3	O4	O5
0	0	0	0	0	0	1	0	0
0	0	0	1	0	0	0	1	0
0	0	1	0	0	0	1	0	0
0	0	1	1	0	0	0	1	0
0	1	0	0	0	0	1	0	0
0	1	0	1	0	0	0	1	0
0	1	1	0	0	0	1	0	0
0	1	1	1	0	0	0	1	0
1	0	0	0	0	0	1	0	0
1	0	0	1	0	0	0	1	0
1	0	1	0	0	0	1	0	0
1	0	1	1	0	0	0	1	0
1	1	0	0	0	0	1	0	0
1	1	0	1	0	0	0	1	0
1	1	1	0	0	0	1	0	0
1	1	1	1	0	0	0	1	0

$$O1 = O2 = O5 = 0$$

$$O3 = \sum m(0, 2, 4, 6, 8, 10, 12, 14) = \sum_{i=0}^{7} m_{2i}$$

$$O4 = \sum m(1, 3, 5, 7, 9, 11, 13, 15) = \sum_{i=0}^{7} m_{2i+1}$$

$$O1 = O2 = O5 = 0$$

$$O3 = \prod M(1, 3, 5, 7, 9, 11, 13, 15) = \prod_{i=0}^{7} M_{2i+1}$$

$$O4 = \prod M(0, 2, 4, 6, 8, 10, 12, 14) = \prod_{i=0}^{7} M_{2i}$$

4-3. Solution.

a.

A1	A0	B1	B0	LT	EQ	GT
0	0	0	0	0	1	0
0	0	0	1	1	0	0
0	0	1	0	1	0	0
0	0	1	1	1	0	0
0	1	0	0	0	0	1
0	1	0	1	0	1	0
0	1	1	0	1	0	0
0	1	1	1	1	0	0
1	0	0	0	0	0	1
1	0	0	1	0	0	1
1	0	1	0	0	1	0
1	0	1	1	1	0	0
1	1	0	0	0	0	1
1	1	0	1	0	0	1
1	1	1	0	0	0	1
1	1	1	1	0	1	0

$$LT = \sum m\,(1, 2, 3, 6, 7, 11)$$

$$EQ = \sum m\,(0, 5, 10, 15)$$

$$GT = \sum m\,(4, 8, 9, 12, 13, 14)$$

$$LT = \prod M\,(0, 4, 5, 8, 9, 10, 12, 13, 14, 15)$$

$$EQ = \prod M\,(1, 2, 3, 4, 6, 7, 8, 9, 11, 12, 13, 14)$$

$$GT = \prod M\,(0, 1, 2, 3, 5, 6, 7, 10, 11, 15)$$

b.

A1	A0	B1	B0	LT_BAR	EQ_BAR	GT_BAR
0	0	0	0	1	0	1
0	0	0	1	0	1	1
0	0	1	0	0	1	1
0	0	1	1	0	1	1
0	1	0	0	1	1	0
0	1	0	1	1	0	1
0	1	1	0	0	1	1
0	1	1	1	0	1	1
1	0	0	0	1	1	0
1	0	0	1	1	1	0
1	0	1	0	1	0	1
1	0	1	1	0	1	1
1	1	0	0	1	1	0
1	1	0	1	1	1	0
1	1	1	0	1	1	0
1	1	1	1	1	0	1

$$\text{LT\_BAR} = \sum m\,(0, 4, 5, 8, 9, 10, 12, 13, 14, 15)$$

$$\text{EQ\_BAR} = \sum m\,(1, 2, 3, 4, 6, 7, 8, 9, 11, 12, 13, 14)$$

$$\text{GT\_BAR} = \sum m\,(0, 1, 2, 3, 5, 6, 7, 10, 11, 15)$$

$$\text{LT\_BAR} = \prod M\,(1, 2, 3, 6, 7, 11)$$

$$\text{EQ\_BAR} = \prod M\,(0, 5, 10, 15)$$

$$\text{GT\_BAR} = \prod M\,(4, 8, 9, 12, 13, 14)$$

4-4. Solution.

a.

X	Y	$X \bullet Y$	$X + (X \bullet Y)$
0	0	0	0
0	1	0	0
1	0	0	1
1	1	1	1

EQUAL

b.

$$X + (X \bullet Y) = (X \bullet 1) + (X \bullet Y)$$ Postulate 2b Identity

$$= X \bullet (1 + Y)$$ Postulate 4b Distributive

$$= X \bullet 1$$ Theorem 2a Identity

$$= X$$ Postulate 2b Identity

4-5. Solution.

a.

X	Y	$\overline{X} \bullet Y$	$X + (\overline{X} \bullet Y)$	$X + Y$
0	0	0	0	0
0	1	1	1	1
1	0	0	1	1
1	1	0	1	1

EQUAL

b.

$$X + (\overline{X} \bullet Y) = (X + \overline{X}) \bullet (X + Y)$$ Postulate 4a Distributive

$$= 1 \bullet (X + Y)$$ Postulate 5a Complement

$$= X + Y$$ Postulate 2b Identity

4-6. Solution.

a.

$$(X+Z) \bullet (X+\overline{X} \bullet Y) \bullet (X+\overline{Z}) = (X+Z) \bullet (X+Y) \bullet (X+\overline{Z}) \quad \text{Theorem 8a Simplification}$$

$$= X \bullet (X+Y) \quad \text{Theorem 6b Adjacency}$$

$$= X \quad \text{Theorem 7b Absorption}$$

b.

$$(A \bullet B \bullet C \bullet D) + (A \bullet D) + (\overline{A} \bullet B \bullet C) + (\overline{A} \bullet D) + (B \bullet C \bullet D) = (A \bullet D) + (\overline{A} \bullet B \bullet C) + (\overline{A} \bullet D) + (B \bullet C \bullet D)$$

Theorem 7a Absorbtion

$$= (\overline{A} \bullet B \bullet C) + D \bullet (\overline{A} + A) + (B \bullet C \bullet D) \quad \text{Postulate 4b Distributive}$$

$$= (\overline{A} \bullet B \bullet C) + D + (B \bullet C \bullet D) \quad \text{Postulates 5b and 2b}$$

$$= (\overline{A} \bullet B \bullet C) + D \quad \text{Theorem 7a Absorbtion}$$

c.

$$(A \bullet B) + (A \bullet B \bullet C) + (\overline{A} \bullet D) + (\overline{B} \bullet D) = (A \bullet B) + (A \bullet B \bullet C) + D \bullet (\overline{A} + \overline{B})$$

Postulate 4b Distributive

$$= (A \bullet B) + (A \bullet B \bullet C) + D \bullet \overline{(A \bullet B)} \quad \text{Theorem 5b DeMorgan}$$

$$= (A \bullet B) + (A \bullet B \bullet C) + D \quad \text{Theorem 8a Simplification}$$

$$= (A \bullet B) + D \quad \text{Theorem 7a Absorbtion}$$

4-7. Solution.

$$\overline{(A \bullet B + \overline{C}) \bullet (E \bullet \overline{F} + \overline{D}) \bullet (G + \overline{H})} = \overline{(A \bullet B + \overline{C})} + \overline{(E \bullet \overline{F} + \overline{D})} + \overline{(G + \overline{H})}$$

$$= (\overline{(A \bullet B)} \bullet C) + \left(\overline{(E \bullet \overline{F})} \bullet D\right) + (\overline{G} \bullet H)$$

$$= ((\overline{A} + \overline{B}) \bullet C) + ((\overline{E} + F) \bullet D) + (\overline{G} \bullet H)$$

$$= (\overline{A} \bullet C) + (\overline{B} \bullet C) + (D \bullet \overline{E}) + (D \bullet F) + (\overline{G} \bullet H)$$

4-8. Solution.

a.

A \ BC	00	01	11	10
0	1	1	0	1
1	0	0	1	1

b.

AB \ CD	00	01	11	10
00	0	0	1	1
01	0	0	1	1
11	1	0	1	1
10	1	1	1	1

c.

AB \ CD	00	01	11	10
00	1	0	0	0
01	0	0	0	1
11	0	0	1	0
10	0	0	0	1

4-9. Solution.

a. ERROR

A \ BC	00	01	11	10
0	0	1	0	1
1	0	1	0	1

b. Implicants: m_1, m_2, m_5, m_6, m_1m_5, and m_2m_6

Prime Implicants: m_1m_5 and m_2m_6

Essential Prime Implicants: m_1m_5 and m_2m_6

Secondary Prime Implicants: None

c. $ERROR = B \oplus C$

4-10. Solution.

a. LT

A1A0 \ B1B0	00	01	11	10
00	0	1	1	1
01	0	0	1	1
11	0	0	0	0
10	0	0	1	0

b. Implicants: m_1, m_2, m_3, m_6, m_7, m_{11}, m_1m_3, m_2m_3, m_6m_7, m_3m_{11}, m_3m_7, m_2m_6, and $m_2m_3m_6m_7$

Prime Implicants: $m_2m_3m_6m_7$, m_1m_3, and m_3m_{11}

Essential Prime Implicants: $m_2m_3m_6m_7$, m_1m_3, and m_3m_{11}

Secondary Prime Implicants: None

c. $LT = (\overline{A1} \bullet \overline{A0} \bullet B0) + (\overline{A0} \bullet B1 \bullet B0) + (\overline{A1} \bullet B1)$

a.

EQ

A1A0 \ B1B0	00	01	11	10
00	1	0	0	0
01	0	1	0	0
11	0	0	1	0
10	0	0	0	1

b. Implicants: m_0, m_5, m_{10}, and m_{15}

Prime Implicants: m_0, m_5, m_{10}, and m_{15}

Essential Prime Implicants: m_0, m_5, m_{10}, and m_{15}

Secondary Prime Implicants: None

c.

$$EQ = (\overline{A1} \bullet \overline{A0} \bullet \overline{B1} \bullet \overline{B0}) + (\overline{A1} \bullet A0 \bullet \overline{B1} \bullet B0) + (A1 \bullet \overline{A0} \bullet B1 \bullet \overline{B0}) + (A1 \bullet A0 \bullet B1 \bullet B0)$$

$$EQ = (A0 \odot B0) \bullet (A1 \odot B1)$$

a.

GT

A1A0 \ B1B0	00	01	11	10
00	0	0	0	0
01	1	0	0	0
11	1	1	0	1
10	1	1	0	0

b. Implicants: m_4, m_8, m_9, m_{12}, m_{13}, m_{14}, $m_{12}m_{14}$, $m_{12}m_{13}$, m_8m_9, m_4m_{12}, m_8m_{12}, m_9m_{13}, and $m_8m_9m_{12}m_{13}$

Prime Implicants: m_4m_{12}, $m_{12}m_{14}$, and $m_8m_9m_{12}m_{13}$

Essential Prime Implicants: m_4m_{12}, $m_{12}m_{14}$, and $m_8m_9m_{12}m_{13}$

Secondary Prime Implicants: None

a. $GT = (A0 \bullet \overline{B1} \bullet \overline{B0}) + (A1 \bullet A0 \bullet \overline{B0}) + (A1 \bullet \overline{B1})$

4-11. Solution.

a.

F

AB \ CD	00	01	11	10
00	1	0	0	1
01	0	0	1	1
11	0	1	0	0
10	1	0	0	1

b. Implicants: m_0, m_2, m_6, m_7, m_8, m_{10}, m_{13}, m_0m_2, m_8m_{10}, m_6m_7, m_0m_8, m_2m_{10}, m_2m_6, and $m_0m_2m_8m_{10}$

Prime Implicants: $m_0m_2m_8m_{10}$, m_2m_6, m_6m_7, and m_{13}

Essential Prime Implicants: $m_0m_2m_8m_{10}$, m_6m_7, and m_{13}

Secondary Prime Implicants: m_2m_6

c. $F = (\overline{B} \bullet \overline{D}) + (\overline{A} \bullet B \bullet C) + (A \bullet B \bullet \overline{C} \bullet D)$

4-12. Solution.

a. F

AB \ CD	00	01	11	10
00	0	0	0	0
01	0	1	1	1
11	1	1	0	1
10	0	0	0	0

b. Implicants: m_5, m_6, m_7, m_{12}, m_{13}, m_{14}, $m_{12}m_{13}$, $m_{12}m_{14}$, m_5m_7, m_6m_7, m_5m_{13}, and m_6m_{14}

Prime Implicants: $m_{12}m_{13}$, $m_{12}m_{14}$, m_5m_7, m_6m_7, m_5m_{13}, and m_6m_{14}

Essential Prime Implicants: None

Secondary Prime Implicants: $m_{12}m_{13}$, $m_{12}m_{14}$, m_5m_7, m_6m_7, m_5m_{13}, and m_6m_{14}

c. $F = (A \bullet B \bullet \overline{C}) + (\overline{A} \bullet B \bullet D) + (B \bullet C \bullet \overline{D})$

4-13. Solution.

a.

F

BC \ DE	00	01	11	10
00	0	0	-	1
01	1	1	1	0
11	0	0	0	0
10	0	0	1	1

A=0

F

BC \ DE	00	01	11	10
00	0	0	0	-
01	0	1	0	0
11	1	0	0	0
10	0	0	0	0

A=1

b. Implicants: m_2, m_3, m_4, m_5, m_7, m_{10}, m_{11}, m_{21}, m_{28}, m_2m_3, m_4m_5, m_5m_7, $m_{10}m_{11}$, m_3m_7, m_3m_{11}, m_2m_{10}, m_5m_{21}, m_2m_{18}, and $m_2m_3m_{10}m_{11}$

Prime Implicants: m_{28}, m_4m_5, m_5m_7, m_3m_7, m_5m_{21}, m_2m_{18}, and $m_2m_3m_{10}m_{11}$

Essential Prime Implicants: m_{28}, m_4m_5, m_5m_{21}, and $m_2m_3m_{10}m_{11}$

Secondary Prime Implicants: m_5m_7, m_3m_7, and m_2m_{18}

c.

$$F(A, B, C, D, E) = (A \bullet B \bullet C \bullet \overline{D} \bullet \overline{E}) + (\overline{A} \bullet \overline{B} \bullet C \bullet \overline{D}) + (\overline{B} \bullet C \bullet \overline{D} \bullet E) + (\overline{A} \bullet \overline{B} \bullet D \bullet E) + (\overline{A} \bullet \overline{C} \bullet D)$$

4-14. Solution.

F

AB \ CD	00	01	11	10
00	0	0	0	0
01	1	1	1	0
11	1	0	1	0
10	1	0	1	0

a. $F = (B \bullet \overline{C} \bullet \overline{D}) + (\overline{A} \bullet B \bullet D) + (A \bullet C \bullet D) + (A \bullet \overline{C} \bullet \overline{D})$

b. $F = (A + B) \bullet (\overline{C} + D) \bullet (\overline{A} + C + \overline{D})$

c. The product-of-sums expression is better because it yields a less complex design.

4-15. Solution.

a.

Implicants	Implicants Pass #1
✔ 0	0,2(2)
✔ 2	2,10(8)
✔ 10	
13	

Prime Implicants	Minterms			
	0	2	10	13
13				Ⓧ
0,2(2)	Ⓧ	X		
2,10(8)		X	Ⓧ	

$$F_1 = (A \bullet B \bullet \overline{C} \bullet D) + (\overline{A} \bullet \overline{B} \bullet \overline{D}) + (\overline{B} \bullet C \bullet \overline{D})$$

b.

Implicants	Implicants Pass #1	Implicants Pass #2
✔ 5	✔ 5,7(2) ✔ 5,13(8)	5,7,13,15(2,8)
✔ 7 ✔ 13	✔ 7,15(8) ✔ 13,15(2)	
✔ 15		

Prime Implicants	Minterms			
	5	7	13	15
5,7,13,15(2,8)	ⓧ	ⓧ	ⓧ	ⓧ

$$F_2 = (B \bullet D)$$

c.

Implicants	Implicants Pass #1	Implicants Pass #2
✔ 1 ✔ 8	✔ 8,10(2) ✔ 8,12(4) 1,9(8) 8,9(1)	8,10,12,14(2,4)
✔ 9 ✔ 10 ✔ 12	✔ 10,14(4) ✔ 12,14(2)	
✔ 14		

Prime Implicants	Minterms 1	8	9	10	12	14
1,9(8)	(X)		X			
8,9(1)		X	X			
8,10,12,14(2,4)		X		(X)	(X)	(X)

$F_3 = (B \bullet \overline{C} \bullet D) + (A \bullet \overline{D})$

4-16. Solution.

a.

Implicants	Implicants Pass #1	Implicants Pass #2
✔ 2 ✔ 4	✔ 2,3(1) 4,5(1) ✔ 2,10(8) 2,18(16)	2,3,10,11(1,8)
✔ 3 ✔ 5 ✔ 10 ✔ 18	3,7(4) 5,7(2) ✔ 3,11(8) ✔ 10,11(1) 5,21(16)	
✔ 7 ✔ 11 ✔ 21 28		

b.

Prime Implicants	Minterms							
	2	4	5	7	10	11	21	28
2,3,10,11(1,8)	X				(X)	(X)		
4,5(1)		(X)	X					
2,18(16)	X							
3,7(4)				X				
5,7(2)			X	X				
5,21(16)			X				(X)	
28								(X)

Prime Implicants	Minterms
	7
2,18(16)	
3,7(4)	X
5,7(2)	X

$$F(A, B, C, D, E) = (A \bullet B \bullet C \bullet \overline{D} \bullet \overline{E}) + (\overline{A} \bullet \overline{B} \bullet C \bullet \overline{D}) + (\overline{B} \bullet C \bullet \overline{D} \bullet E) + (\overline{A} \bullet \overline{B} \bullet D \bullet E) + (\overline{A} \bullet \overline{C} \bullet D)$$

4-17. Solution.

Individual Minimization

a. F_1

Implicants	Implicants Pass #1	Implicants Pass #2
✔ 2 4 ✔ 8	✔ 2,3(1) ✔ 8,9(1) ✔ 2,10(8) ✔ 8,10(2)	2,3,10,11(1,8) 8,9,10,11(1,2)
✔ 3 ✔ 9 ✔ 10	✔ 3,11(8) ✔ 9,11(2) ✔ 10,11(1)	
✔ 11		

F_2

Implicants	Implicants Pass #1
4 ✔ 8	8,9(1)
✔ 3 ✔ 9	3,7(4)
✔ 7	

b. F_1

Prime Implicants	Minterms						
	2	3	4	8	9	10	11
2,3,10,11(1,8)	Ⓧ	Ⓧ				X	X
8,9,10,11(1,2)				Ⓧ	Ⓧ	X	X
4			Ⓧ				

F_2

Prime Implicants	**Minterms**				
	3	**4**	**7**	**8**	**9**
8,9(1)				ⓧ	ⓧ
3,7(4)	ⓧ		ⓧ		
4		ⓧ			

$$F_1 = (\bar{A} \bullet B \bullet \bar{C} \bullet \bar{D}) + (\bar{B} \bullet C) + (A \bullet \bar{B})$$

$$F_2 = (\bar{A} \bullet B \bullet \bar{C} \bullet \bar{D}) + (A \bullet \bar{B} \bullet \bar{C}) + (\bar{A} \bullet C \bullet D)$$

Collective Minimization

a. F_1

Implicants	**Implicants Pass #1**	**Implicants Pass #2**
✔ 2 4 ✔ 8	✔ 2,3(1) ✔ 8,9(1) ✔ 2,10(8) ✔ 8,10(2)	2,3,10,11(1,8) 8,9,10,11(1,2)
✔ 3 ✔ 9 ✔ 10	✔ 3,11(8) ✔ 9,11(2) ✔ 10,11(1)	
✔ 11		

F_2

Implicants	Implicants Pass #1
4 ✔ 8	8,9(1)
✔ 3 ✔ 9	3,7(4)
✔ 7	

F_{12}

Implicants	Implicants Pass #1
4 ✔ 8	8,9(1)
3 ✔ 9	

b.

	Prime Implicants	Minterms F_1							Minterms F_2				
		2	**3**	**4**	**8**	**9**	**10**	**11**	**3**	**4**	**7**	**8**	**9**
F_1	2,3,10,11(1,8)	(X)	X				X	X					
	8,9,10,11(1,2)				X	X	X	X					
	4			X									
F_2	8,9(1)											X	X
	3,7(4)								X		(X)		
	4									X			
F_{12}	3		X						X				
	4			X						X			
	8,9(1)				X	X						X	X

Function	Prime Implicants	Minterms F_1			Minterms F_2		
		4	8	9	4	8	9
F_1	8,9,10,11(1,2)		X	X			
	4	X					
F_2	8,9(1)					X	X
	4				X		
F_{12}	3						
	4	X			X		
	8,9(1)		X	X		X	X

$$F_1 = (\bar{B} \bullet C) + (\bar{A} \bullet B \bullet \bar{C} \bullet \bar{D}) + (A \bullet \bar{B} \bullet \bar{C})$$

$$F_2 = (\bar{A} \bullet C \bullet D) + (\bar{A} \bullet B \bullet \bar{C} \bullet \bar{D}) + (A \bullet \bar{B} \bullet \bar{C})$$

c. The individual and collective minimizations yield different results. The collective minimization yields a less complex design because the products terms $(\bar{A} \bullet B \bullet \bar{C} \bullet \bar{D})$ and $(A \bullet \bar{B} \bullet \bar{C})$ are common to both functions.

4-18. Solution.

a.

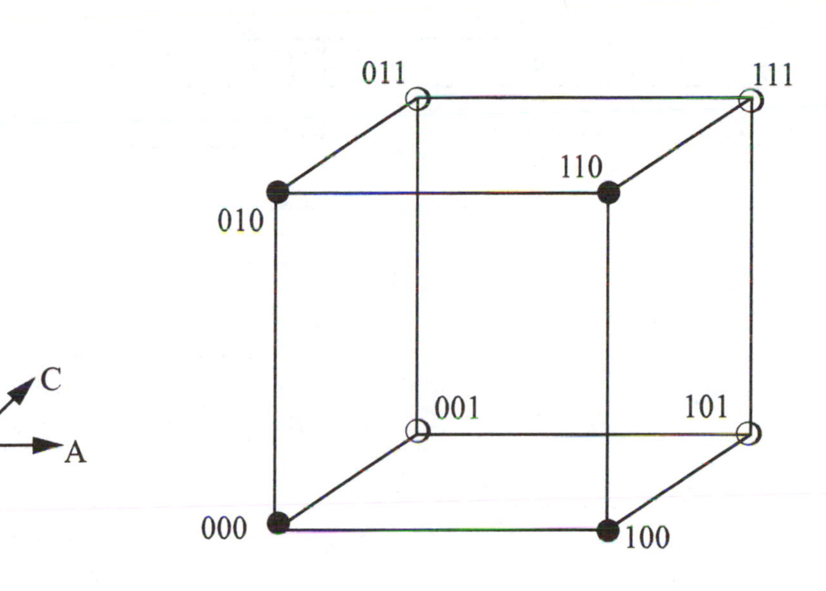

b.

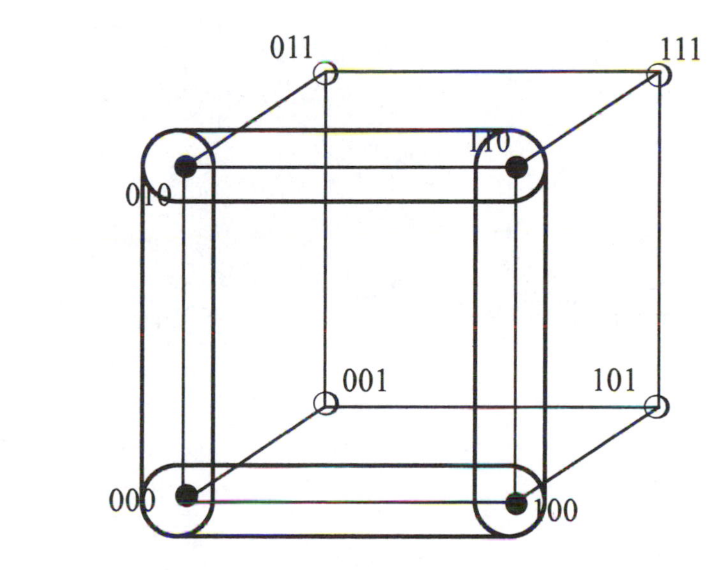

c. $F = \overline{C}$

d.

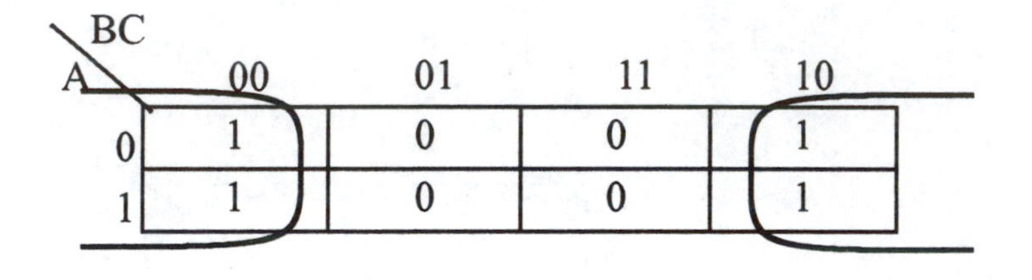

Chapter 5. Combinational Design - Implementation

5-1. Solution.

AB \ CD	00	01	11	10
00	0	1	0	0
01	0	1	0	1
11	0	1	0	0
10	0	1	1	0

a. $F = (\overline{C} \bullet D) + (A \bullet \overline{B} \bullet D) + (\overline{A} \bullet B \bullet C \bullet \overline{D})$

b. $F = (C + D) \bullet (\overline{A} + D) \bullet (A + B + \overline{C}) \bullet (\overline{B} + \overline{C} + \overline{D})$

c. The **nand-nand** implementation is better because it yields a lower gate count.

5-2. Solution.

AB \ CD	00	01	11	10
00	0	0	1	0
01	0	0	1	1
11	1	1	-	-
10	0	1	1	0

a. $F = (C \bullet D) + (A \bullet B) + (A \bullet D) + (B \bullet C)$

b. $F = (A + C) \bullet (B + D)$

c. The **nor-nor** implementation is better because it yields a lower gate count.

5-3. Solution.

AB \ CD	00	01	11	10
00	0	0	1	0
01	0	0	1	0
11	1	1	0	1
10	0	0	1	0

a. $F = (A \bullet B \bullet \overline{C}) + (\overline{A} \bullet C \bullet D) + (\overline{B} \bullet C \bullet D) + (A \bullet B \bullet \overline{D})$

b. $F = (A \bullet B) \oplus (C \bullet D)$

c. The **xor-and** implementation is better because it yields a lower gate count.

5-4. Solution.

$$F = (A \bullet B \bullet \overline{C}) + (\overline{A} \bullet C \bullet D) + (\overline{B} \bullet C \bullet D) + (A \bullet B \bullet \overline{D})$$

$$F = (A \bullet B) \bullet (\overline{C} + \overline{D}) + (C \bullet D) \bullet (\overline{A} + \overline{B})$$

5-5. Solution.

a.

b. The multiplexers can generate clocks signals 180° out of phase.

5-6. Solution.

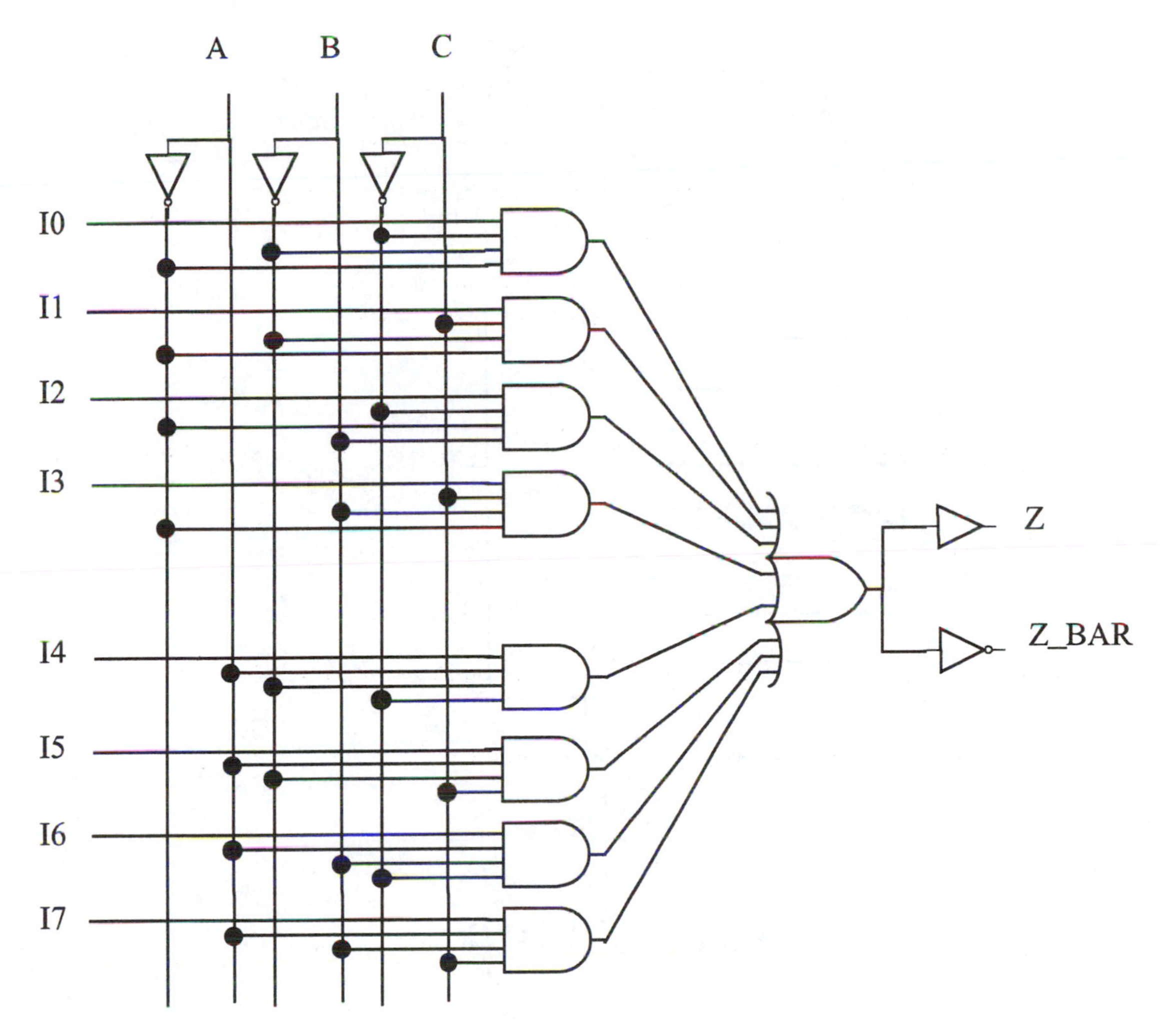

5-7. Solution.

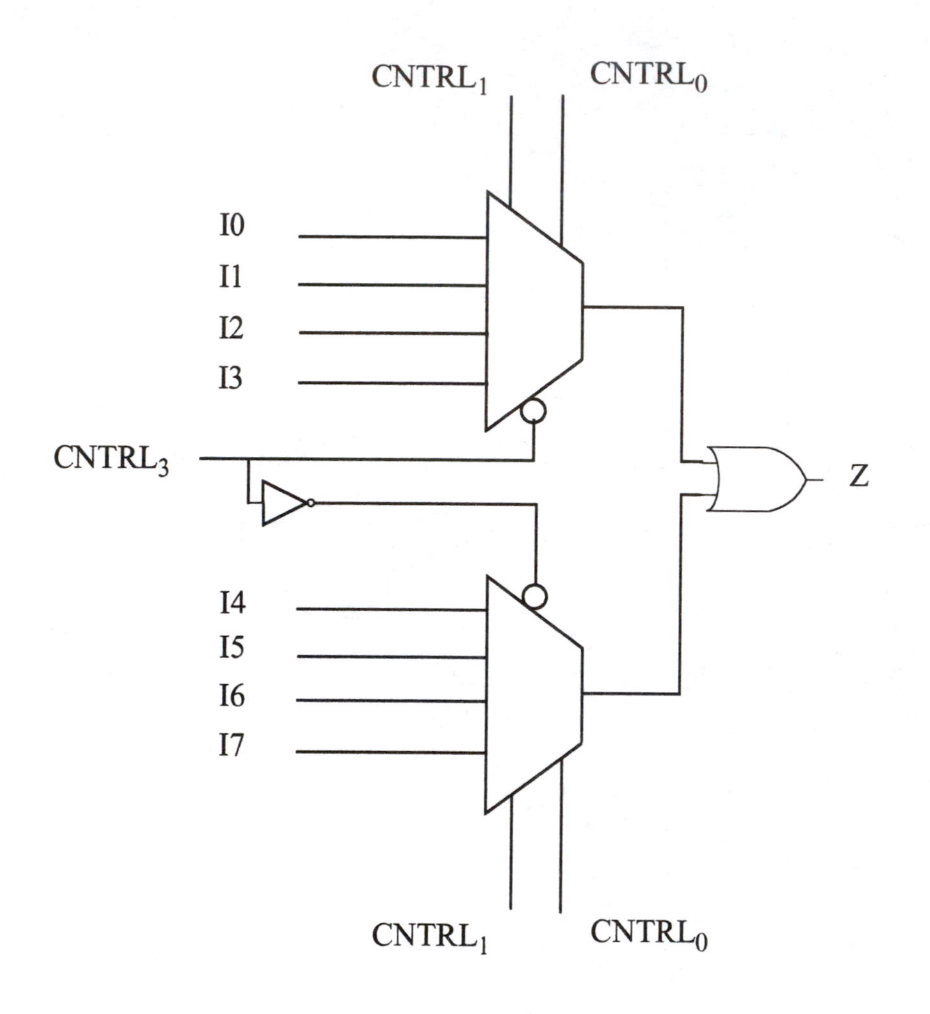

5-8. Solution.

Decimal Number	A	B	C	D	PRIME
	0	0	0	0	-
	0	0	0	1	-
	0	0	1	0	-
0	0	0	1	1	1
1	0	1	0	0	1
2	0	1	0	1	1
3	0	1	1	0	1
4	0	1	1	1	0
5	1	0	0	0	1
6	1	0	0	1	0
7	1	0	1	0	1
8	1	0	1	1	0
9	1	1	0	0	0
	1	1	0	1	-
	1	1	1	0	-
	1	1	1	1	-

a.

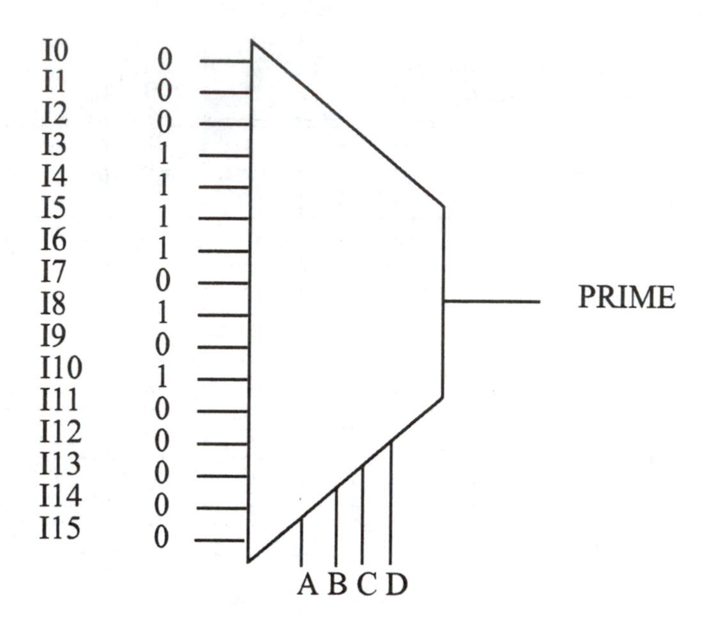
I0 0
I1 0
I2 0
I3 1
I4 1
I5 1
I6 1
I7 0
I8 1
I9 0
I10 1
I11 0
I12 0
I13 0
I14 0
I15 0
PRIME
A B C D

b.

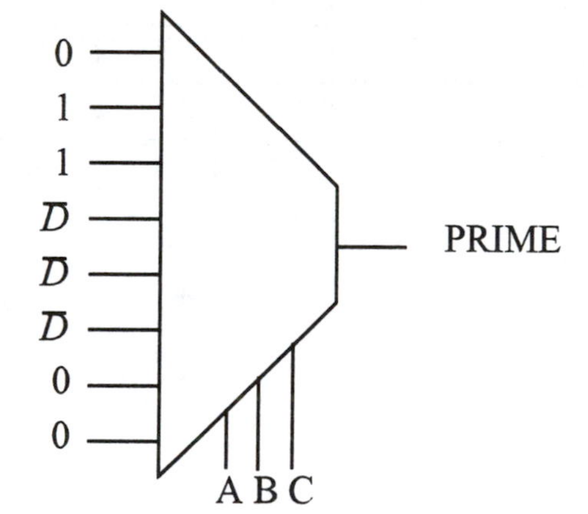
0
1
1
$\overline{D}$
$\overline{D}$
$\overline{D}$
0
0
PRIME
A B C

c.

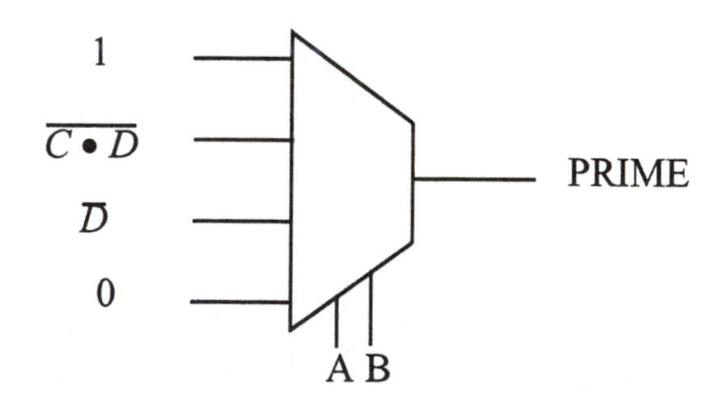
1
$\overline{C \bullet D}$
$\overline{D}$
0
PRIME
A B

5-9. Solution.

a.

F (AB \ CD)	00	01	11	10
00	0	0	0	0
01	0	1	0	1
11	0	0	0	0
10	1	0	1	0

$$F(A=0, B=0) = 0$$
$$F(A=0, B=1) = C \oplus D = G(C, D)$$
$$F(A=1, B=0) = \overline{C \oplus D} = \overline{G(C, D)}$$
$$F(A=1, B=1) = 0$$

b.

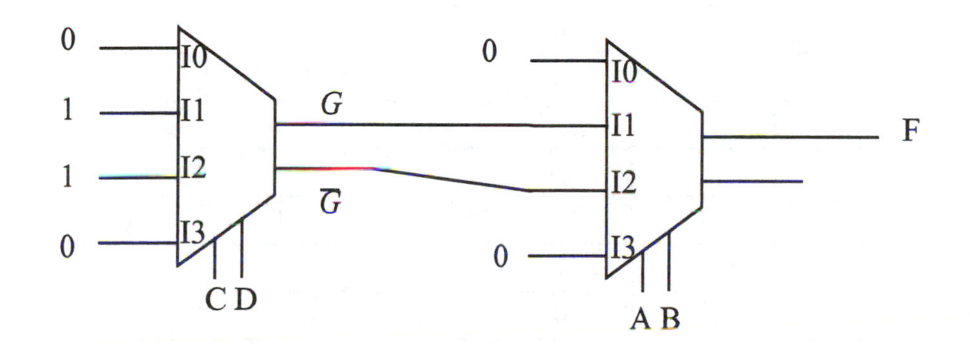

5-10. Solution.

a.

A	B	C	D	E	Z
0	0	0	0	0	0
0	0	0	0	1	0
0	0	0	1	0	1
0	0	0	1	1	1
0	0	1	0	0	0
0	0	1	0	1	0
0	0	1	1	0	0
0	0	1	1	1	0
0	1	0	0	0	0
0	1	0	0	1	1
0	1	0	1	0	1
0	1	0	1	1	1
0	1	1	0	0	1
0	1	1	0	1	1
0	1	1	1	0	1
0	1	1	1	1	1
1	0	0	0	0	0
1	0	0	0	1	1
1	0	0	1	0	1
1	0	0	1	1	1
1	0	1	0	0	1
1	0	1	0	1	1
1	0	1	1	0	1
1	0	1	1	1	1
1	1	0	0	0	0
1	1	0	0	1	1
1	1	0	1	0	1
1	1	0	1	1	1
1	1	1	0	0	1

A	B	C	D	E	Z
1	1	1	0	1	1
1	1	1	1	0	1
1	1	1	1	1	1

Implicants	Implicants Pass #1	Implicants Pass #2	Implicants Pass #3
✔ 2	✔ 2,3(1) ✔ 2,10(8)	✔ 2,3,10,11(1,8) ✔ 2,10,18,26(8,16) ✔ 2,3,18,19(1,16)	2,3,10,11,18,19,26,27(1,8,16)
✔ 3 ✔ 9 ✔ 10 ✔ 12 ✔ 17 ✔ 18 ✔ 20	✔ 3,11(8) ✔ 9,11(2) ✔ 9,13(4) ✔ 9,25(16) ✔ 10,11(1) ✔ 10,14(4) ✔ 10,26(16) ✔ 12,13(1) ✔ 12,14(2) ✔ 12,28(16) ✔ 17,21(4) ✔ 17,25(8) ✔ 18,22(4) ✔ 18,26(8) ✔ 20,22(2) ✔ 20,28(8) ✔ 3,19(16) ✔ 17,19(2) ✔ 18,19(1)	✔ 9,11,13,15(2,4) ✔ 9,11,25,27(2,16) ✔ 9,13,25,29(4,16) ✔ 10,11,14,15(1,4) ✔ 10,11,26,27(1,16) ✔ 10,14,26,20(4,16) ✔ 12,13,14,15(1,2) ✔ 12,13,28,29(1,16) ✔ 12,14,28,30(2,16) ✔ 17,21,25,29(4,8) ✔ 18,22,26,30(4,8) ✔ 20,21,22,23(1,2) ✔ 20,22,28,30(2,8) ✔ 20,21,28,29(1,8) ✔ 3,11,19,27(8,16) ✔ 17,19,21,23(2,4) ✔ 17,19,25,27(2,8) ✔ 18,19,22,23(1,4) ✔ 18,19,26,27(1,8)	9,11,13,15,25,27,29,31(2,4,16) 10,11,14,15,26,27,30,31(1,4,16) 12,13,14,15,28,29,30,31(1,2,16) 20,21,22,23,28,29,30,31(1,2,8) 17,19,21,23,25,27,29,31(2,4,8) 18,19,22,23,26,27,30,31(1,4,8)

Implicants	Implicants Pass #1	Implicants Pass #2	Implicants Pass #3
✔ 11	✔ 11,15(4)	✔ 11,15,27,31(4,16)	
✔ 13	✔ 11,27(16)	✔ 13,15,29,31(2,16)	
✔ 14	✔ 13,15(2)	✔ 14,15,30,31(1,16)	
✔ 19	✔ 13,29(16)	✔ 21,23,29,31(2,8)	
✔ 21	✔ 14,15(1)	✔ 22,23,30,31(1,8)	
✔ 22	✔ 14,30(16)	✔ 25,27,29,31(2,4)	
✔ 25	✔ 21,23(2)	✔ 26,27,30,31(1,4)	
✔ 26	✔ 21,29(8)	✔ 28,29,30,31(1,2)	
✔ 28	✔ 22,23(1)	✔ 19,23,27,31(4,8)	
	✔ 22,30(8)		
	✔ 25,27(2)		
	✔ 25,29(4)		
	✔ 26,27(1)		
	✔ 26,30(4)		
	✔ 28,29(1)		
	✔ 28,30(2)		
	✔ 19,23(4)		
	✔ 19,27(8)		
✔ 15	✔ 15,31(16)		
✔ 23	✔ 23,31(8)		
✔ 27	✔ 27,31(4)		
✔ 29	✔ 29,31(2)		
✔ 30	✔ 30,31(1)		
✔ 31			

$A \bullet C =$ 20, 21, 22, 23, 28, 29, 30, 31 (1, 2, 8)

$A \bullet D =$ 18, 19, 22, 23, 26, 27, 30, 31 (1, 4, 8)

$A \bullet E =$ 17, 19, 21, 23, 25, 27, 29, 31 (2, 4, 8)

$B \bullet C =$ 12, 13, 14, 15, 28, 29, 30, 31 (1, 2, 16)

$B \bullet D =$ 10, 11, 14, 15, 26, 27, 30, 31 (1, 4, 16)

$B \bullet E =$ 9, 11, 13, 15, 25, 27, 29, 31 (2, 4, 16)

$\overline{C} \bullet D =$ 2, 3, 10, 11, 18, 19, 26, 27 (1, 8, 16)

b.

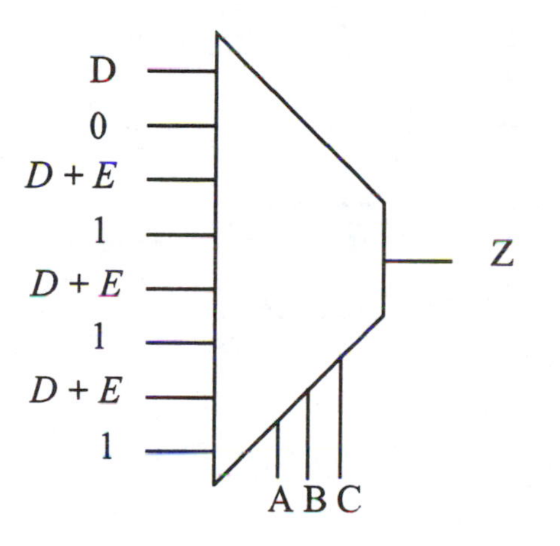

5-11. Solution.

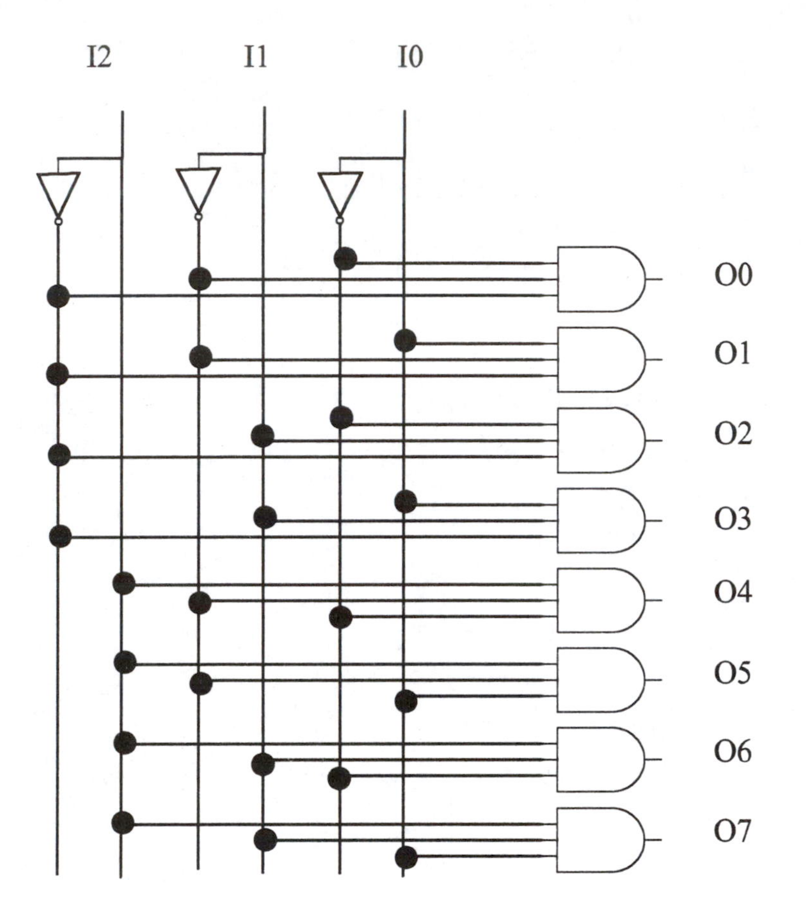

5-12. Solution.

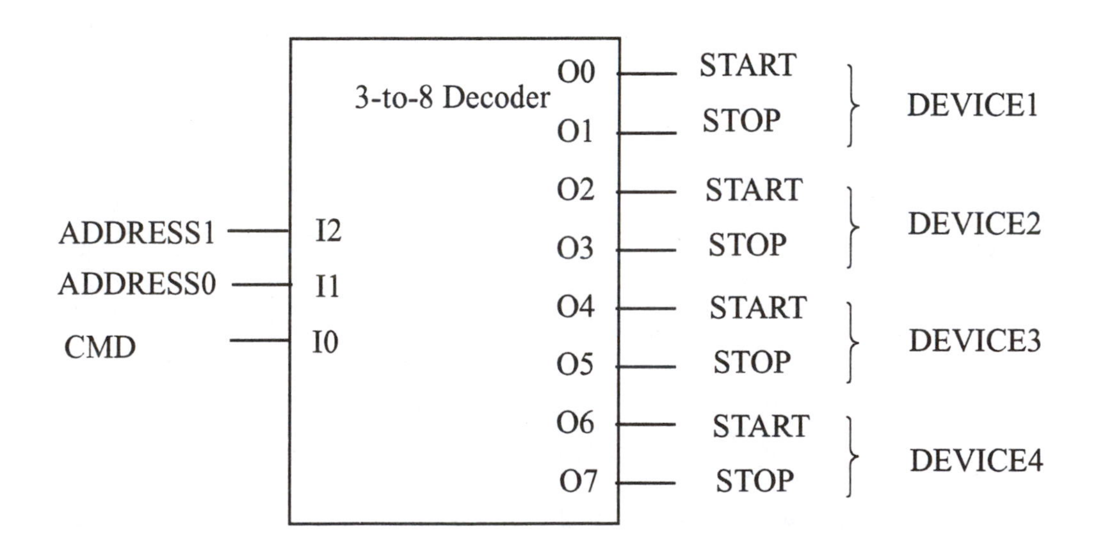

5-13. Solution.

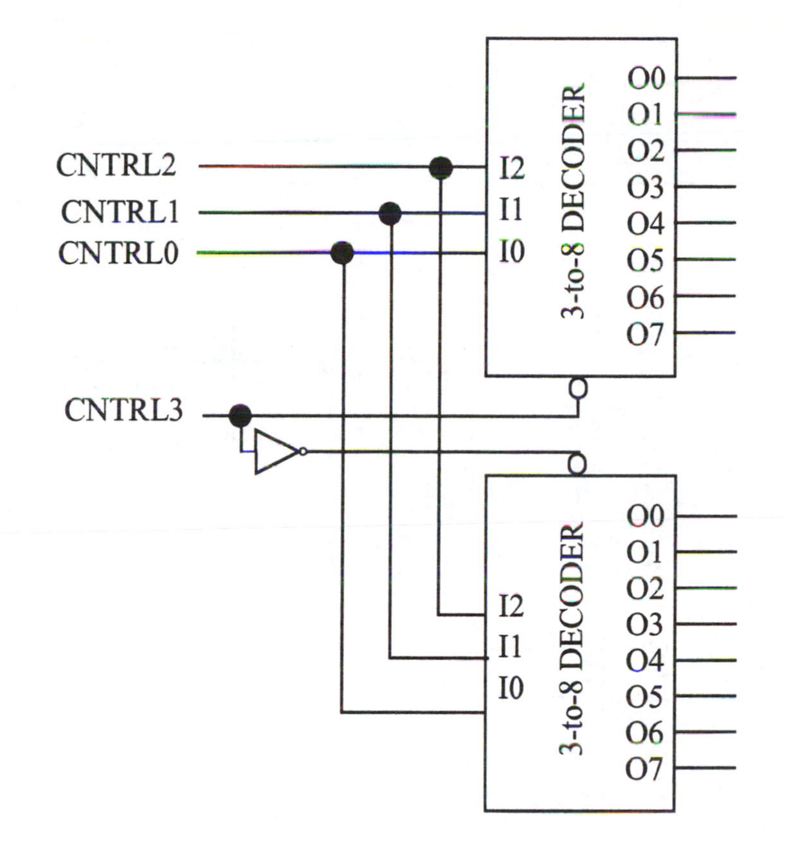

5-14. Solution.

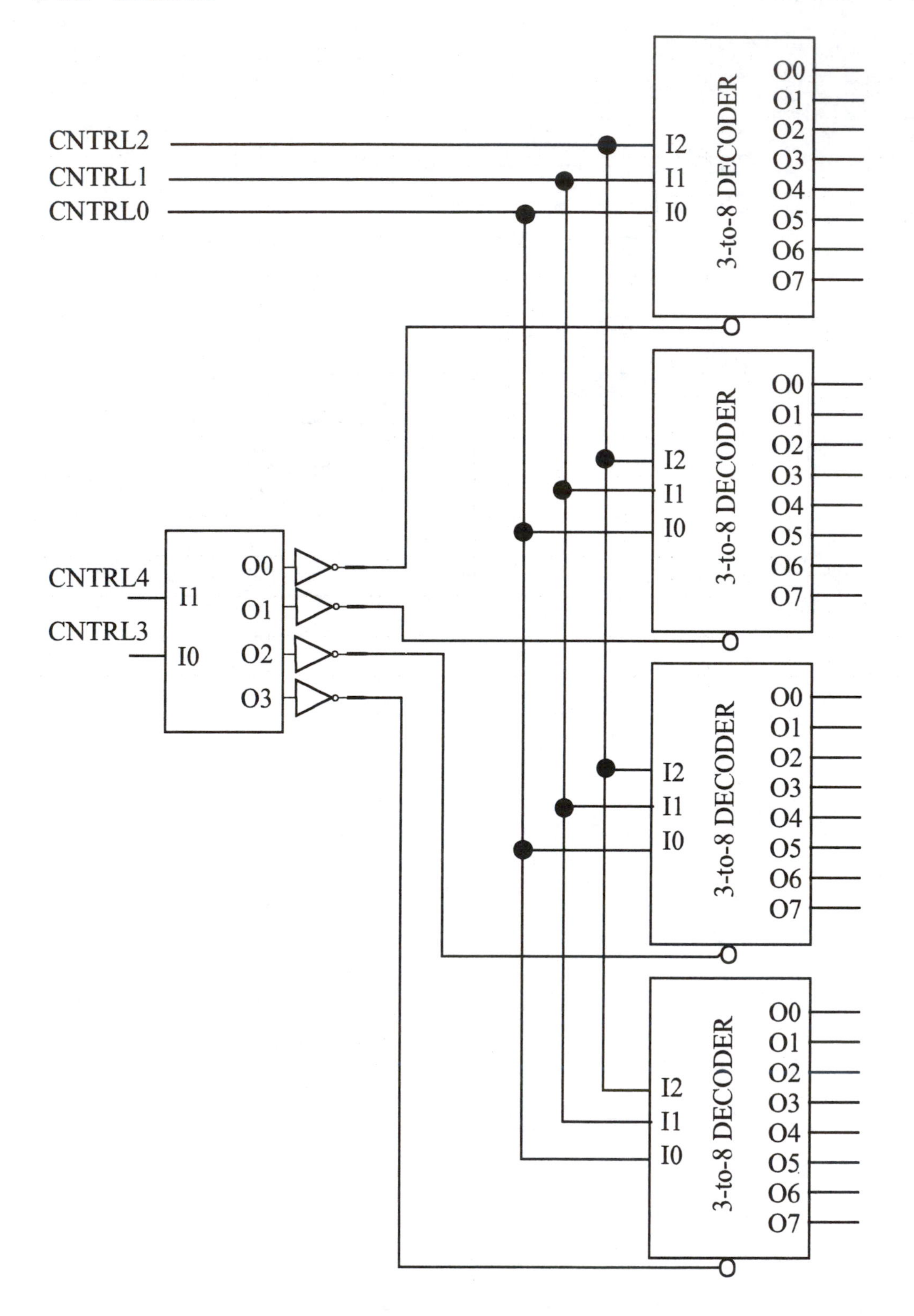

5-15. Solution.

A	B	C	D	C4	C3	C2	C1	C0
0	0	0	0	1	1	0	0	0
0	0	0	1	0	0	0	1	1
0	0	1	0	0	0	1	0	1
0	0	1	1	0	0	1	1	0
0	1	0	0	0	1	0	0	1
0	1	0	1	0	1	0	1	0
0	1	1	0	0	1	1	0	0
0	1	1	1	1	0	0	0	1
1	0	0	0	1	0	0	1	0
1	0	0	1	1	0	1	0	0
1	0	1	0	-	-	-	-	-
1	0	1	1	-	-	-	-	-
1	1	0	0	-	-	-	-	-
1	1	0	1	-	-	-	-	-
1	1	1	0	-	-	-	-	-
1	1	1	1	-	-	-	-	-

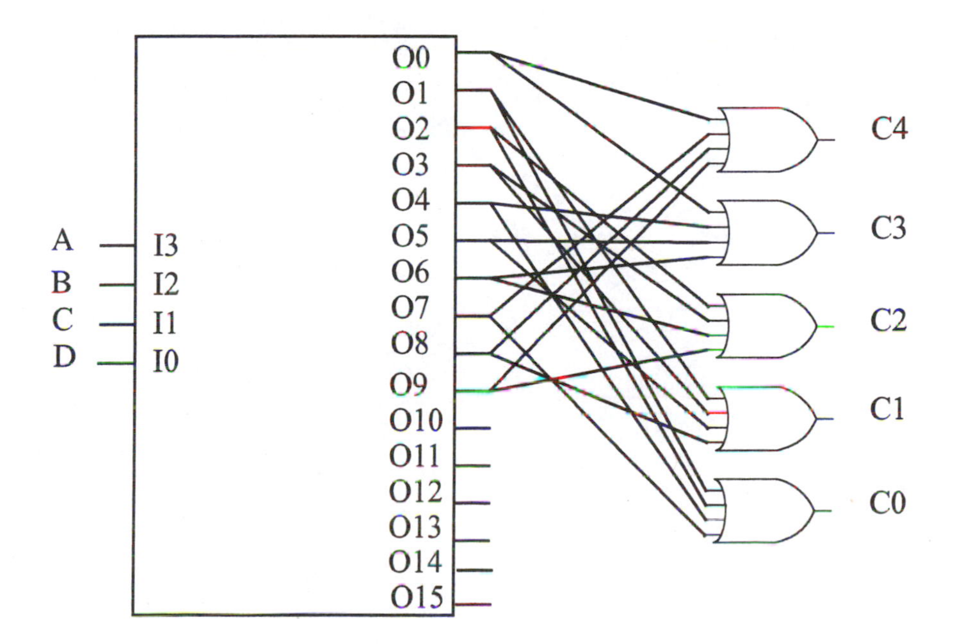

5-16. Solution.

a.

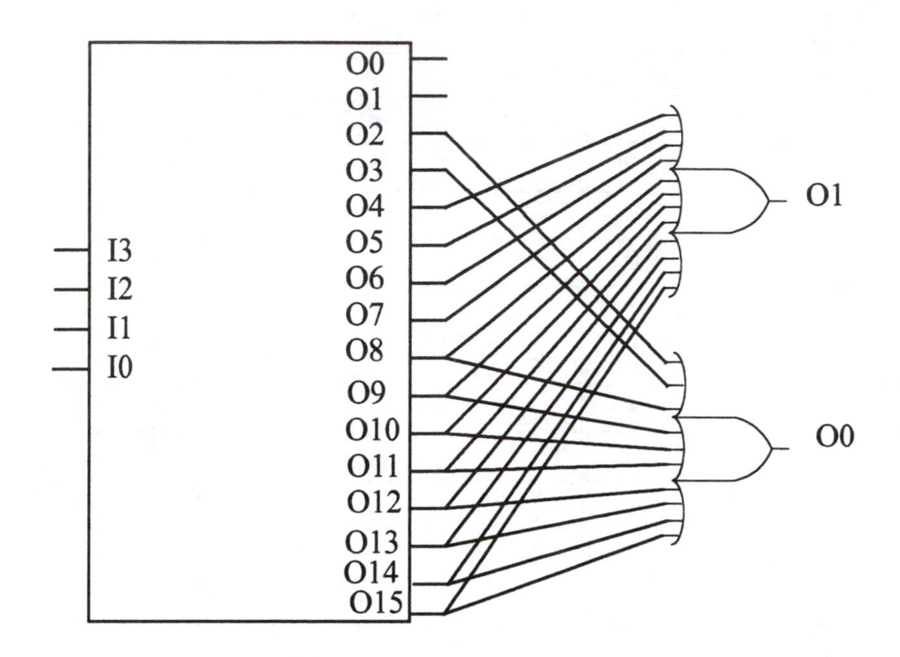

5-17. Solution.

a.

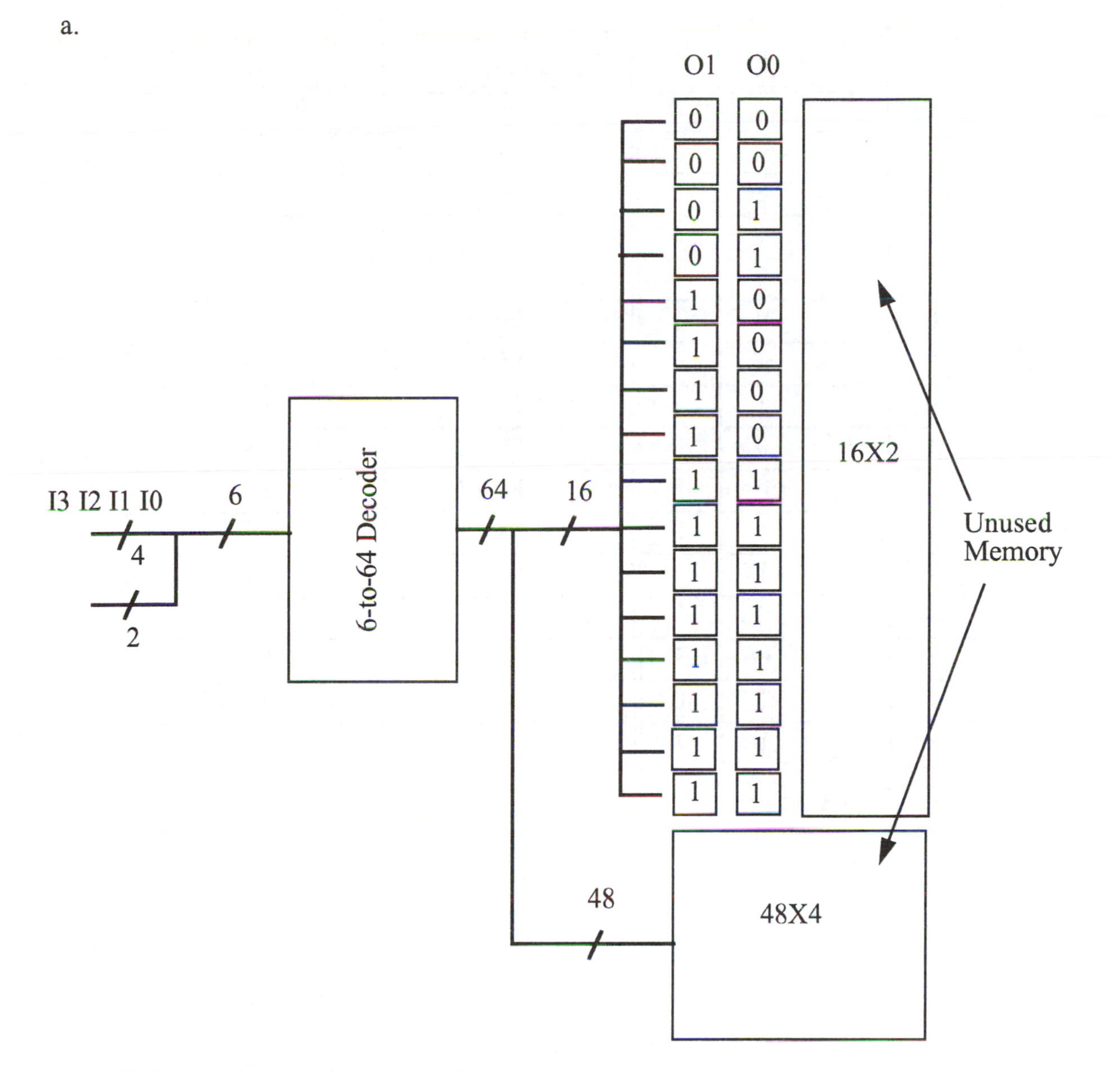

b. 16X2 and 4080X4 unused blocks yields16,352 unused storage locations out of 16,384 or 0.2% utilization.

5-18. Solution.

B_3	B_2	B_1	B_0	High-Order Digit Hexadecimal	Low-Order Digit Hexadecimal
0	0	0	0	F0	F0
0	0	0	1	F0	F1
0	0	1	0	F0	F2
0	0	1	1	F0	F3
0	1	0	0	F0	F4
0	1	0	1	F0	F5
0	1	1	0	F0	F6
0	1	1	1	F0	F7
1	0	0	0	F0	F8
1	0	0	1	F0	F9
1	0	1	0	F1	F0
1	0	1	1	F1	F1
1	1	0	0	F1	F2
1	1	0	1	F1	F3
1	1	1	0	F1	F4
1	1	1	1	F1	F5

8-bit memory locations are addressed 0000 to 1111, top to bottom.

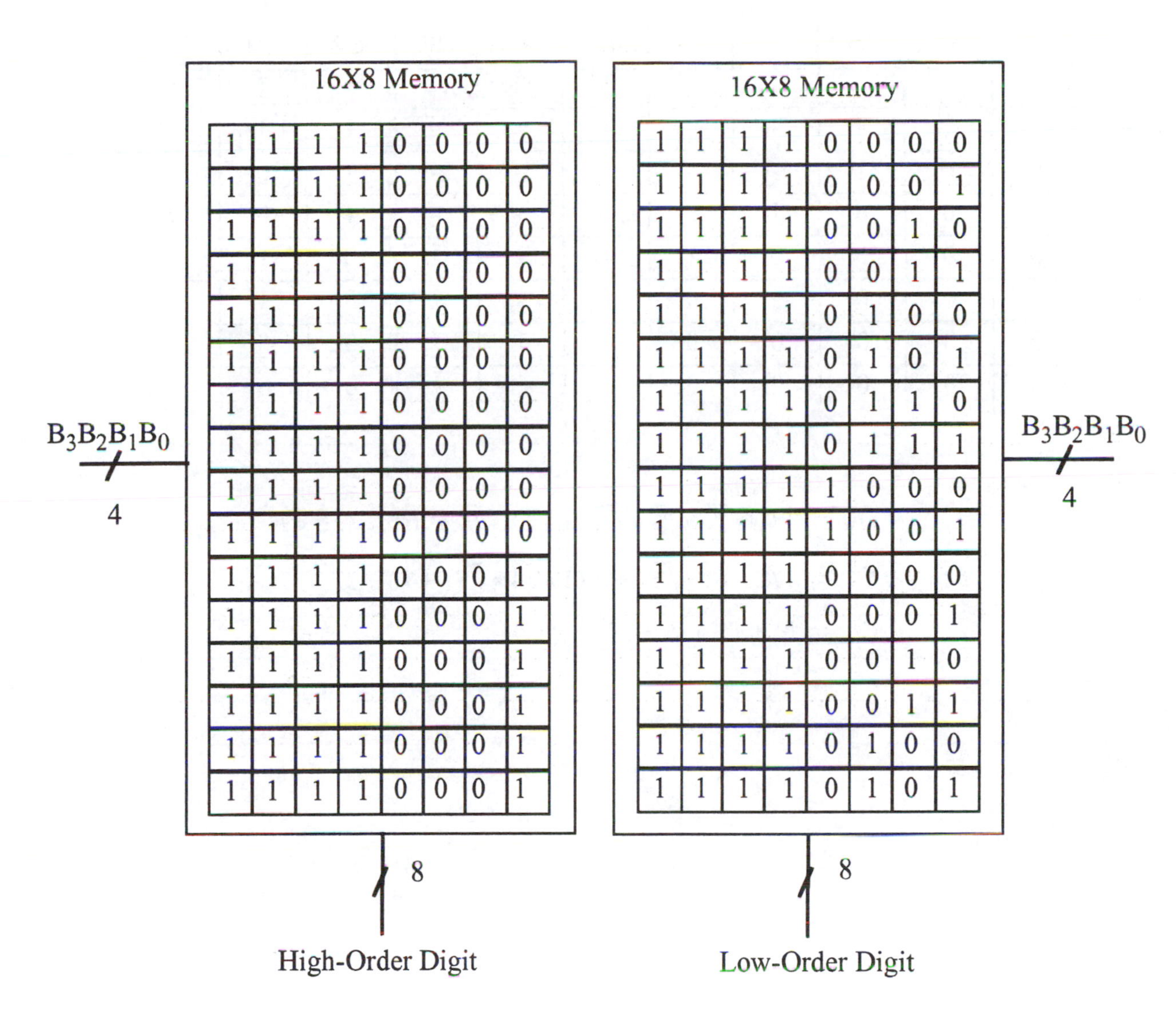

5-19. Solution.

I7	I6	I5	I4	I3	I2	I1	I0	O2	O1	O0
1	-	-	-	-	-	-	-	1	1	1
0	1	-	-	-	-	-	-	1	1	0
0	0	1	-	-	-	-	-	1	0	1
0	0	0	1	-	-	-	-	1	0	0
0	0	0	0	1	-	-	-	0	1	1
0	0	0	0	0	1	-	-	0	1	0
0	0	0	0	0	0	1	-	0	0	1
0	0	0	0	0	0	0	1	0	0	0

$$O0 = (I1 \bullet \overline{I2} \bullet \overline{I4} \bullet \overline{I6}) + (I3 \bullet \overline{I4} \bullet \overline{I6}) + (I5 \bullet \overline{I6}) + I7$$

$$O1 = (I2 \bullet \overline{I4} \bullet \overline{I5}) + (I3 \bullet \overline{I4} \bullet \overline{I5}) + I6 + I7$$

$$O2 = I4 + I5 + I6 + I7$$

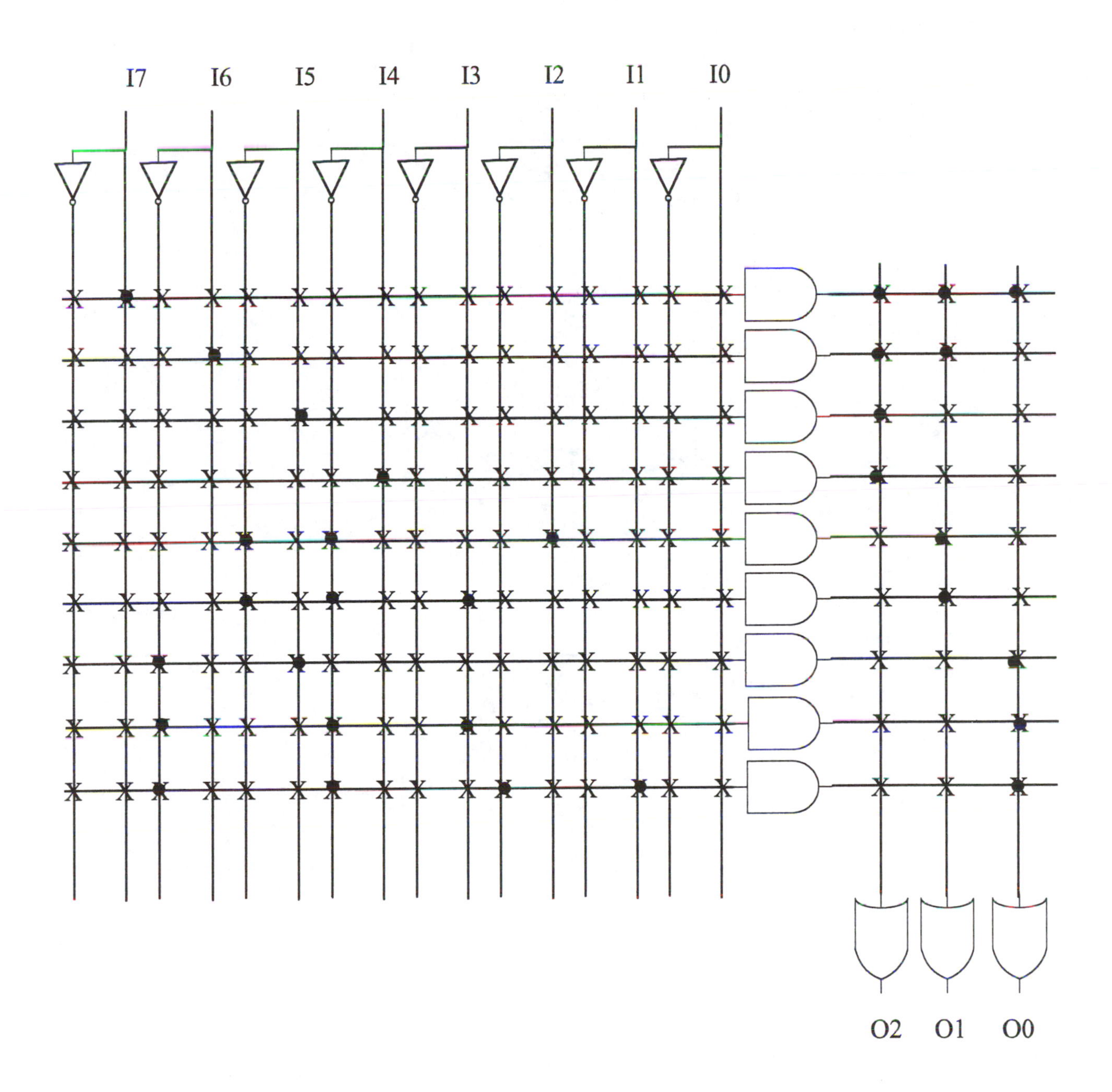

5-20. Solution.

$$SUM = ((A \oplus B) \bullet \overline{CIN}) + (\overline{(A \oplus B)} \bullet CIN)$$

$$CARRY = ((A \oplus B) \bullet CIN) + (\overline{(A \oplus B)} \bullet A)$$

5-21. Solution.

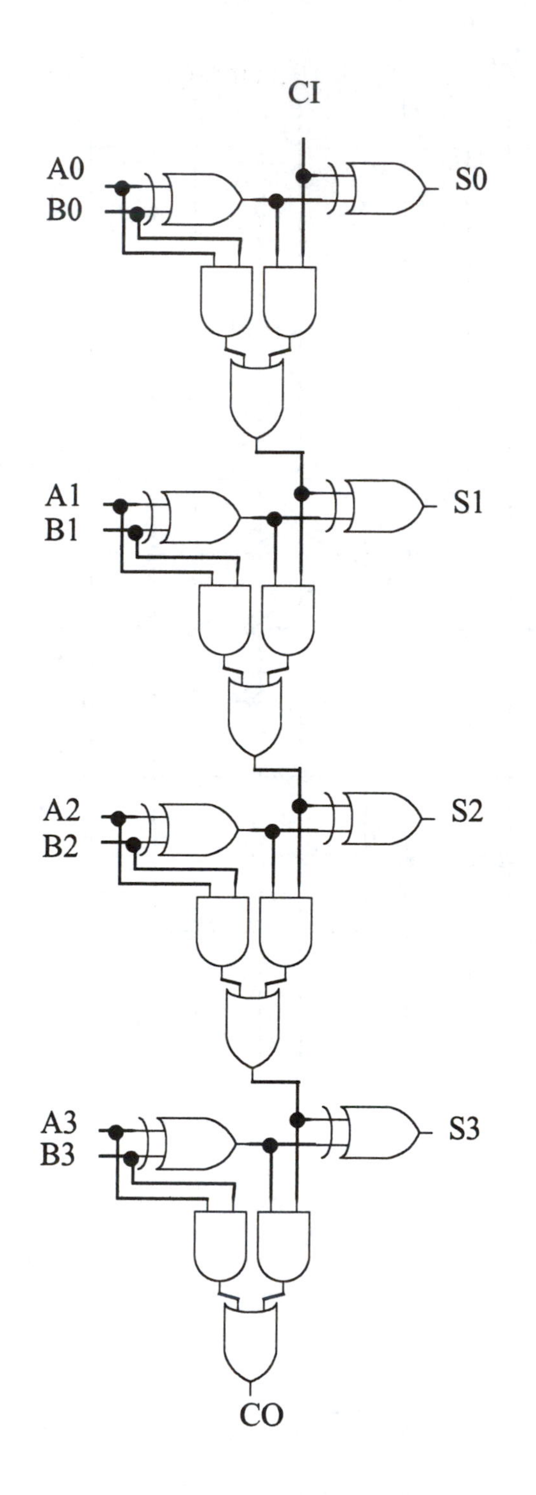

5-22. Solution.

a. Borrow-In: BI; MINUEND: M; SUBTRAHEND: S; DIFFERENCE: D; and Borrow-Out: BO.

$$D = M \oplus S \oplus BI$$

$$BO = (BI \bullet \overline{M}) + (BI \bullet S) + (\overline{M} \bullet S)$$

b.

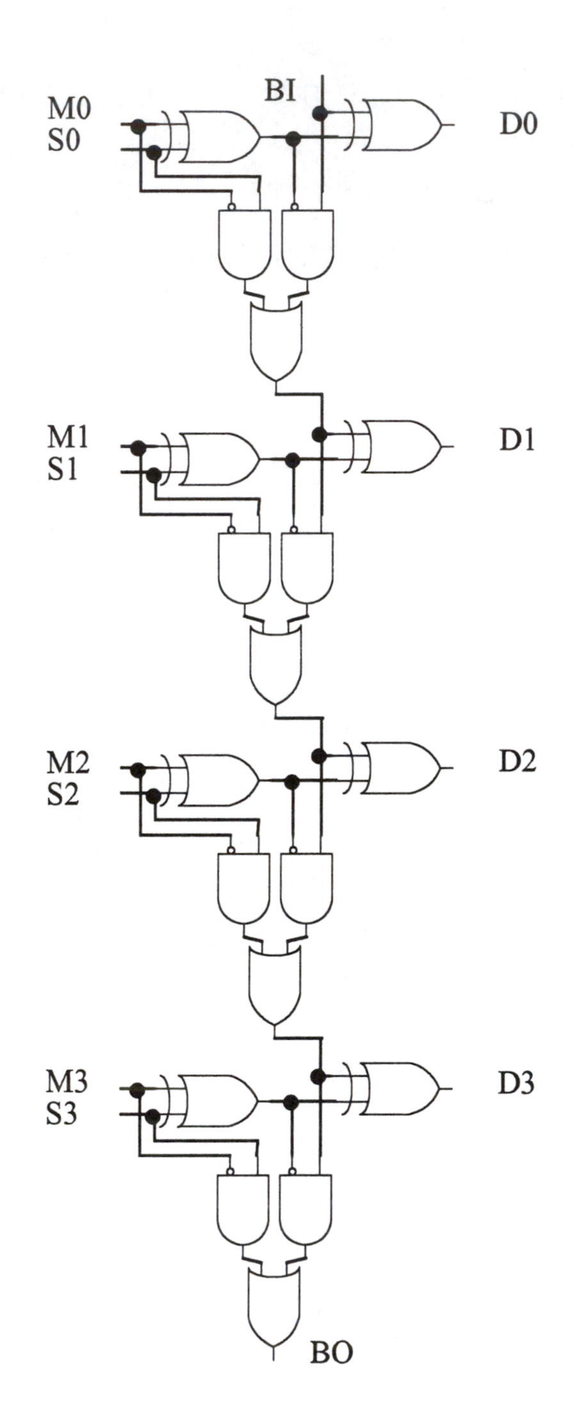

c.

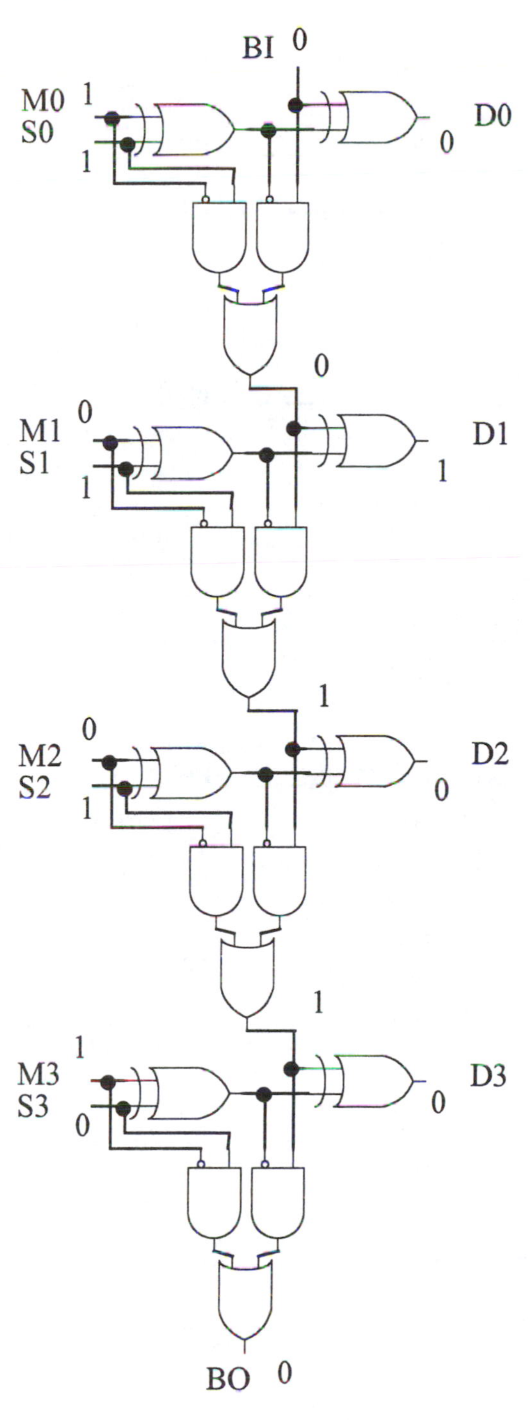

$9_{10} - 7_{10}$

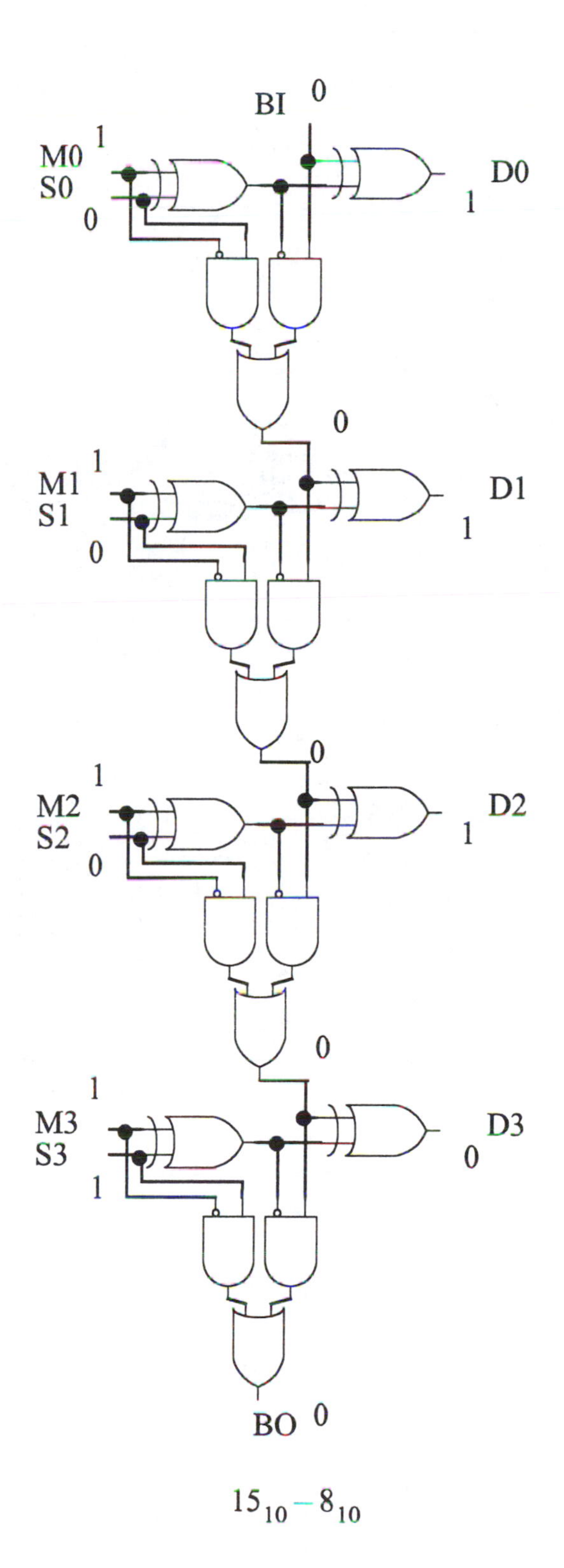

$15_{10} - 8_{10}$

5-23. Solution.

a.

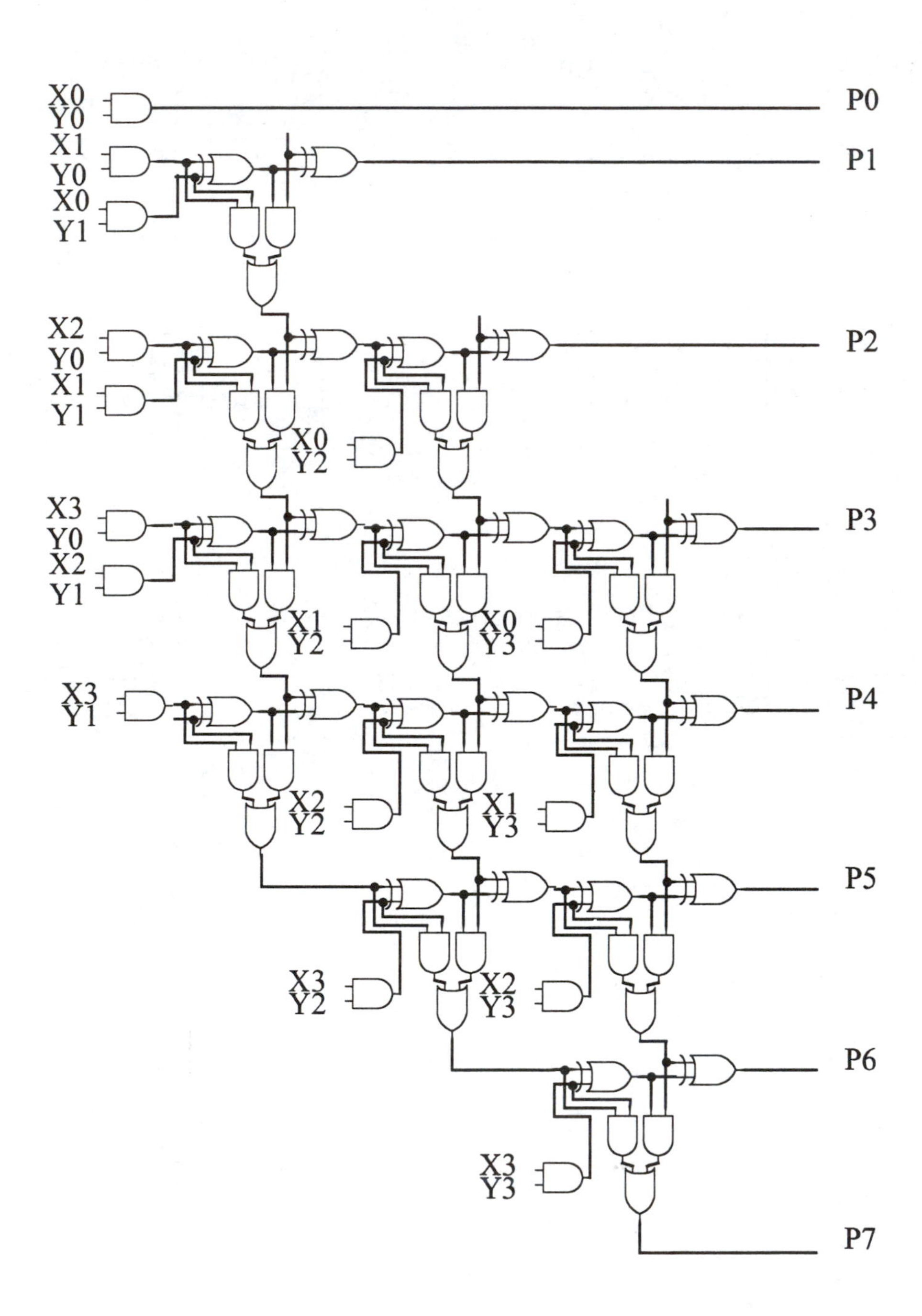

b.

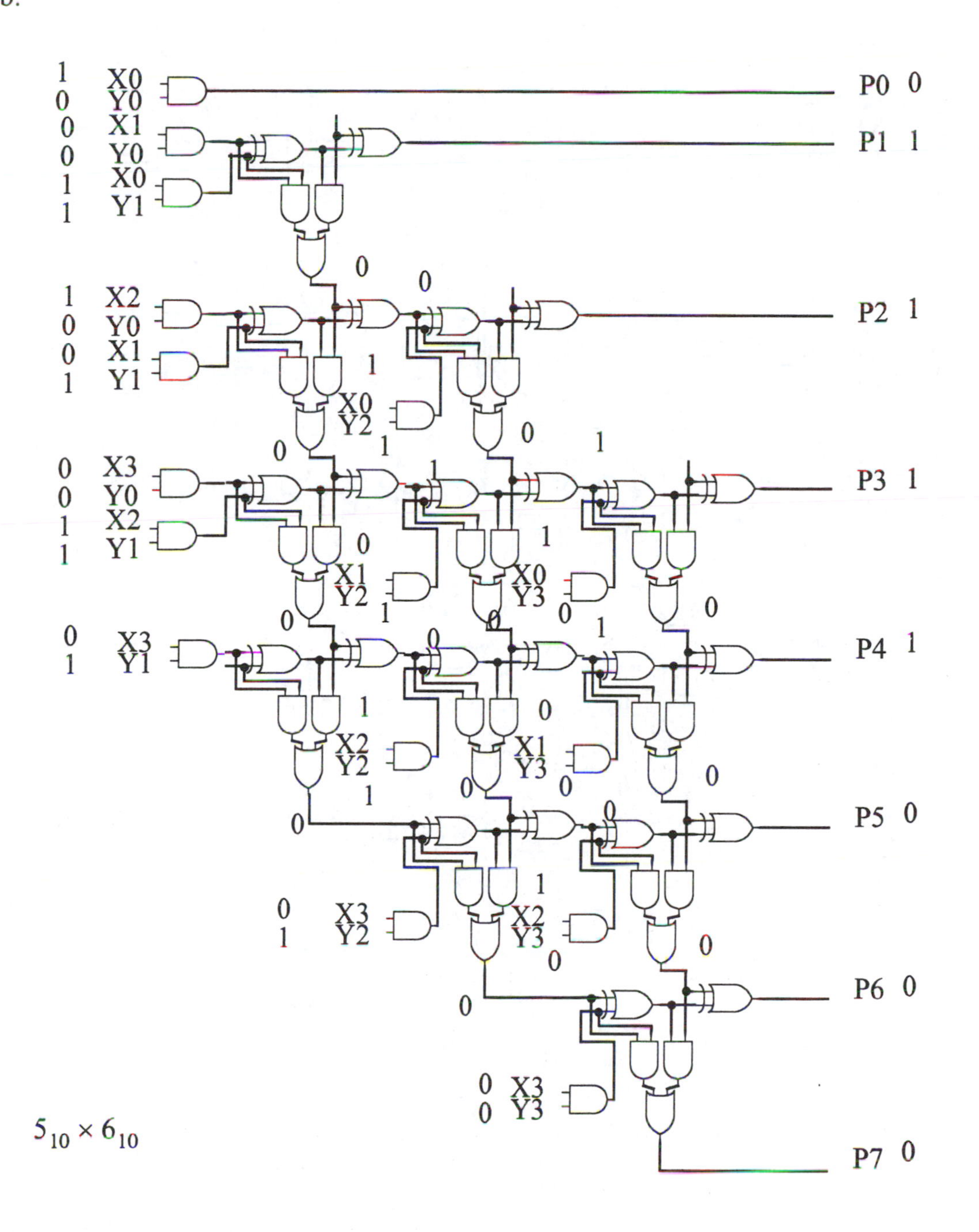
1 X0
0 Y0
0 X1
0 Y0
1 X0
1 Y1
P0 0
P1 1
1 X2
0 Y0
0 X1
1 Y1
X0
Y2
P2 1
0 X3
0 Y0
1 X2
1 Y1
X1
Y2
X0
Y3
P3 1
0 X3
1 Y1
X2
Y2
X1
Y3
P4 1
P5 0
0 X3
1 Y2
X2
Y3
P6 0
0 X3
0 Y3
P7 0
$5_{10} \times 6_{10}$

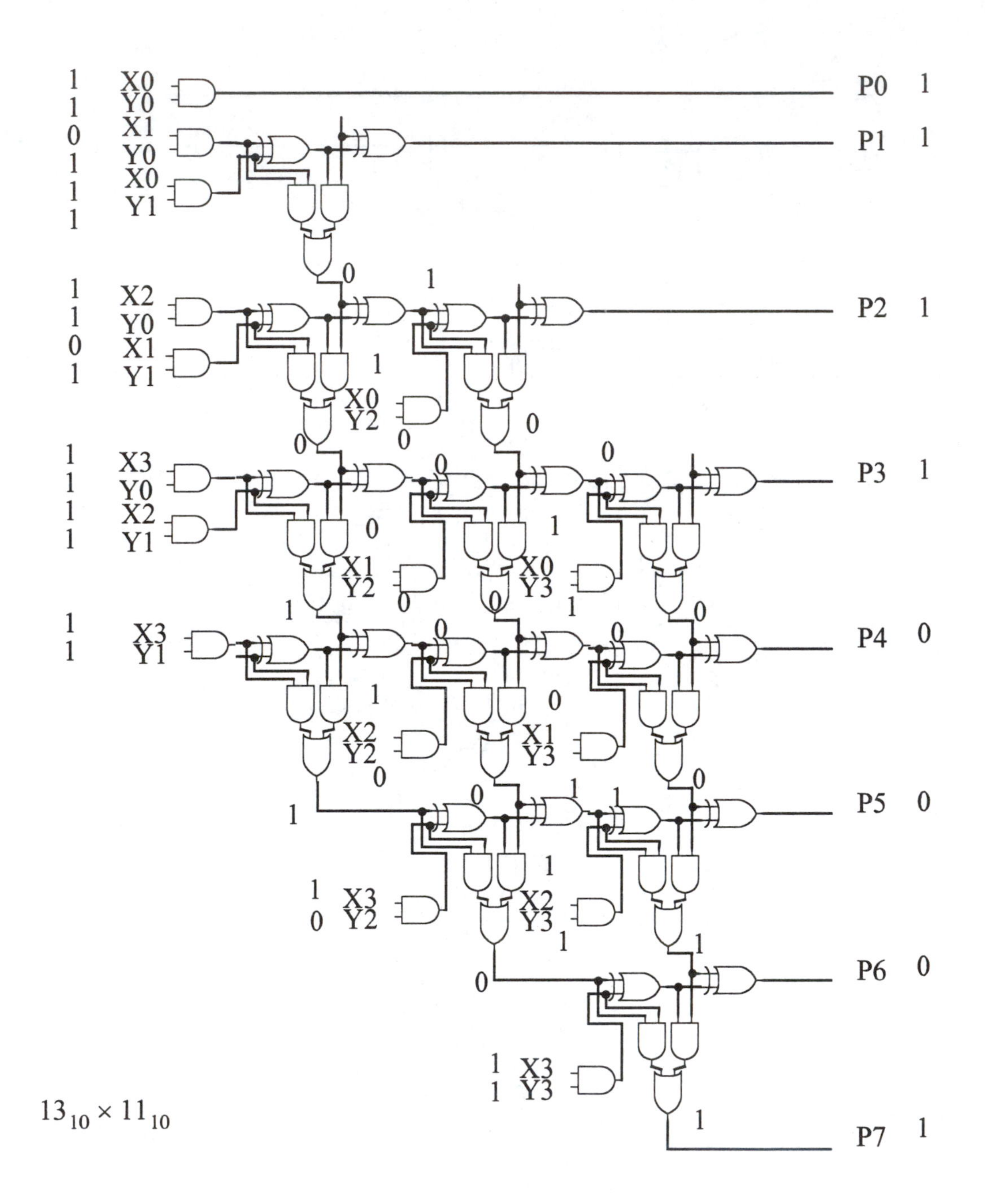

$13_{10} \times 11_{10}$

Chapter 6. Logic Families

6-1. Solution.

a.

A	B	Z
0	0	0
0	1	1
1	0	1
1	1	0

$Z = A \oplus B$

b.

A	B	Z
1	1	1
1	0	0
0	1	0
0	0	1

$Z = \overline{A \oplus B}$

6-2. Solution.

a.

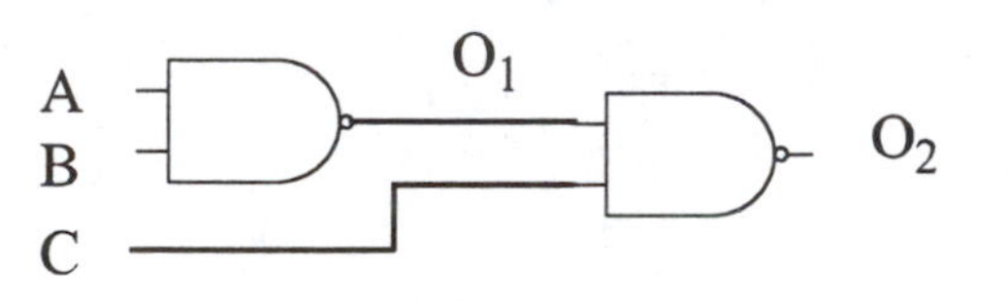

b. O1 should be L and O2 should be H.

c.

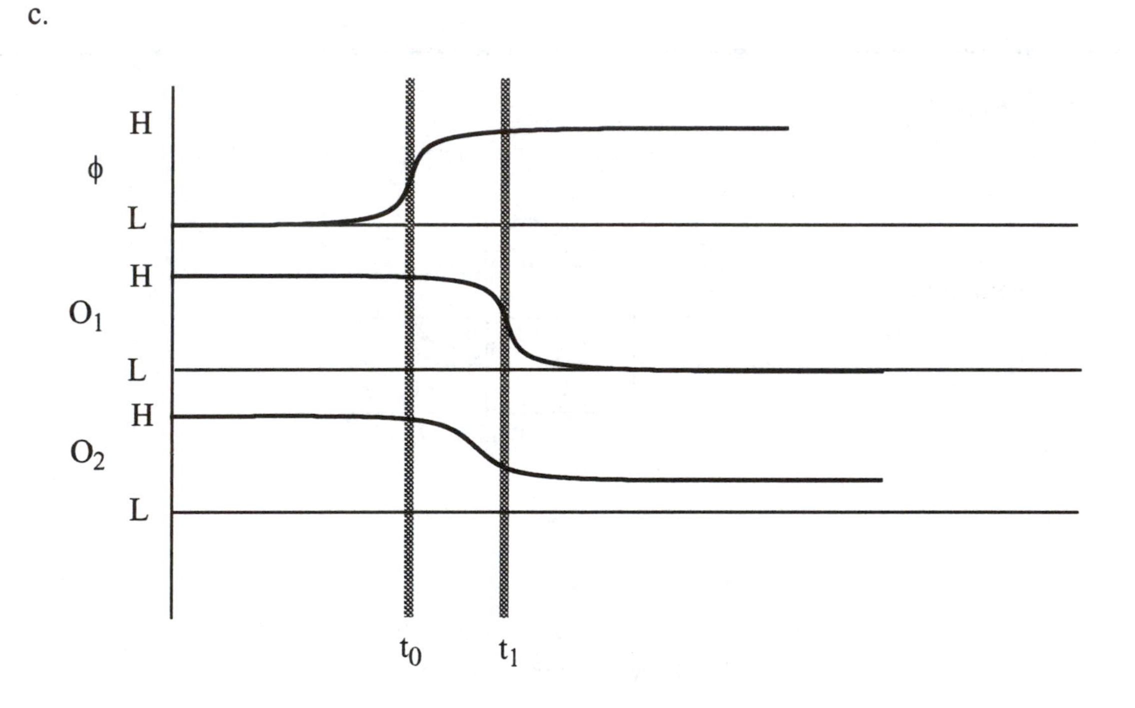

The low-to-high transition on ϕ signifies the end of the precharge phase and the start of the evaluation phase. The transistors Q_2, Q_3, and Q_4 turn on to discharge O_1, but this does not happen instantaneously. Thus, O_1 remains high for a short period and Q_6, Q_7, and Q_8 begin to discharge incorrectly O_2. At time t_1, the transistors Q_2, Q_3, and Q_4 have finally discharged O_1 enough to turn off Q_7 and stop the discharge of O_2. However, the period t_1-t_0 may be long enough to discharge O_2 to a value no longer considered a logic 1.

6-3. Solution.

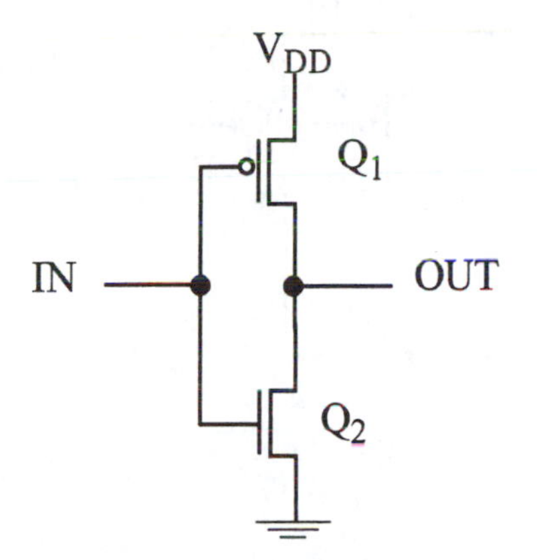

When IN is L, Q1 turns on and Q_2 turns off. Q_1 charges OUT to H. When IN is H, Q_1 turns off and Q_2 turns on. Q_2 discharges OUT to L.

6-4. Solution.

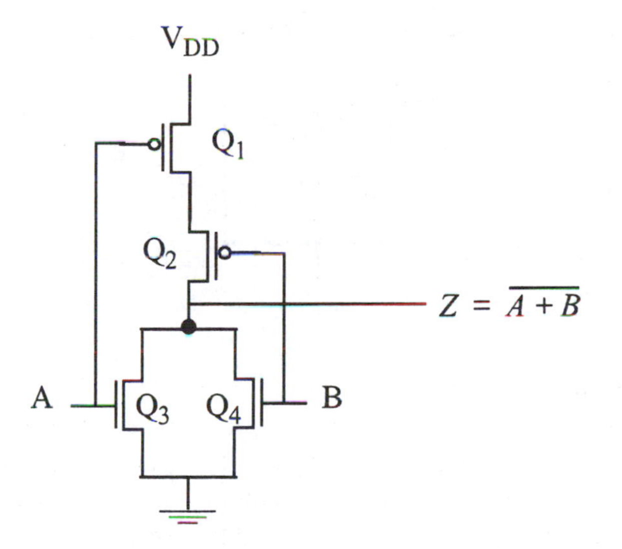

A=L and B=L: Q_3 and Q_4 are turned off. Q_1 and Q_2 turn on and charge Z to H.

A=L and B=H: Q_2 and Q_3 are turned off. Q_1 and Q_4 turn on and Q_4 discharges Z to L.

A=H and B=L: Q_1 and Q_4 are turned off. Q_2 and Q_3 turn on and Q_3 discharges Z to L.

A=H and B=H: Q_1 and Q_2 are turned off. Q_3 and Q_4 turn on and discharge Z to L.

6-5. **Solution.**

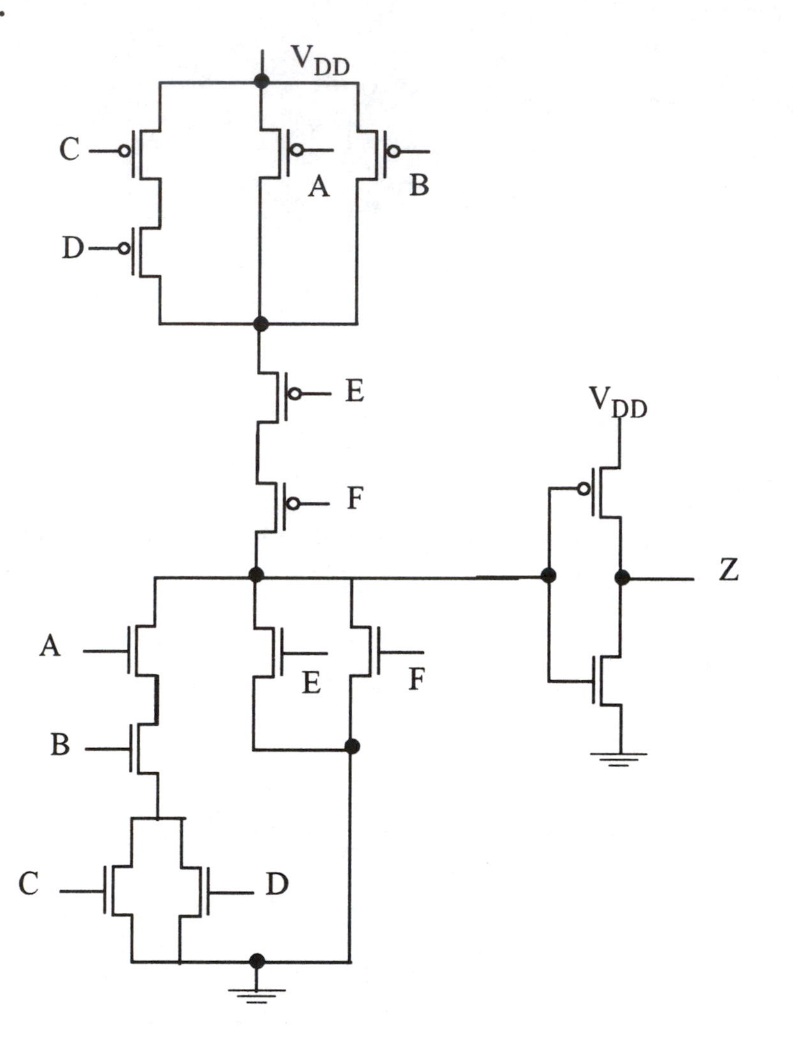

6-6. **Solution.** $Z = A \oplus B$

6-7. **Solution.**

a.

A	B	X
L	L	H
L	H	L
H	L	L
H	H	L

$X = \overline{A + B}$ with positive logic convention and $X = \overline{A \bullet B}$ with negative logic convention.

A=L and B=L: Q_1 and Q_2 are on which turn Q_3 and Q_4 off. Q_6 turns off and Q_5 turns on to charge X to H.

A=L and B=H: Q_1 is on which turns Q_3 off. Q_2 is off which turns Q_4 on. Q_5 turns off and Q_6 turns on to discharge X to L.

A=H and B=L: Q_2 is on which turns Q_4 off. Q_1 is off which turns Q_3 on. Q_5 turns off and Q_6 turns on to discharge X to L

A=H and B=H: Q_1 is off which turns Q_3 on. Q_2 is off which turns Q_4 on. Q_5 turns off and Q_6 turns on to discharge X to L

b.

A	B	C	D	X
L	L	L	L	H
L	L	L	H	H
L	L	H	L	H
L	L	H	H	L
L	H	L	L	H
L	H	L	H	H
L	H	H	L	H
L	H	H	H	L
H	L	L	L	H
H	L	L	H	H
H	L	H	L	H
H	L	H	H	L
H	H	L	L	L
H	H	L	H	L
H	H	H	L	L
H	H	H	H	L

$X = \overline{(A \bullet B) + (C \bullet D)}$ with positive logic convention and $X = \overline{(A + B) \bullet (C + D)}$ with negative logic convention.

A=L, B=L, C=L and D=L: Q_1 and Q_2 are on which turns Q_3 and Q_4 off. Q_6 turns off and Q_5 turns on to charge X to H.

A=L, B=L, C=L and D=H: Q_1 and Q_2 are on which turns Q_3 and Q_4 off. Q_6 turns off and Q_5 turns on to charge X to H.

A=L, B=L, C=H and D=L: Q_1 and Q_2 are on which turns Q_3 and Q_4 off. Q_6 turns off and Q_5 turns on to charge X to H.

A=L, B=L, C=H and D=H: Q_1 is on which turns Q_3 off. Q_2 is off which turns Q_4 on. Q_5 turns off and Q_6 turns on to discharge X to L.

A=L, B=H, C=L and D=L: Q_1 and Q_2 are on which turns Q_3 and Q_4 off. Q_6 turns off and Q_5 turns on to charge X to H.

A=L, B=H, C=L and D=H: Q_1 and Q_2 are on which turns Q_3 and Q_4 off. Q_6 turns off and Q_5 turns on to charge X to H.

A=L, B=H, C=H and D=L: Q_1 and Q_2 are on which turns Q_3 and Q_4 off. Q_6 turns off and Q_5 turns on to charge X to H.

A=L, B=H, C=H and D=H: Q_1 is on which turns Q_3 off. Q_2 is off which turns Q_4 on. Q_5 turns off and Q_6 turns on to discharge X to L.

A=H, B=L, C=L and D=L: Q_1 and Q_2 are on which turns Q_3 and Q_4 off. Q_6 turns off and Q_5 turns on to charge X to H.

A=H, B=L, C=L and D=H: Q_1 and Q_2 are on which turns Q_3 and Q_4 off. Q_6 turns off and Q_5 turns on to charge X to H.

A=H, B=L, C=H and D=L: Q_1 and Q_2 are on which turns Q_3 and Q_4 off. Q_6 turns off and Q_5 turns on to charge X to H.

A=H, B=L, C=H and D=H: Q_1 is on which turns Q_3 off. Q_2 is off which turns Q_4 on. Q_5 turns off and Q_6 turns on to discharge X to L.

A=H, B=H, C=L and D=L: Q_2 is on which turns Q_4 off. Q_1 is off which turns Q_3 on. Q_5 turns off and Q_6 turns on to discharge X to L.

A=H, B=H, C=L and D=H: Q_2 is on which turns Q_4 off. Q_1 is off which turns Q_3 on. Q_5 turns off and Q_6 turns on to discharge X to L.

A=H, B=H, C=H and D=L: Q_2 is on which turns Q_4 off. Q_1 is off which turns Q_3 on. Q_5 turns off and Q_6 turns on to discharge X to L.

A=H, B=H, C=H and D=H: Q_1 and Q2 are off which turns Q_3 and Q_4 on. Q_5 turns off and Q_6 turns on to discharge X to L.

6-8. Solution.

a.

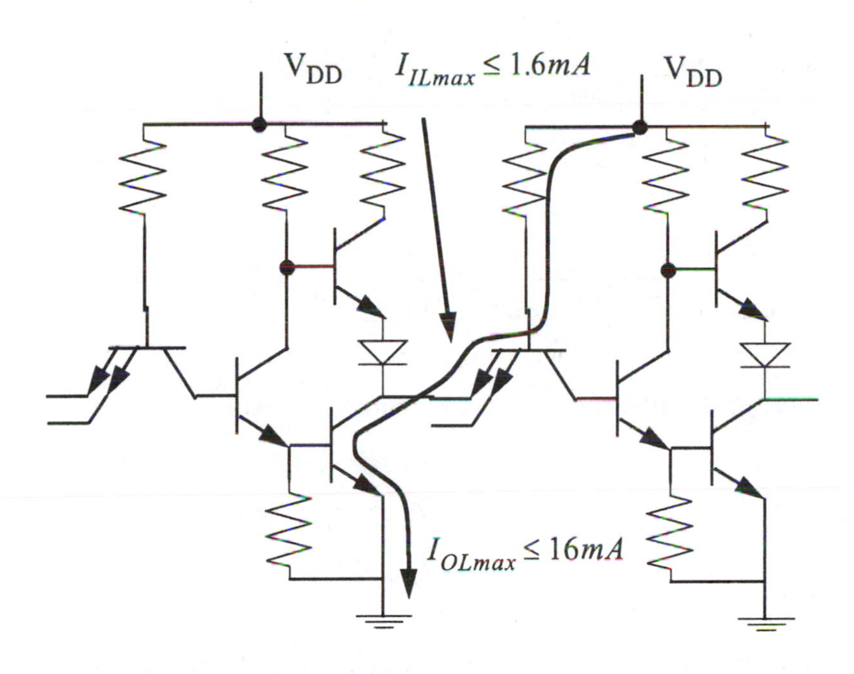

b.

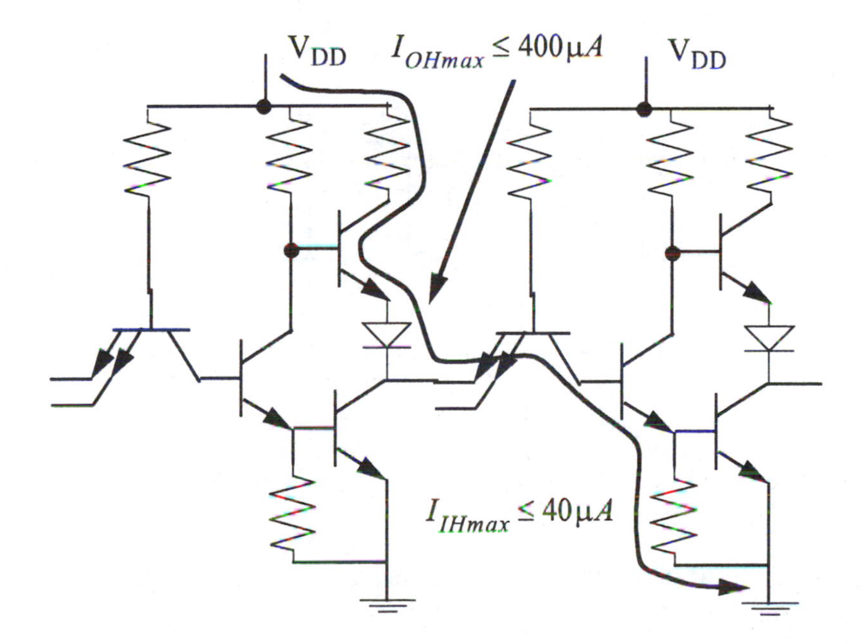

6-9. Solution.

a.

$$R \geq \frac{V_{DD} - V_{OLmax}}{I_{OLmax} - (2 \cdot I_{ILmax})} = \frac{5 - 0.4}{16mA - (2 \cdot 1.6mA)} = 359\Omega$$

b.

$$R \leq \frac{V_{DD} - V_{OHmin}}{(2 \cdot I_{OHmax}) + (2 \cdot I_{IHmax})} = \frac{5 - 2.5}{(2 \cdot 250\mu A) + (2 \cdot 40\mu A)} = 4310\Omega$$

c. Larger values of R allow more load TTL gates because

$$N = \frac{I_{OLmax} - \frac{V_{DD} - V_{OLmax}}{R}}{I_{ILmax}}$$

Smaller values of R allow higher currents to charge and discharge nodes which increases switching speed.

6-10. Solution.

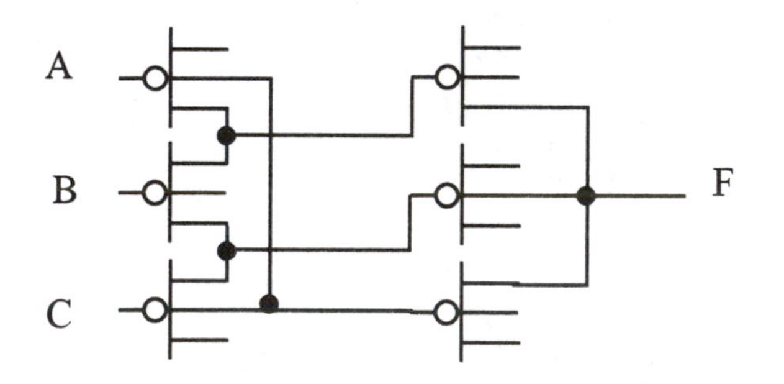

6-11. Solution.

The BiCMOS circuit performs a logic complement operation. When INPUT=H, Q1 turns off, which turns off base current to Q_2 and turns Q_2 off. Also, Q_3 turns on, which turns on base current to Q_4. Q_4 turns on and discharges the OUTPUT to a L. When INPUT=L, Q_3 turns off, which turns off base current to Q_4 and turns Q_4 off. Also, Q_1 turns on, which turns on base current to Q_2. Q_2 turns on and charges the OUTPUT to a H.

6-12. Solution.

TTL driving CMOS

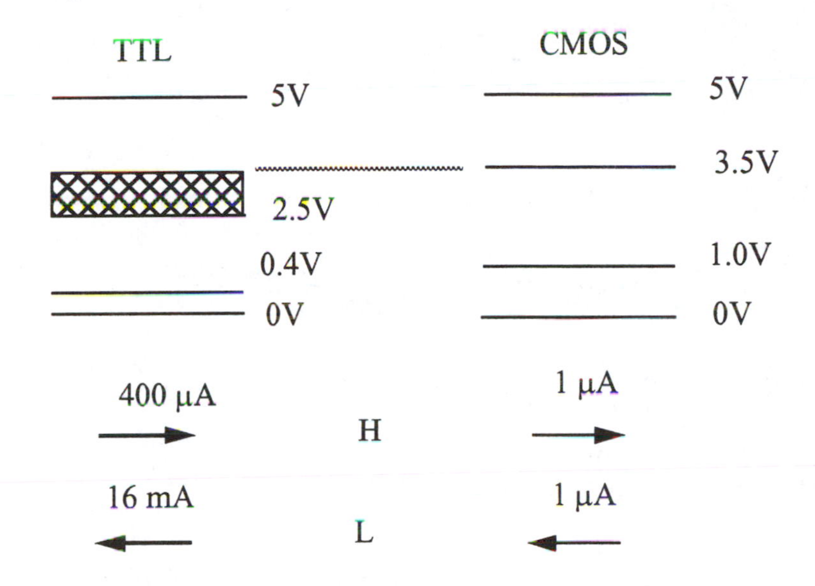

No. Fanin and fanout currents are compatible, but the voltage levels are not compatible. The TTL output can yield a H output that will not be recognized as a H CMOS input because $NM_{HIGH} = V_{OHmin}^{TTL} - V_{IHmin}^{CMOS} = 2.5 - 3.5 = -1.0V$.

CMOS driving TTL

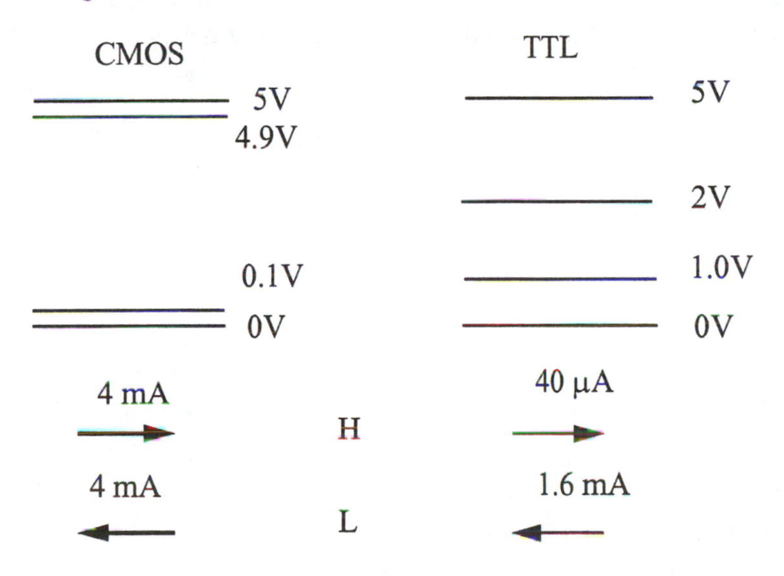

Yes. Fanin and fanout currents are compatible and voltage levels are compatible. However, the fanin/fanout currents limit the fanout of CMOS driving TTL to two loads.

6-13. Solution.

$$NM_{LOW}^{TTL} = V_{ILmax}^{TTL} - V_{OLmax}^{TTL} = 0.8V - 0.4V = 0.4V$$

$$NM_{HIGH}^{TTL} = V_{OHmin}^{TTL} - V_{IHmin}^{TTL} = 2.5V - 2V = 0.5V$$

$$NM_{LOW}^{STTL} = V_{ILmax}^{STTL} - V_{OLmax}^{STTL} = 0.8V - 0.5V = 0.3V$$

$$NM_{HIGH}^{STTL} = V_{OHmin}^{STTL} - V_{IHmin}^{STTL} = 2.7V - 2V = 0.7V$$

$$NM_{LOW}^{CMOS} = V_{ILmax}^{CMOS} - V_{OLmax}^{CMOS} = 1.0V - 0.1V = 0.9V$$

$$NM_{HIGH}^{CMOS} = V_{OHmin}^{CMOS} - V_{IHmin}^{CMOS} = 4.9V - 3.5V = 1.4V$$

$$NM_{LOW}^{ECL} = V_{ILmax}^{ECL} - V_{OLmax}^{ECL} = -1.48V - -1.63V = 0.15V$$

$$NM_{HIGH}^{ECL} = V_{OHmin}^{ECL} - V_{IHmin}^{ECL} = -0.98V - -1.13V = 0.15V$$

Based on noise margins, the logic families rank best to worst: CMOS, STTL, TTL, ECL. The larger the noise margins, the more noise immunity, the better the logic family.

6-14. Solution.

- Higher voltage supply levels,
- Lower temperatures,
- Lower circuit resistances,
- Lower circuit capacitances,
- Nonsaturating transistors, and
- Small voltage swings.

6-15. Solution.

$$P_{DYNAMIC} = C \times V^2 \times F \times G = 15 \times 10^{-12} \times 5^2 \times 100 \times 10^3 \times 80K = 3W$$

$$P_{DYNAMIC} = C \times V^2 \times F \times G = 15 \times 10^{-12} \times 5^2 \times 500 \times 10^3 \times 80K = 15W$$

$$P_{DYNAMIC} = C \times V^2 \times F \times G = 15 \times 10^{-12} \times 5^2 \times 1 \times 10^6 \times 80K = 30W$$

6-16. Solution.

No. Like totem pole configured TTL, static CMOS having active pull-up and pull-down transistors will fight for control of the wired output node and create a low resistance path between power and ground.

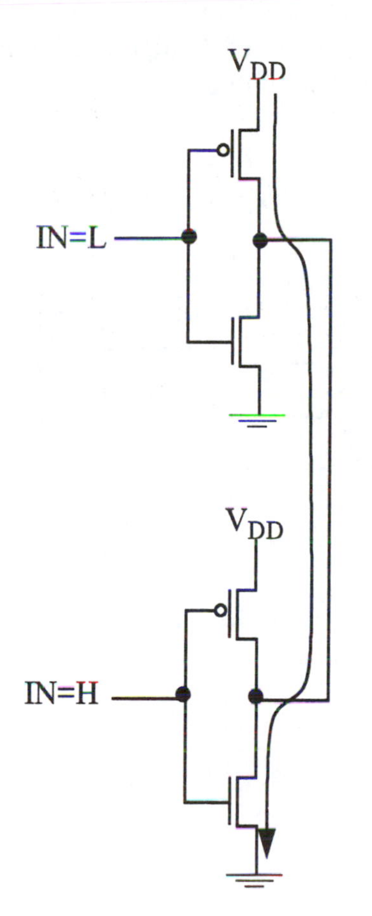

6-17. Solution.

Tri-state inverter with A being the active-1 enable control signal and B being the data signal.

A=L and B=L: Q_1 turns on, which turns off Q_2. Q_2 turns off Q_4 and the L on A turns off Q_3. Thus, both active pull-up and pull-down are turned off.

A=L and B=H: Q1 turns on, which turns off Q_2. Q_2 turns off Q_4 and the L on A turns off Q_3. Thus, both active pull-up and pull-down are turned off.

A=H and B=L: Q_1 turns on, which turns off Q_2. Q_2 turns off Q_4 and turns on Q_3, which charges Z to a H. The diode D_1 is reversed biased and blocks the ability of A to control Q_3.

A=H and B=H: Q1 turns off, which turns on Q_2. Q_2 turns off Q_3 and turns on Q_4, which discharges Z to a L. The diode D_1 is reversed biased and blocks the ability of A to control Q_3.

Bus arbitration. Provides means to realize wired-logic properties with the advantages of totem pole (active pull-up and push-down) TTL outputs.

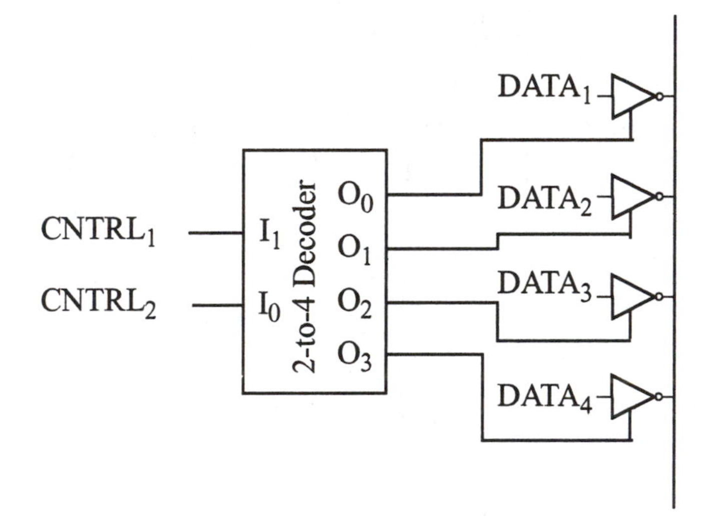

Chapter 7. Integrated Circuits

7-1. Solution.

a.

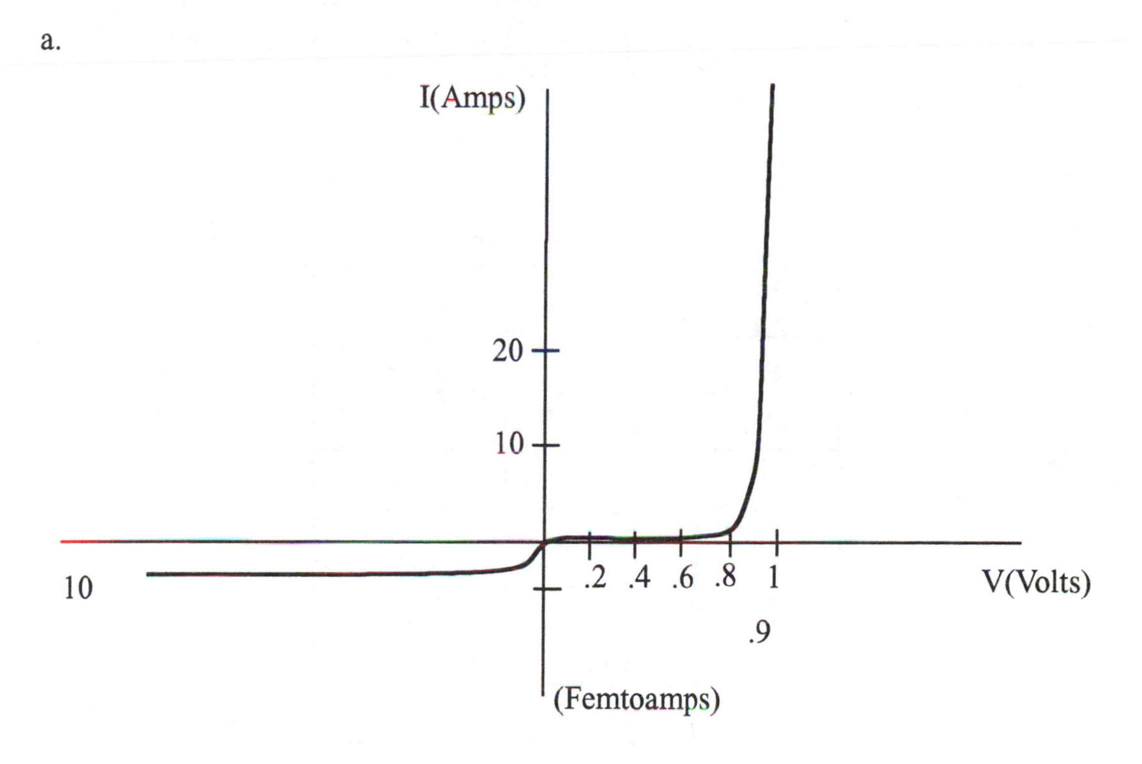

b. $V_t \approx 0.6V$

c. I_S is the diode current under reverse bias conditions.

7-2. Solution.

a.

$$^oF = \frac{9}{5} \cdot {^oC} + 32$$

$$\Delta^oC = \frac{5}{9} \cdot \Delta^oF = \frac{5}{9} \bullet 20 = 11.1$$

$$\Delta^oK = \Delta^oC$$

$$I_{AVG} = 1\times10^{-14}\left(e^{V\bullet 1.6\times10^{-19}/1.38\times10^{-23}\bullet 300} - 1\right)$$

$$I_{HOT} = 2\left(\frac{11.1}{10}\right)(1\times10^{-14})\left(e^{V\bullet 1.6\times10^{-19}/1.38\times10^{-23}\bullet 311.1} - 1\right)$$

V (Volts)	I (Amps)	
	$T = 80^oF$	$T = 100^oF$
-0.4	-1×10^{-14}	-2.2×10^{-14}
-0.2	-1×10^{-14}	-2.2×10^{-14}
-0	-1×10^{-14}	-2.2×10^{-14}
0.2	4.6×10^{-13}	1.0×10^{-12}
0.4	5.1×10^{-8}	1.1×10^{-7}
0.6	1.1×10^{-4}	2.5×10^{-4}
0.8	0.27	0.59
1.0	608.6	1338

b. It is important to consider heat dissipation to prevent increases in ambient temperatures, which can increase diode currents. Increased diode currents can, in turn, cause intermittent circuit failures due to voltage shifts and/or permanent circuit failures due to device damage.

7-3. Solution.

$$\left(\frac{1.38\times10^{-23}\bullet 300}{1.6\times10^{-19}}\right)\bullet ln(\frac{1\times10^{-3}}{1\times10^{-14}}+1) = 0.65V$$

7-4. Solution.

a.

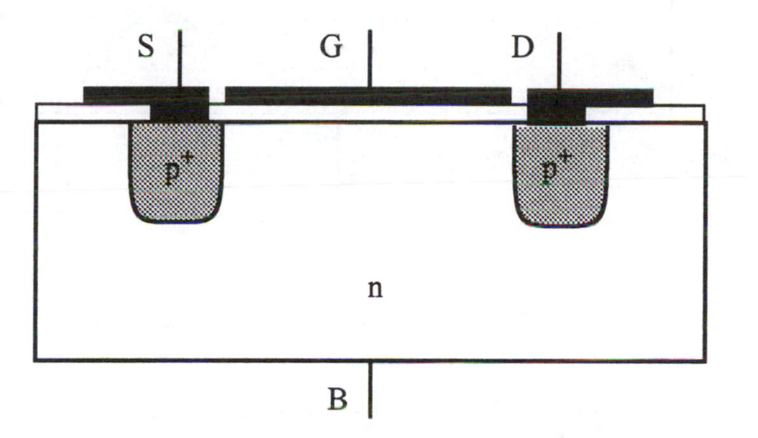

b.

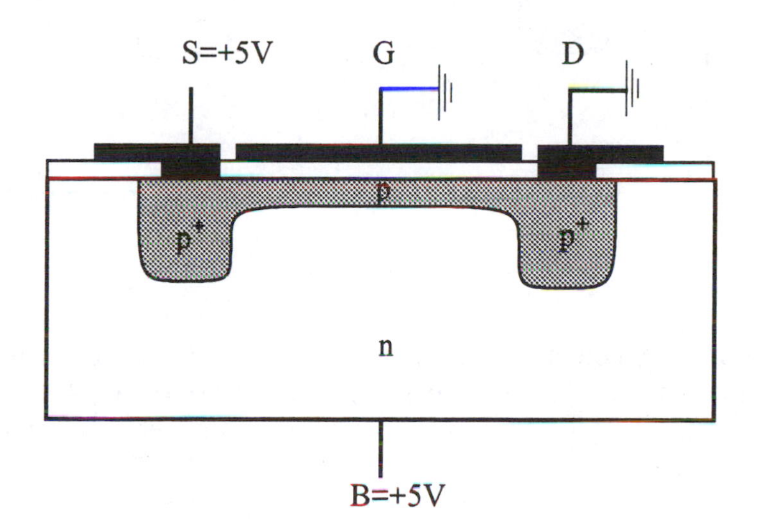

Applying a negative potential to the gate with respect to the source attracts positively-charged holes into the channel. The conductivity of the channel increases and the connection between the source and drain is created.

7-5. Solution.

a.

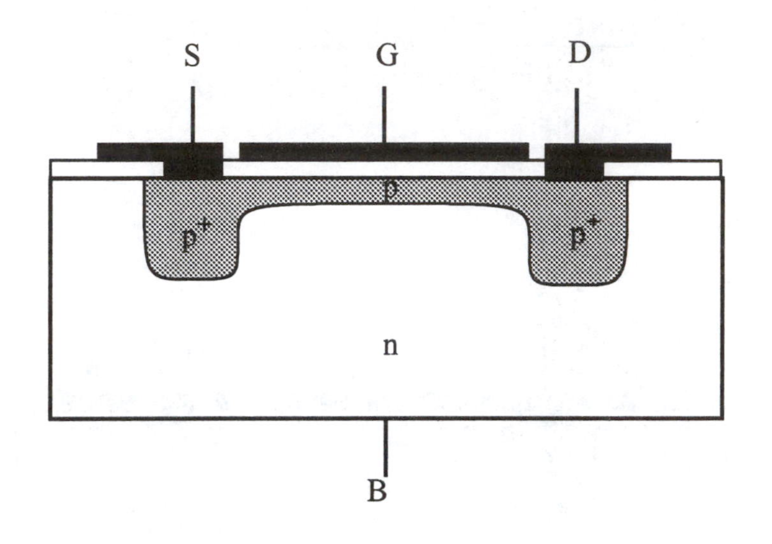

b.

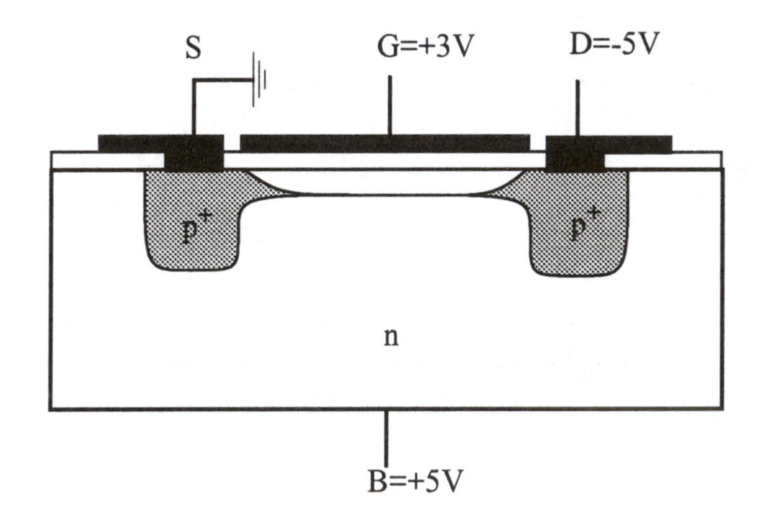

Applying a positive potential to the gate with respect to the source attracts negatively charged electrons into the channel, which depletes the channel of holes. The conductivity of the channel decreases and the connection between the source and drain is destroyed.

7-6. Solution.

a.

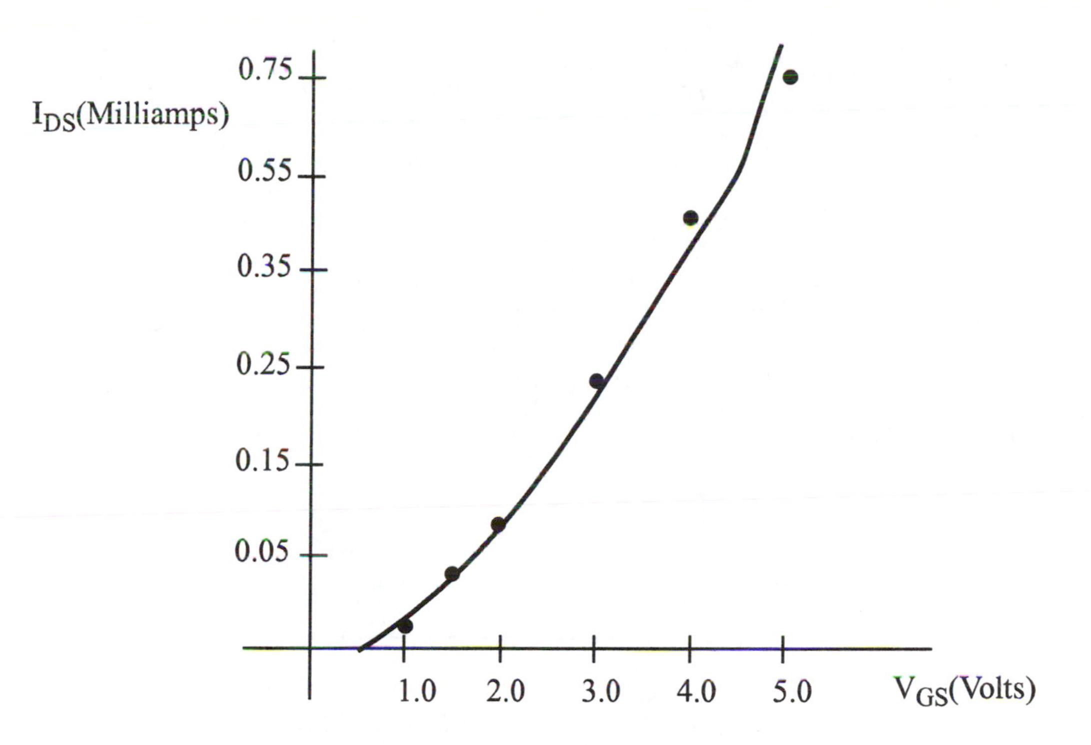

b.

- Reduce temperature to reduce electron scattering to increase mobility.
- Reduce lattice imperfections to reduce electron scattering to increase mobility.
- Increase dielectric constant of insulator between gate electrode and conducting channel.
- Decrease thickness of insulator between gate electrode and conducting channel.
- Increase width of transistor.
- Decrease length of transistor.
- Increase gate voltage.
- Decrease substrate doping to decrease threshold voltage.
- Decrease substrate bias to decrease threshold voltage.

7-7. Solution.

a.

$$L = \left(\frac{R}{R_s}\right) \bullet W = \left(\frac{300}{25}\right) \bullet 5\mu m = 60\mu m$$

b.

$$L = \left(\frac{R}{R_s}\right) \bullet W = \left(\frac{300}{50}\right) \bullet 5\mu m = 30\mu m$$

7-8. Solution.

a.

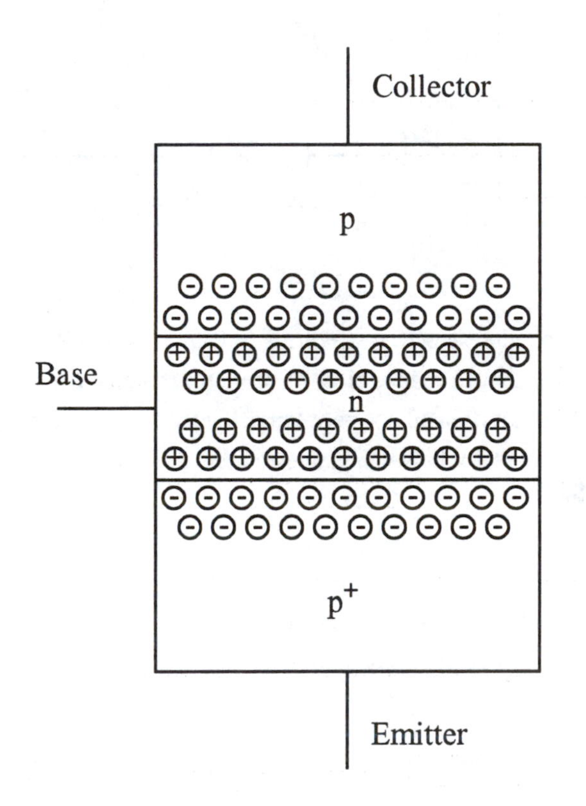

b.

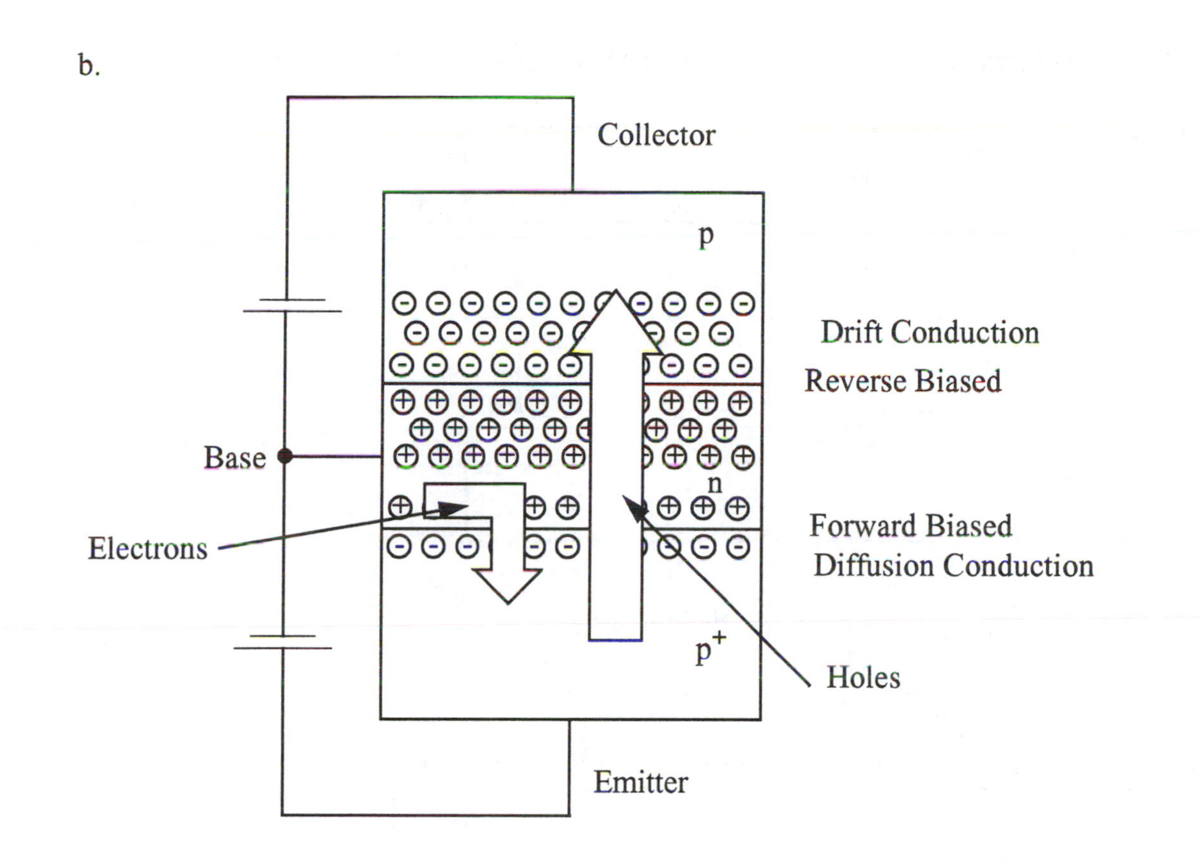

7-9. Solution.

Deposit layer of Si_3N_4, define area for p-well, grow field oxide, strip away remaining Si_3N_4 and create p-well.

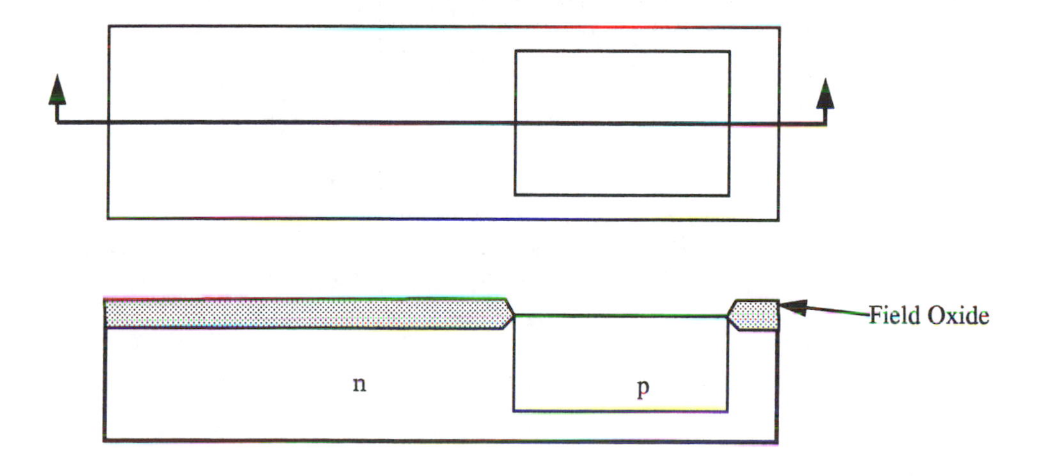

Strip field oxide away to define active PMOS transistor and grow gate oxide.

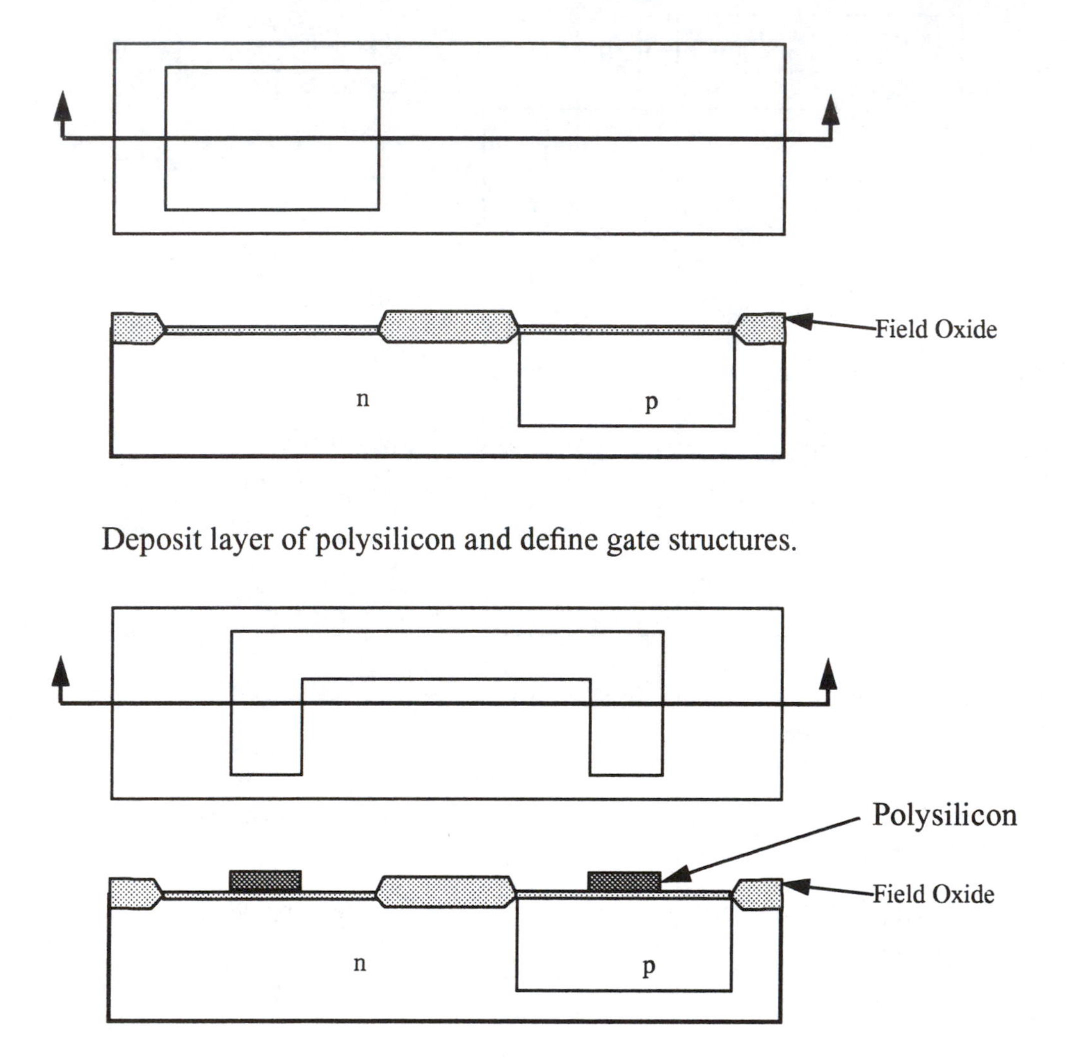

Deposit layer of polysilicon and define gate structures.

Define p^+ source and drain areas.

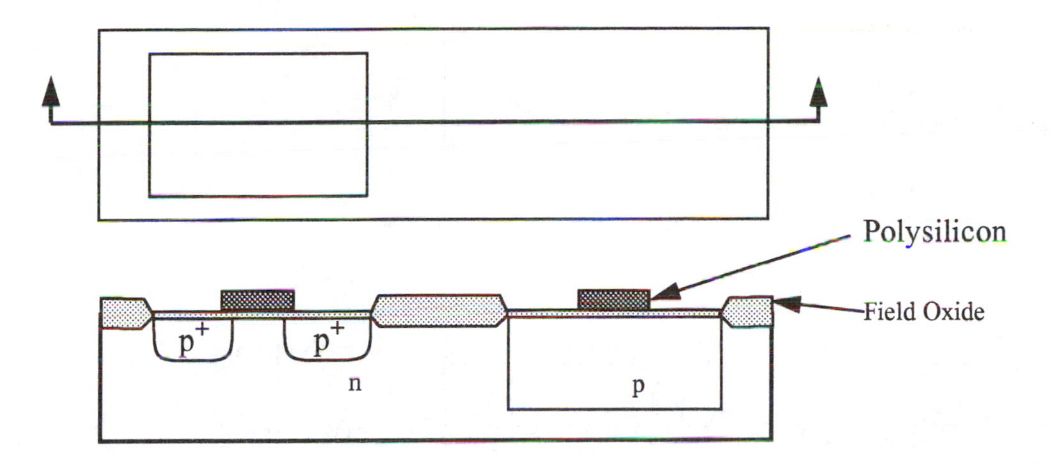

Define n^+ source and drain areas.

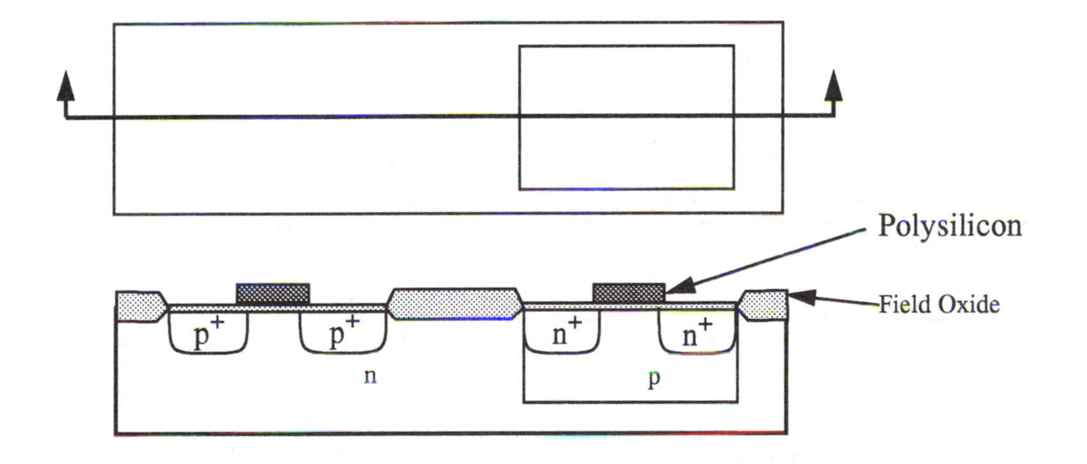

Deposit layer of SiO_2 and define contacts.

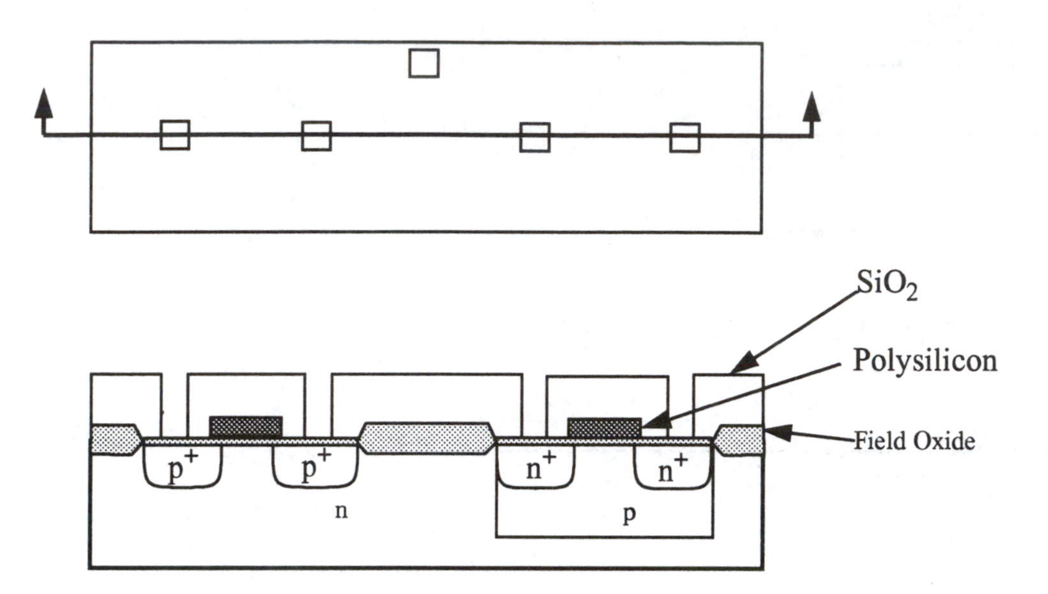

Define pattern for metallization.

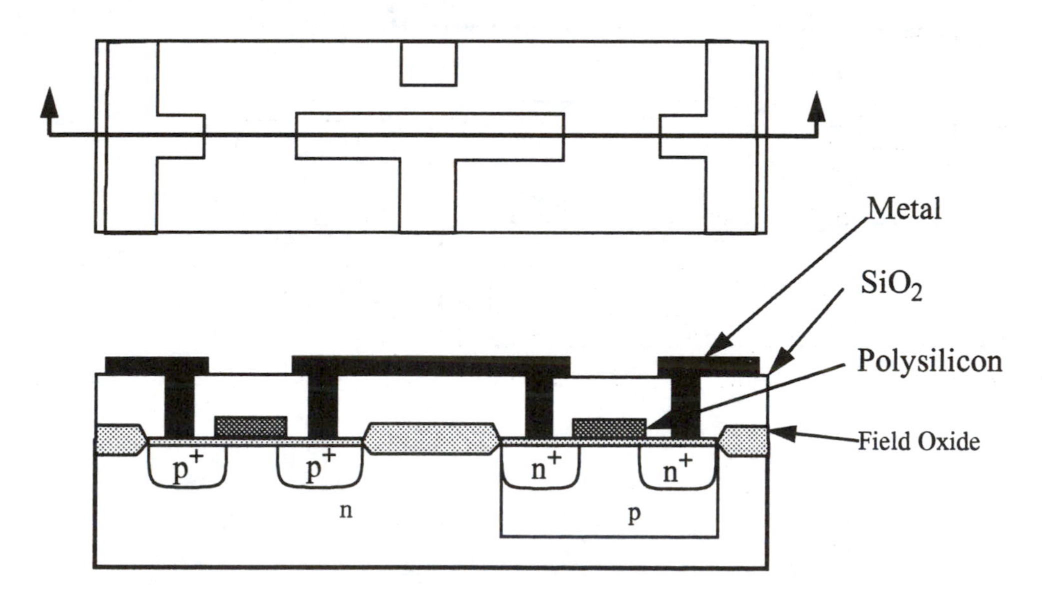

7-10. Solution.

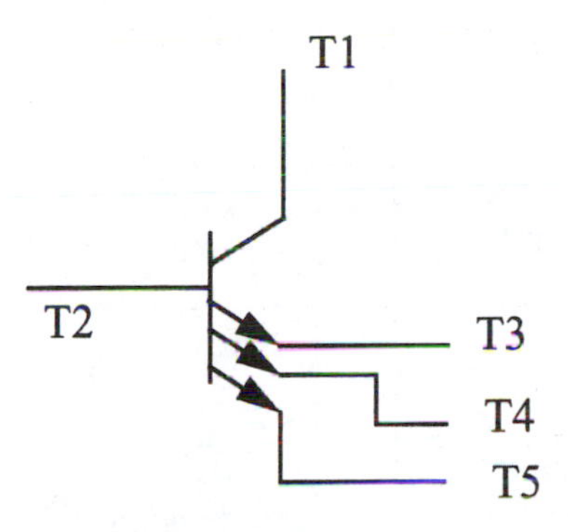

NPN bipolar transistor having three emitters. TTL uses multiple emitter transistors to perform logic conjunction.

7-11. Solution.

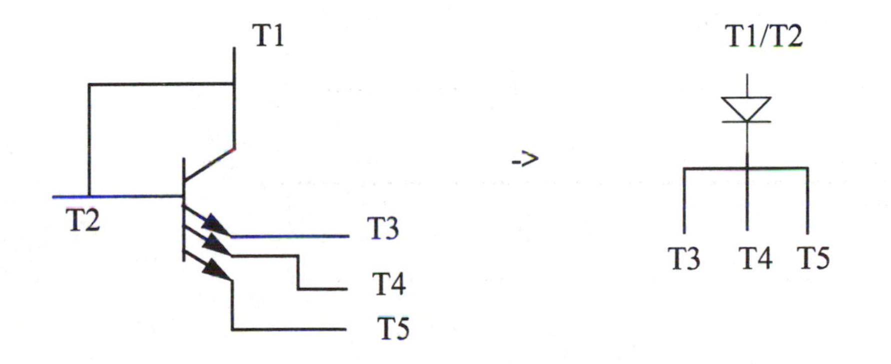

Diode with T1 and T2 being anode connections and T3, T4, and T5 being cathode connections.

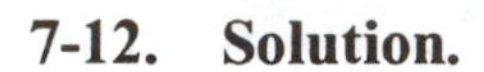
7-12. Solution.

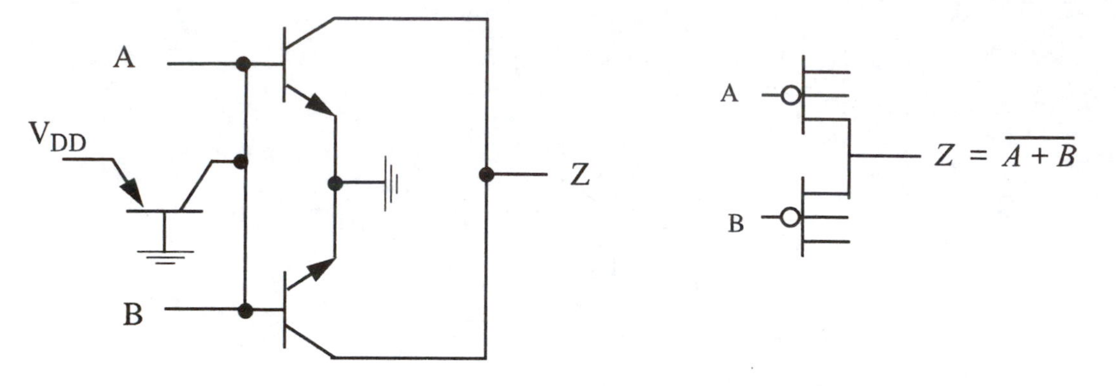

The logic family is I^2L and the circuit operation is a logic **nor**.

7-13. Solution.

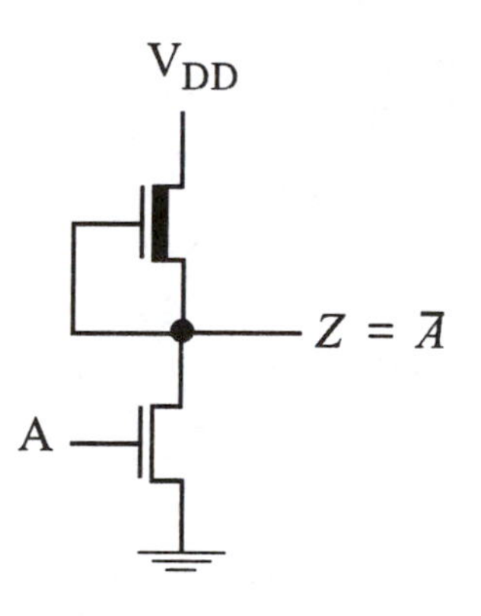

The logic family is static NMOS and the circuit operation is a logic **not**.

7-14. Solution.

a. Standard Products.

Advantages. Short design time. Low design costs.

Disadvantages. Low performance. Low programmability.

b. Programmable Logic Arrays

Advantages. Low-Moderate design time. Low-Moderate design costs. High programmability.

Disadvantages. Low-Moderate performance.

c. Gate Arrays

Advantages. Moderate design time. Moderate design costs. Moderate programmability.

Disadvantages. Moderate performance.

d. Full Custom

Advantages. High performance.

Disadvantages. High design time. High design costs. Low programmability.

Chapter 8. Sequential Systems - Definition & Analysis

8-1. Solution.

a.

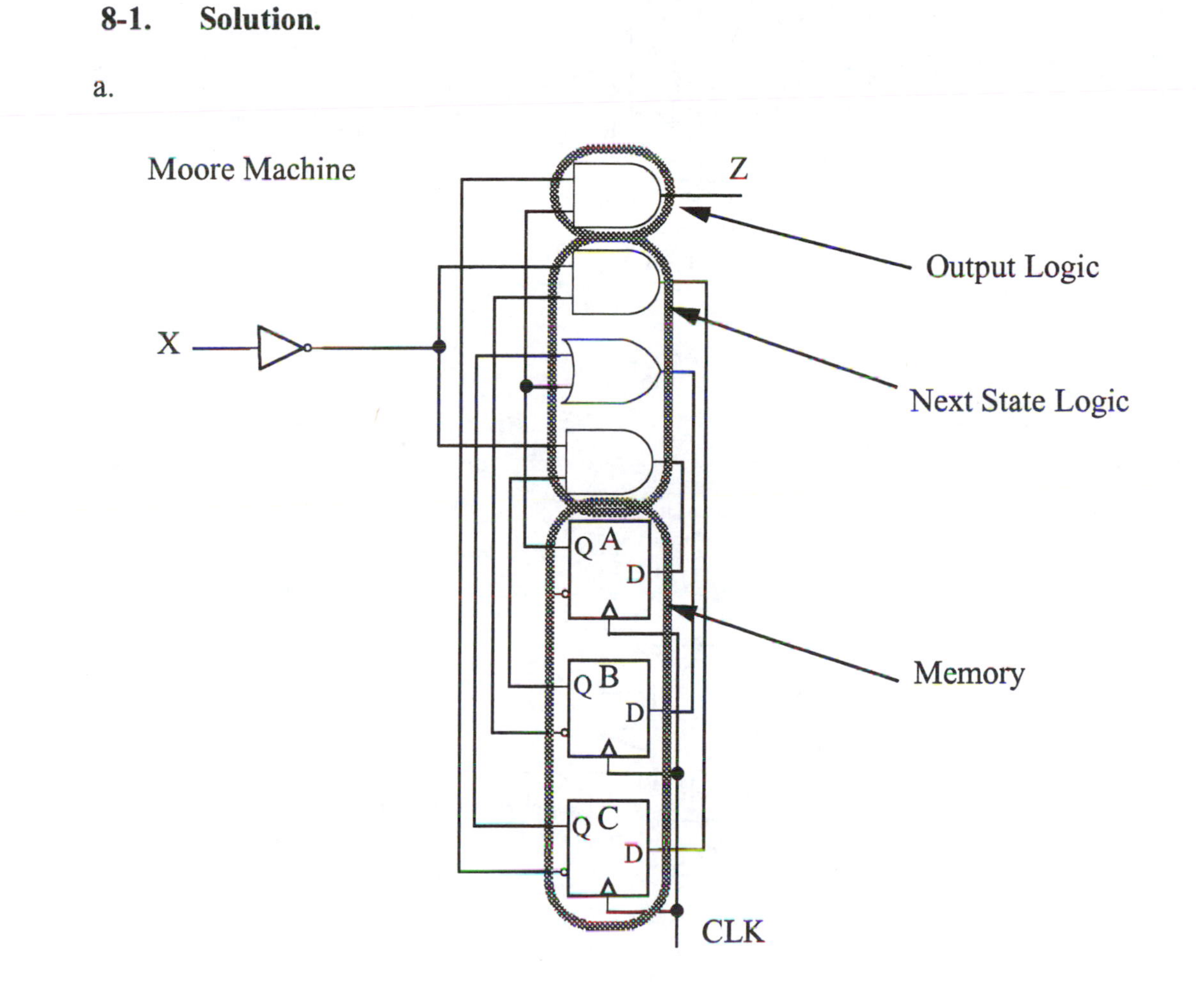

b.

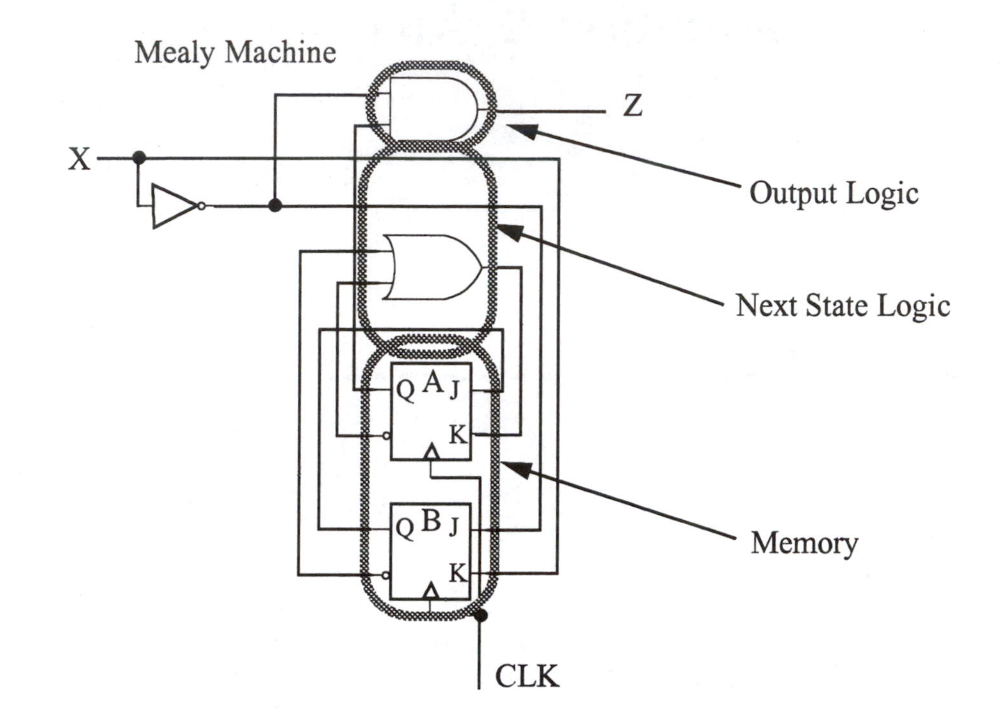

8-2. Solution.

a.

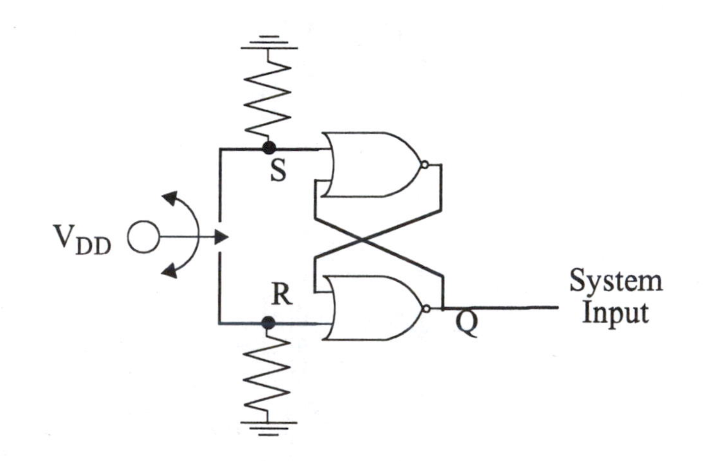

b. The resistors tie-off the S and R inputs to 0 when no input is applied. To set the system input to 1, the switch is connected to the S input. The switch sets S to 1 and the bottom resistor keeps R at 0, which sets the SR latch and the system input to 1. If the input switch bounces, the S input will receive an initial series of 1's and 0's and the SR latch will receive an initial series "set" (S = 1 and R=0) and "no change" (S=0 and R=0) commands. This initial series of "set" and "no change" commands will always leave the SR latch holding a 1, so the switch is debounced for set. To set the system input to 0, the switch is connected to the R input. The switch sets R to 1 and the top resistor keeps S at 0, which resets the SR latch and the system input to 0. If the input switch bounces, the R input will receive an initial series of 1's and 0's and the SR latch will receive an initial series "reset" (S =0 and R=1) and "no change" (S=0 and R=0) commands. This initial series of "reset" and "no change" commands will always leave the SR latch holding a 0, so the switch is debounced for reset.

8-3. Solution.

a.

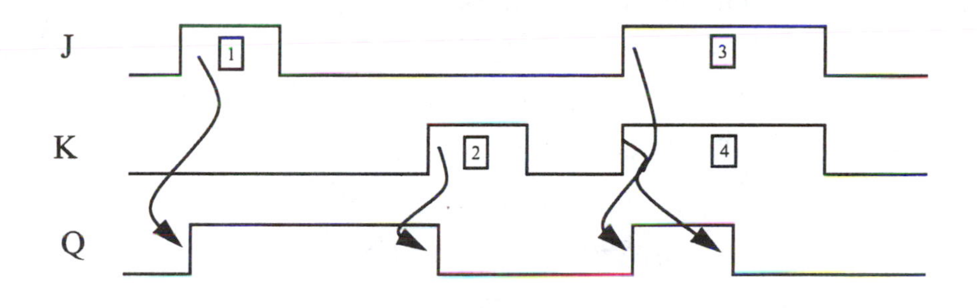

b. The JK latch oscillates, continually changing between 0 and 1.

8-4. Solution.

X	Y	Q	Q^+	Q_BAR^+
0	0	0	1	0
0	0	1	1	0
0	1	0	1	0
0	1	1	1	0
1	0	0	0	1
1	0	1	1	0
1	1	0	1	0
1	1	1	1	0

When X=0, Q is forced to 1, which forces Q_BAR to 0. When Y=1, Q_BAR is forced to 0, which forces Q to 1. The input state of X=1 and Y=0 is a stable, no change state.

8-5. Solution.

a.

J	$\overline{K}$	Q^+
0	0	0
0	1	Q
1	0	$\overline{Q}$
1	1	1

b. A $J\overline{K}$ flip-flop readily realizes a D flip-flop

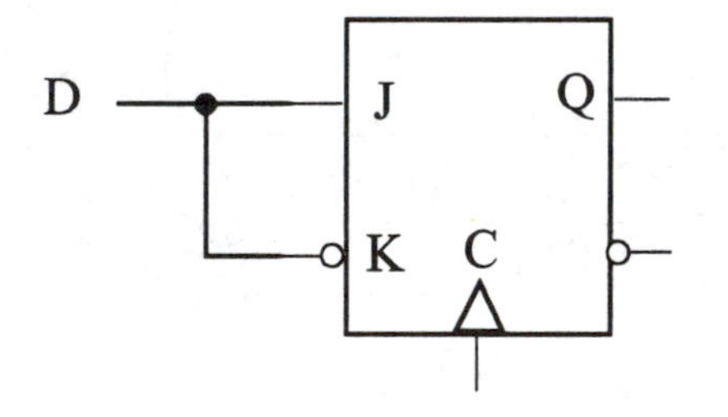

8-6. Solution.

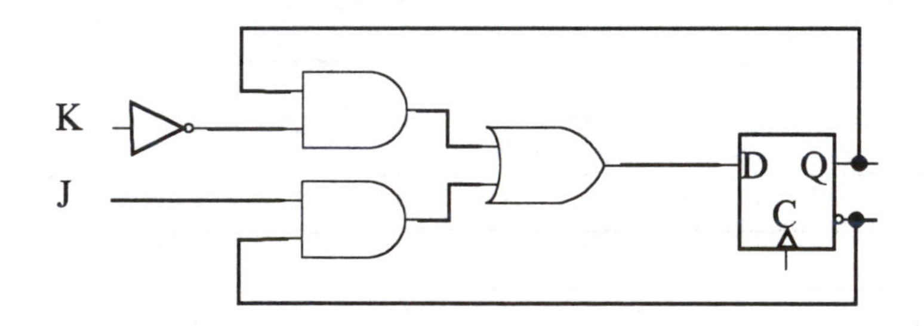

8-7. Solution.

a.

M	C	Q	Q^+	Q_BAR^+	Latch	Action
0	0	0	0	1	T	No Change
0	0	1	1	0		
0	1	0	1	0		Toggle
0	1	1	0	1		
1	0	0	0	1	D	Reset
1	0	1	0	1		
1	1	0	1	0		Set
1	1	1	1	0		

b.

Q^+: M \ CQ	00	01	11	10
0	0	1	0	1
1	0	0	1	1

c.

$$Q^+ = (M \bullet C) + (C \bullet \overline{Q}) + (\overline{M} \bullet \overline{C} \bullet Q)$$

8-8. Solution.

a.

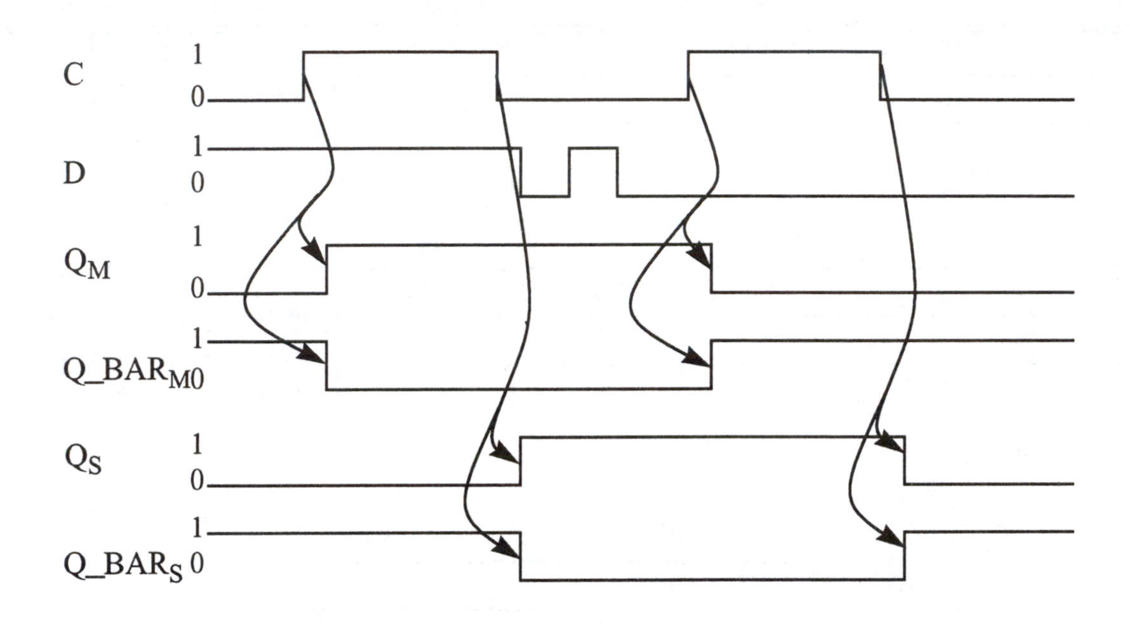

b.

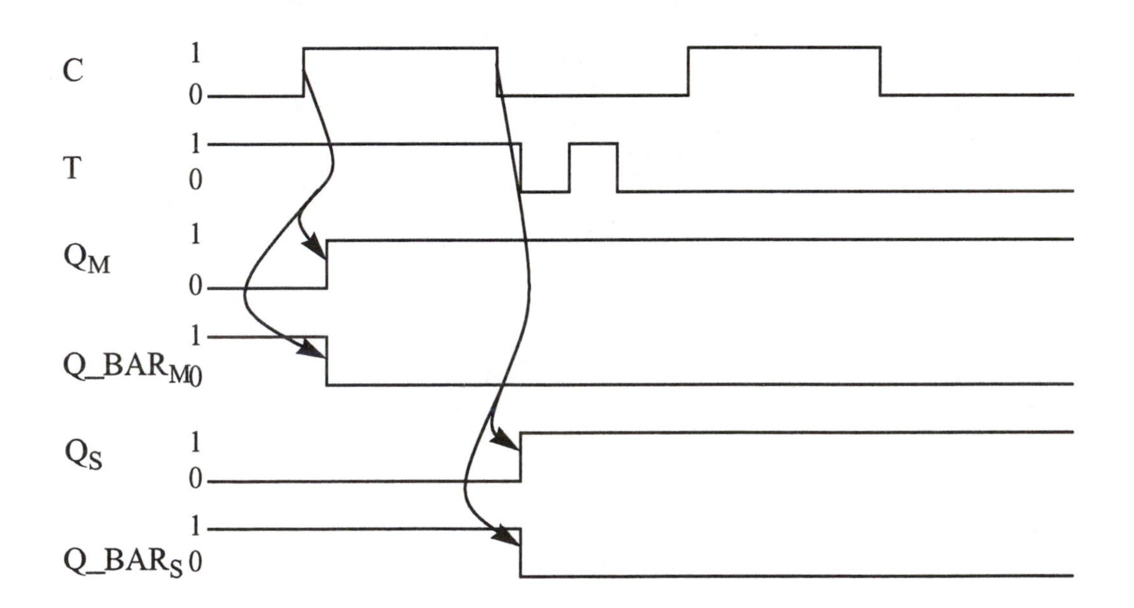

8-9. Solution.

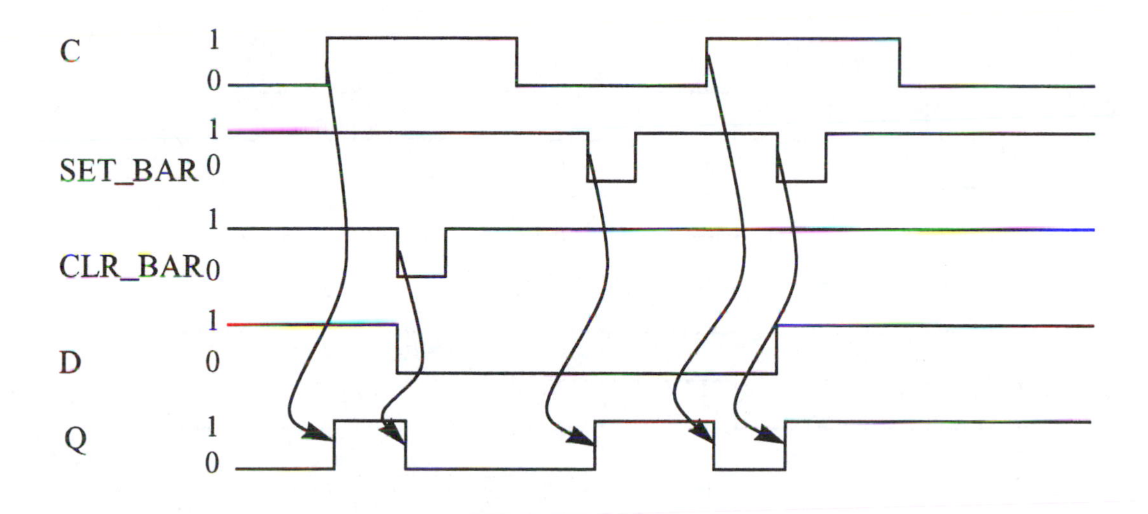

8-10. Solution.

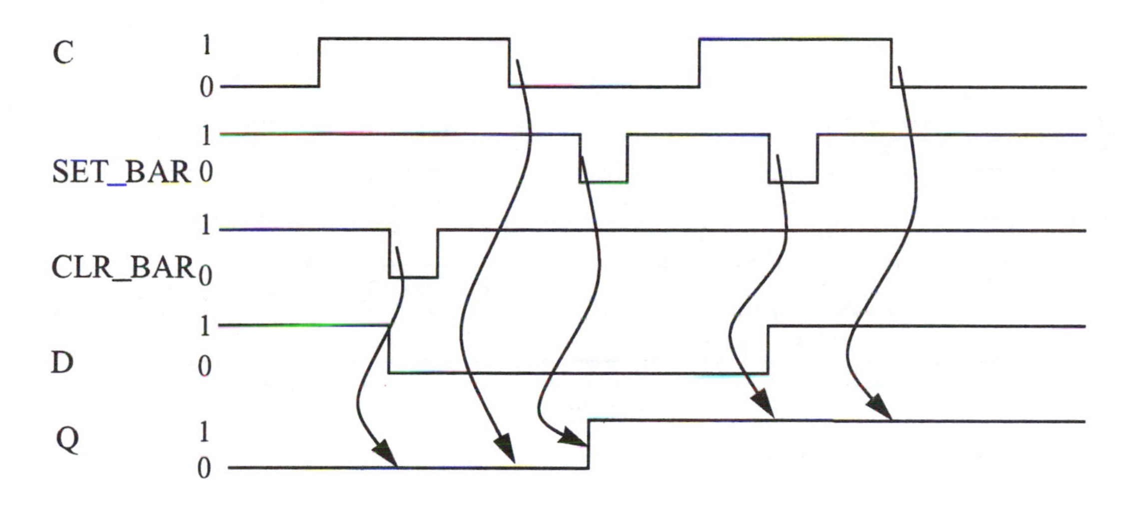

8-11. Solution.

a.

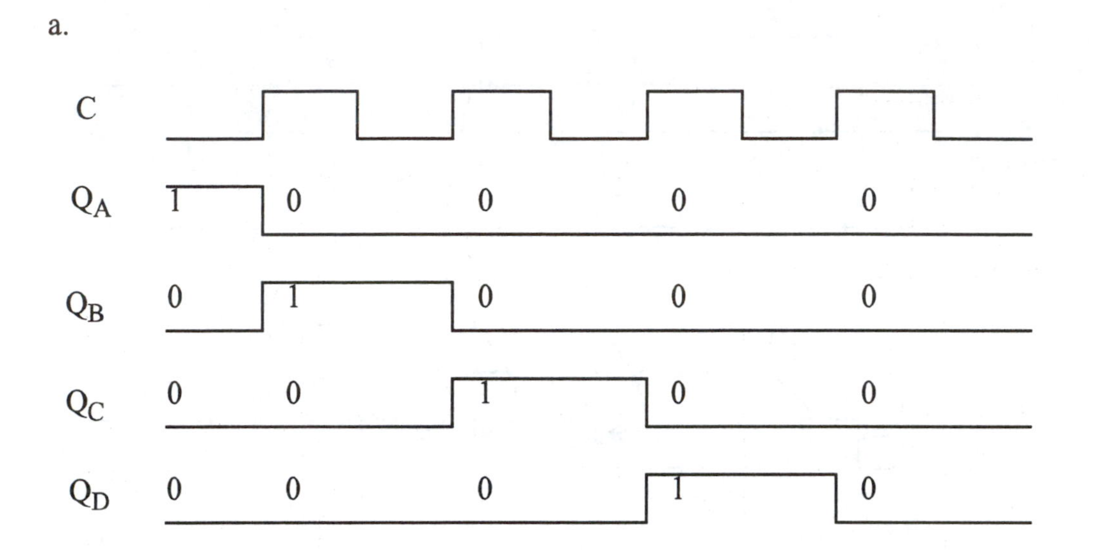

Applications: Divide-by-Two, Multiply-by-Two, and Signal Delay

b.

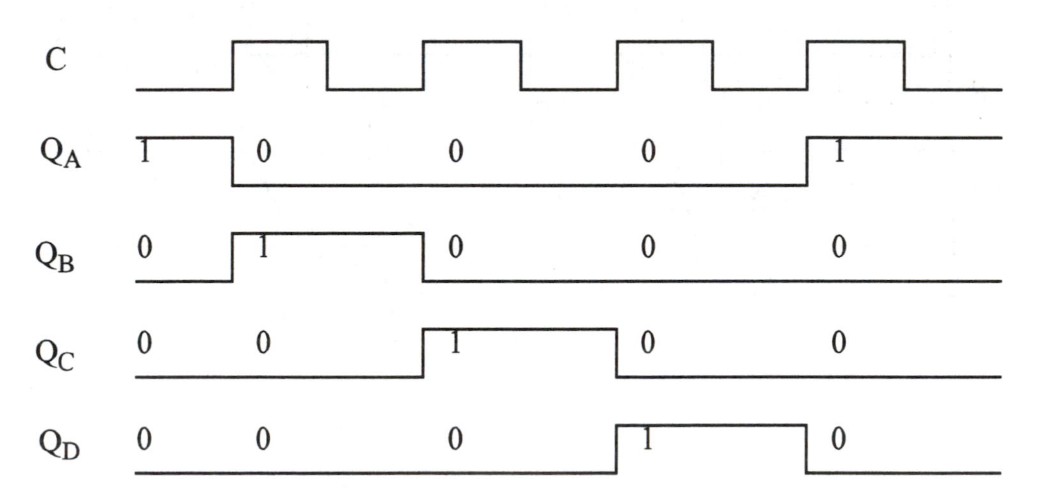

Applications: One-at-a-time sequence control and polling; the output of each flip-flop drives an enable signal.

8-12. Solution.

a.

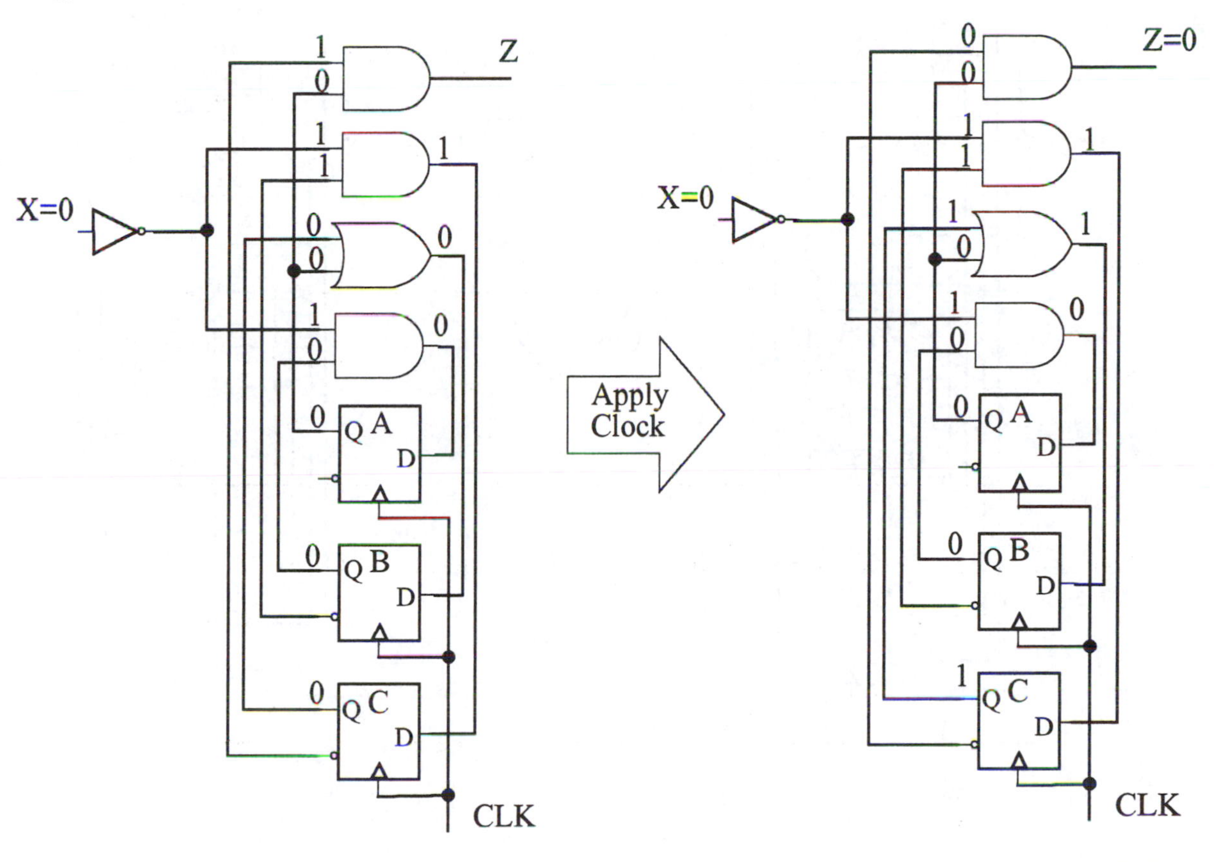

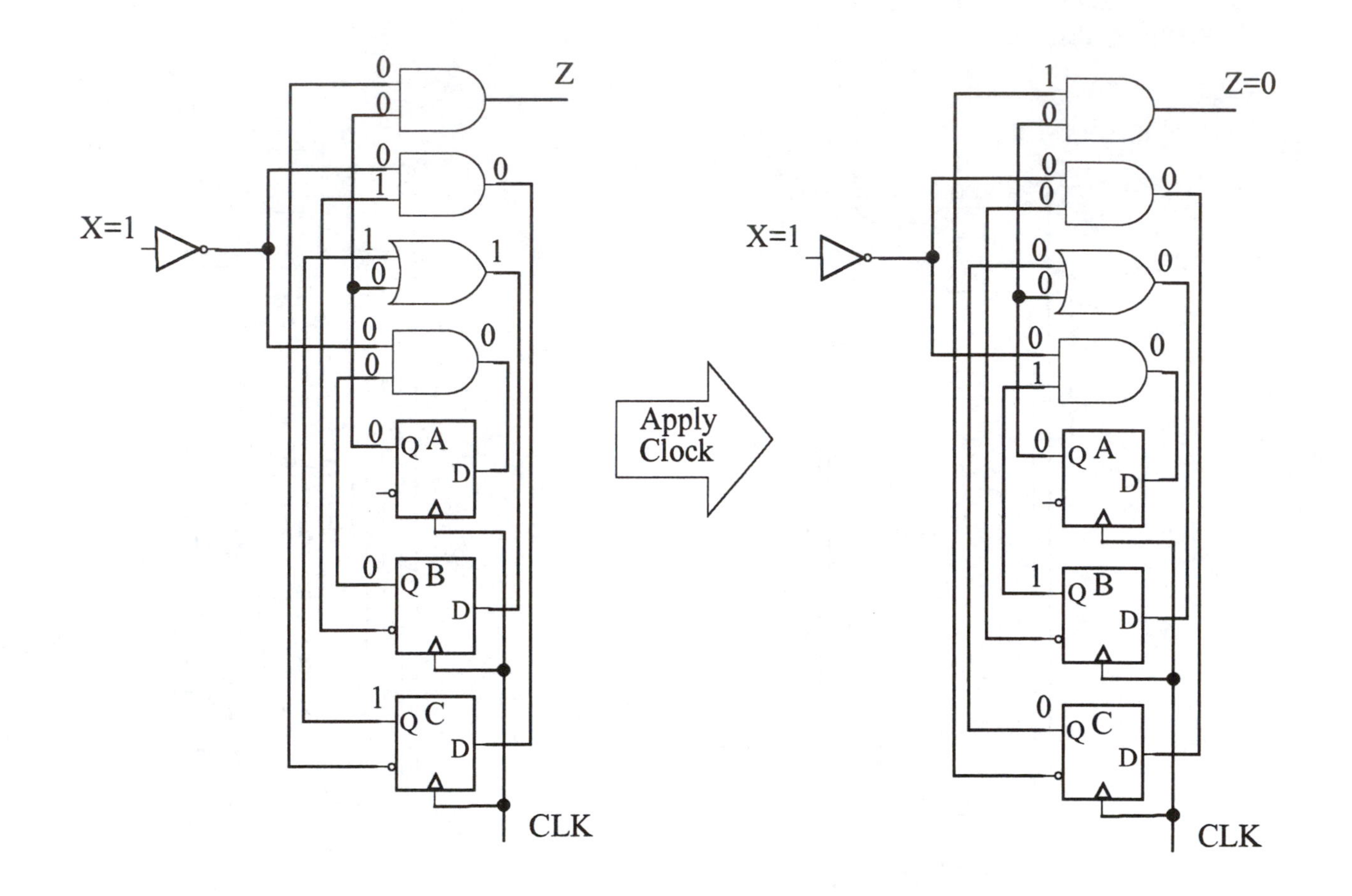

X=1
Z
Q A
D
Q B
D
Q C
D
CLK
Apply Clock
X=1
Z=0
Q A
D
Q B
D
Q C
D
CLK

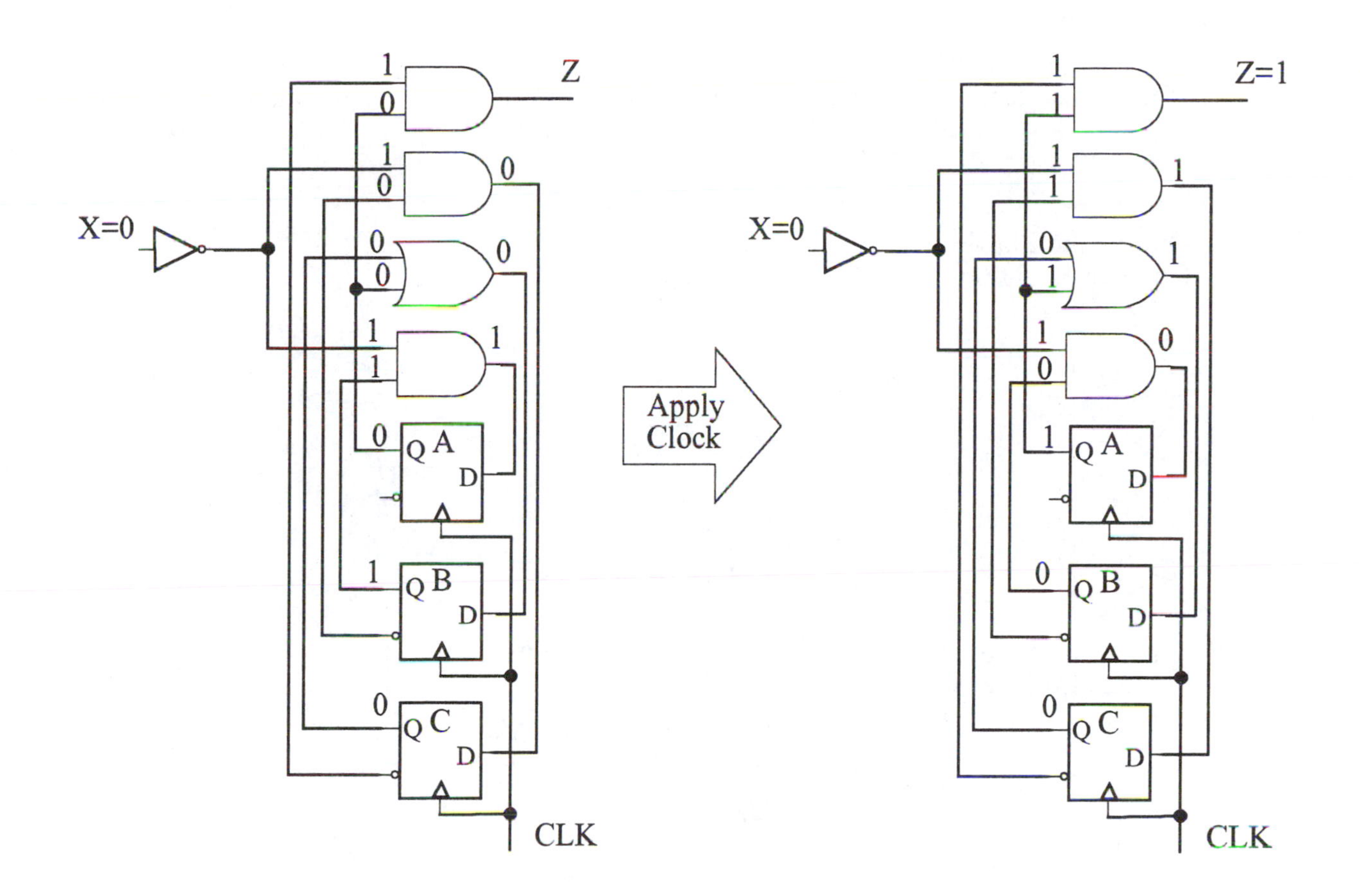

b.

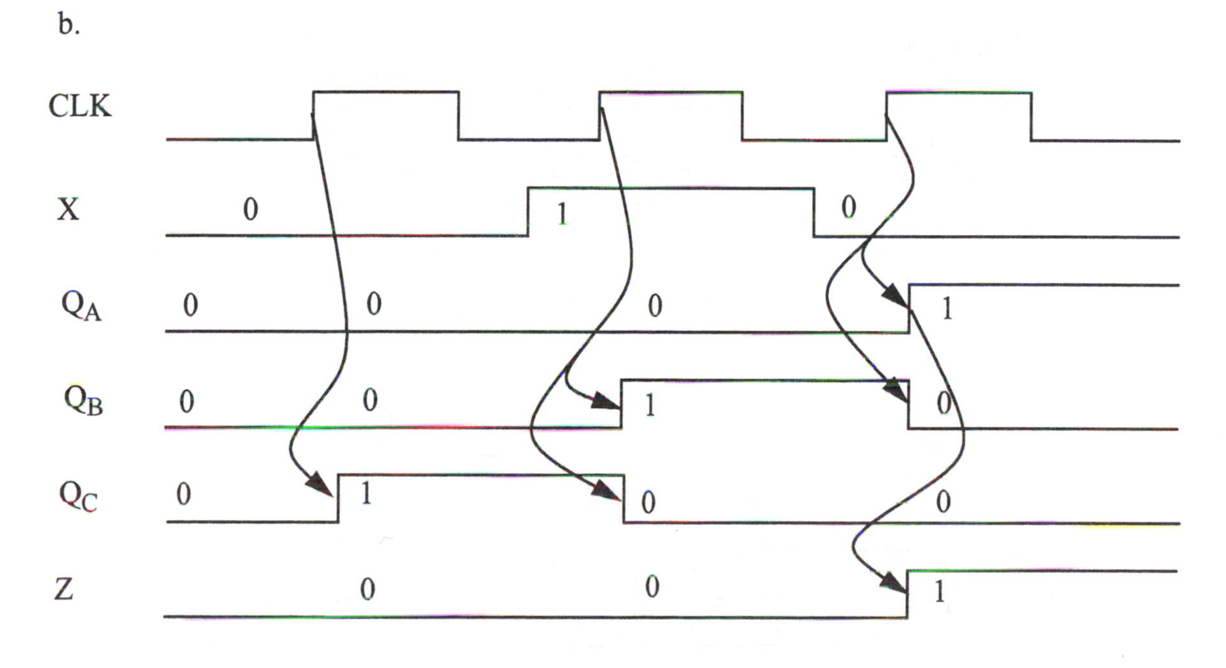

8-13. Solution.

Moore Machine

a.

Output and Flip-Flop Input Boolean Equations

$$Z = Q_A \bullet \overline{Q}_C$$

$$D_A = \overline{X} \bullet Q_B$$

$$D_B = Q_A + Q_C$$

$$D_C = \overline{X} \bullet \overline{Q}_B$$

Flip-Flop Next State Boolean Equations

$$Q_A^+ = D_A = \overline{X} \bullet Q_B$$

$$Q_B^+ = D_B = Q_A + Q_C$$

$$Q_C^+ = D_C = \overline{X} \bullet \overline{Q}_B$$

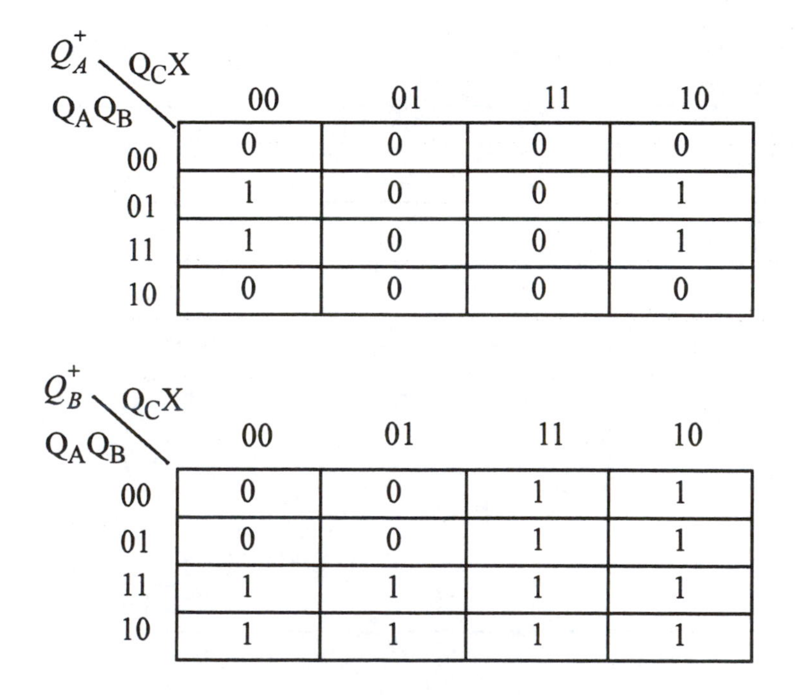

Q_A^+ Q_AQ_B \ Q_CX	00	01	11	10
00	0	0	0	0
01	1	0	0	1
11	1	0	0	1
10	0	0	0	0

Q_B^+ Q_AQ_B \ Q_CX	00	01	11	10
00	0	0	1	1
01	0	0	1	1
11	1	1	1	1
10	1	1	1	1

Q_C^+ (Q_AQ_B \ Q_CX)	00	01	11	10
00	1	0	0	1
01	0	0	0	0
11	0	0	0	0
10	1	0	0	1

Z (Q_A \ Q_BQ_C)	00	01	11	10
0	0	0	0	0
1	1	0	0	1

State Table

$Q_A Q_B Q_C$	$Q_A^+ Q_B^+ Q_C^+$		Z
	$X = 0$	$X = 1$	
000	001	000	0
001	011	010	0
010	100	000	0
011	110	010	0
100	011	010	1
101	011	010	0
110	110	010	1
111	110	010	0

State Diagram

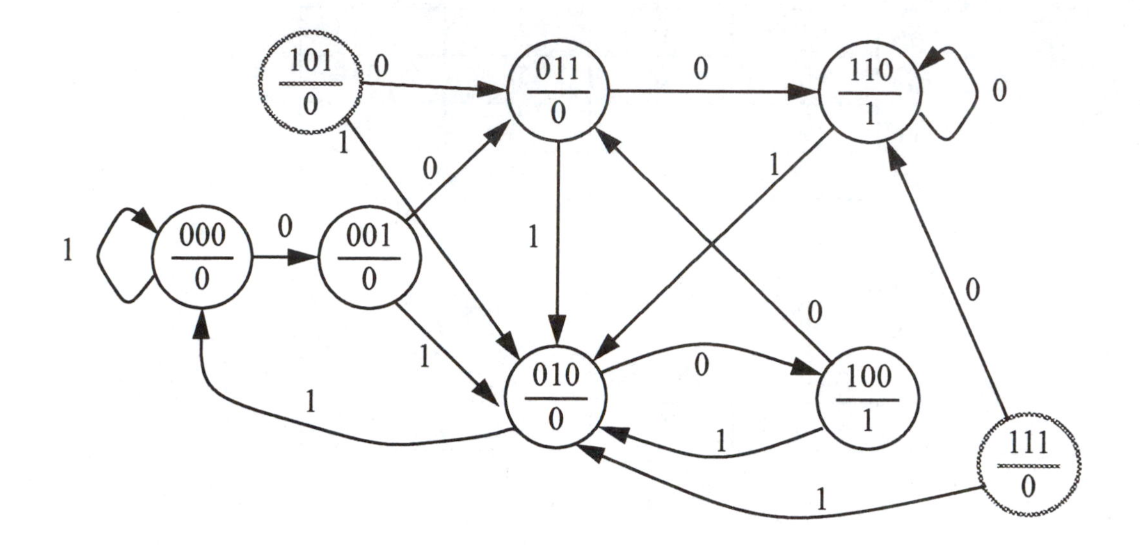

b.

$$X = 010\ldots$$
$$Z = 001\ldots$$

Mealy Machine

a.

Output and Flip-Flop Input Boolean Equations

$$Z = Q_A \bullet \overline{X}$$
$$J_A = Q_B$$
$$K_A = \overline{Q_A \bullet Q_B}$$
$$J_B = \overline{X}$$
$$K_B = X$$

Flip-Flop Next State Boolean Equations

$$Q_A^+ = (J_A \bullet \overline{Q}_A) + (\overline{K}_A \bullet Q_A) = (Q_B \bullet \overline{Q}_A) + (Q_A \bullet Q_B \bullet Q_A) = Q_B$$
$$Q_B^+ = (J_B \bullet \overline{Q}_B) + (\overline{K}_B \bullet Q_B) = (\overline{X} \bullet \overline{Q}_B) + (\overline{X} \bullet Q_B) = \overline{X}$$

Q_A^+: Q_A \ Q_BX	00	01	11	10
0	0	0	1	1
1	0	0	1	1

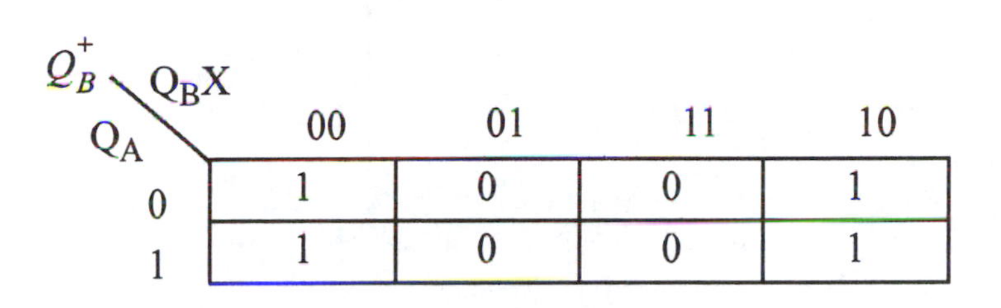

Q_B^+: Q_A \ Q_BX	00	01	11	10
0	1	0	0	1
1	1	0	0	1

Z: Q_A \ Q_BX	00	01	11	10
0	0	0	0	0
1	1	0	0	1

State Table

$Q_A Q_B$	$Q_A^+ Q_B^+$		Z	
	$X = 0$	$X = 1$	$X = 0$	$X = 1$
00	01	00	0	0
01	11	10	0	0
10	01	00	1	0
11	11	10	1	0

State Diagram

X/Z

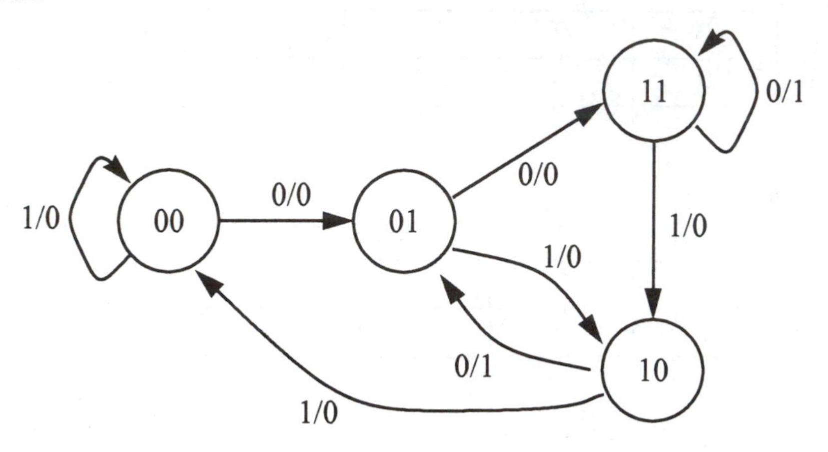

b.

$$X = 010\ldots$$
$$Z = 001\ldots$$

c. Both finite state machines detect the nonresetting input sequence 010. Moreover, the finite state machines are equivalent; they both detect the nonresetting sequences 010 or 000. The Mealy machine yields one less flip-flop. The complexities of the combinational logic of the Moore and Mealy machines are roughly equivalent for this example.

8-14. Solution.

a. Mealy machine.

b.

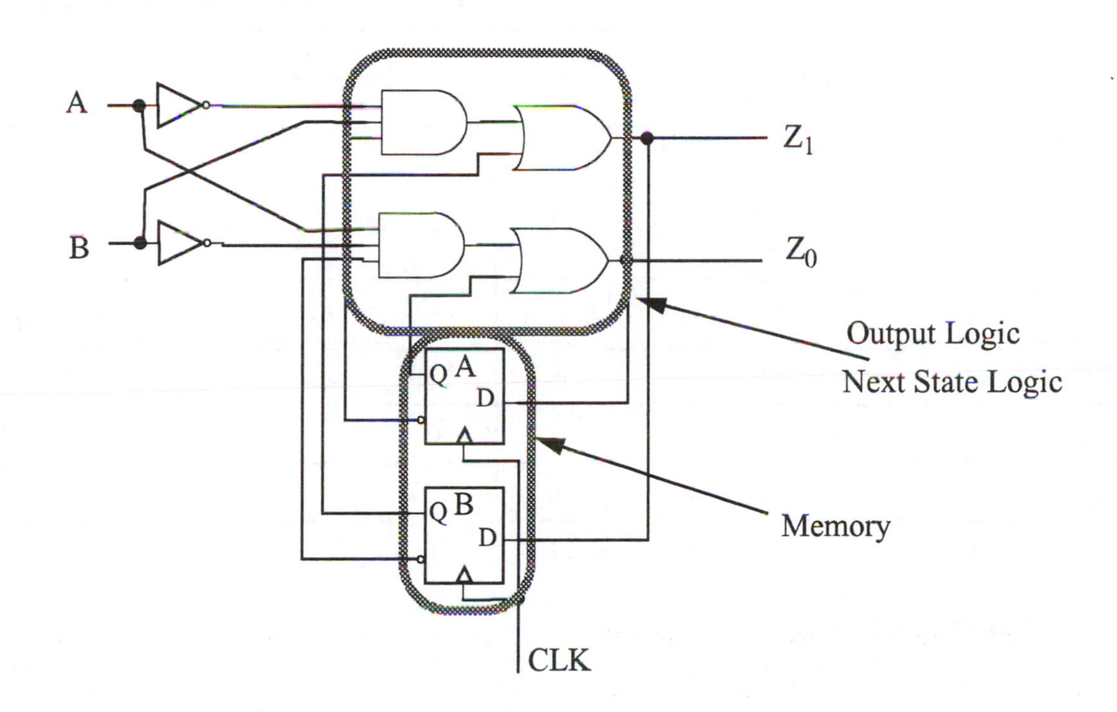

Combinational logic serves as both output logic and next state logic.

c.

Output and Flip-Flop Input Boolean Equations

$$Q_B^+ = Z_1 = D_B = (\overline{A} \bullet B \bullet \overline{Q}_A) + Q_B$$

$$Q_A^+ = Z_0 = D_A = (A \bullet \overline{B} \bullet \overline{Q}_B) + Q_A$$

Q_A^+ AB / Q_AQ_B	00	01	11	10
00	0	0	0	1
01	0	0	0	0
11	1	1	1	1
10	1	1	1	1

Q_B^+ \ AB, Q_AQ_B	00	01	11	10
00	0	1	0	0
01	1	1	1	1
11	1	1	1	1
10	0	0	0	0

State Table

$Q_A Q_B$	$Q_A^+ Q_B^+$				$Z_0 Z_1$			
	$AB = 00$	$AB = 01$	$AB = 10$	$AB = 11$	$AB = 00$	$AB = 01$	$AB = 10$	$AB = 11$
00	00	01	10	00	00	01	10	00
01	01	01	01	01	01	01	01	01
10	10	10	10	10	10	10	10	10
11	11	11	11	11	11	11	11	11

State Diagram

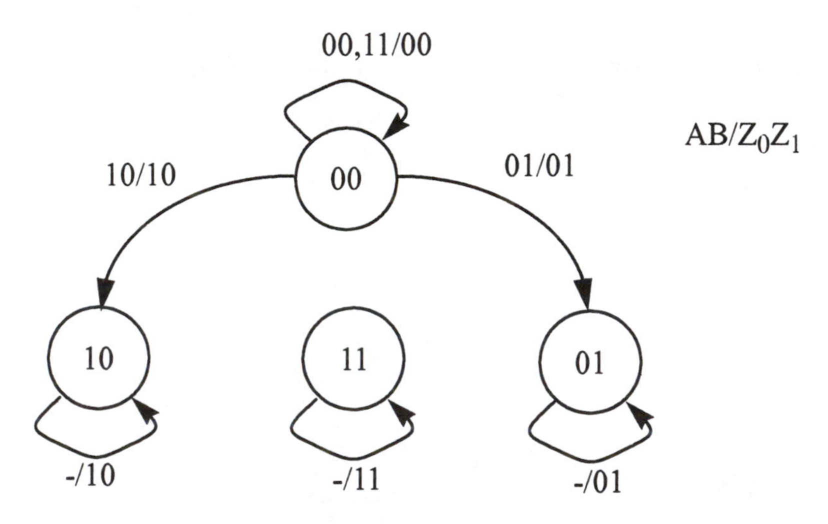

d. State 11 is an unused state. If the Mealy machine accidentally enters state 11, it will stay in state 11, yield a constant output of 11 and never return to the desired state sequencing.

e. The Mealy machine is a sequential comparator. The output denotes the relative magnitude of A and B: 00 denotes A=B, 01 denotes $A<B$, and 10 denotes $A>B$. Starting in state 00, if the A and B bits are equal, then the operands are equal (output is 00) and Mealy machine stays in state 00. If A=0 and B=1, then $A<B$ (output is 01) and the Mealy machine transitions to state 01. If A=1 and B=0, then $A>B$ (output is 10) and the Mealy machine transitions to state 10. Once a pair of bits differ and the relative magnitude is determined, any further bits will not change the relative ordering because the operands are applied most significant bits first. Thus, once the Mealy machine reaches state 01 or 10, it stays in these states for the remainder of the bit vectors. State 11 is an unused state and output 11 is an unused output.

8-15. Solution.

a.

Output and Flip-Flop Input Boolean Equations

$$Z = Q_A \bullet \overline{Q}_C$$

$$T_A = (Q_B \bullet Q_C \bullet X) + (\overline{Q}_B \bullet Q_C \bullet \overline{X}) + Q_A$$

$$T_B = (Q_A \bullet Q_C \bullet \overline{X}) + Q_B$$

$$T_C = (\overline{Q}_C \bullet X) + Q_B$$

b.

Flip-Flop Output Boolean Equations

$$Q_A^+ = (T_A \bullet \overline{Q}_A) + (\overline{T}_A \bullet Q_A) = (\overline{Q}_A \bullet Q_B \bullet Q_C \bullet X) + (\overline{Q}_A \bullet \overline{Q}_B \bullet Q_C \bullet \overline{X})$$

$$Q_B^+ = (T_B \bullet \overline{Q}_B) + (\overline{T}_B \bullet Q_B) = (Q_A \bullet \overline{Q}_B \bullet Q_C \bullet \overline{X})$$

$$Q_C^+ = (T_C \bullet \overline{Q}_C) + (\overline{T}_C \bullet \overline{Q}_C) = (Q_B \bullet \overline{Q}_C) + (\overline{Q}_C \bullet X) + (\overline{Q}_B \bullet Q_C) + (\overline{Q}_B \bullet Q_C \bullet \overline{X})$$

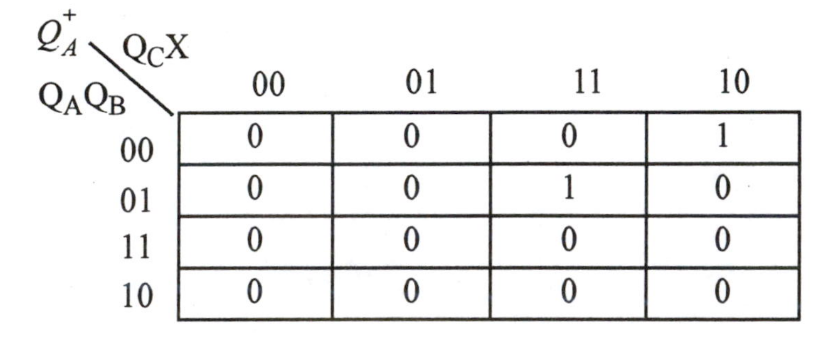

Q_A^+ — Q_AQ_B \ Q_CX	00	01	11	10
00	0	0	0	1
01	0	0	1	0
11	0	0	0	0
10	0	0	0	0

Q_B^+ (rows Q_AQ_B, columns Q_CX)

Q_AQ_B \ Q_CX	00	01	11	10
00	0	0	0	0
01	0	0	0	0
11	0	0	0	0
10	0	0	0	1

Q_C^+ (rows Q_AQ_B, columns Q_CX)

Q_AQ_B \ Q_CX	00	01	11	10
00	0	1	1	1
01	1	1	0	0
11	1	1	0	0
10	0	1	1	1

State Table

$Q_A Q_B Q_C$	$Q_A^+ Q_B^+ Q_C^+$		Z
	$X = 0$	$X = 1$	
000	000	001	0
001	101	001	0
010	001	001	0
011	000	100	0
100	000	001	1
101	011	001	0
110	001	001	1
111	000	000	0

State Diagram

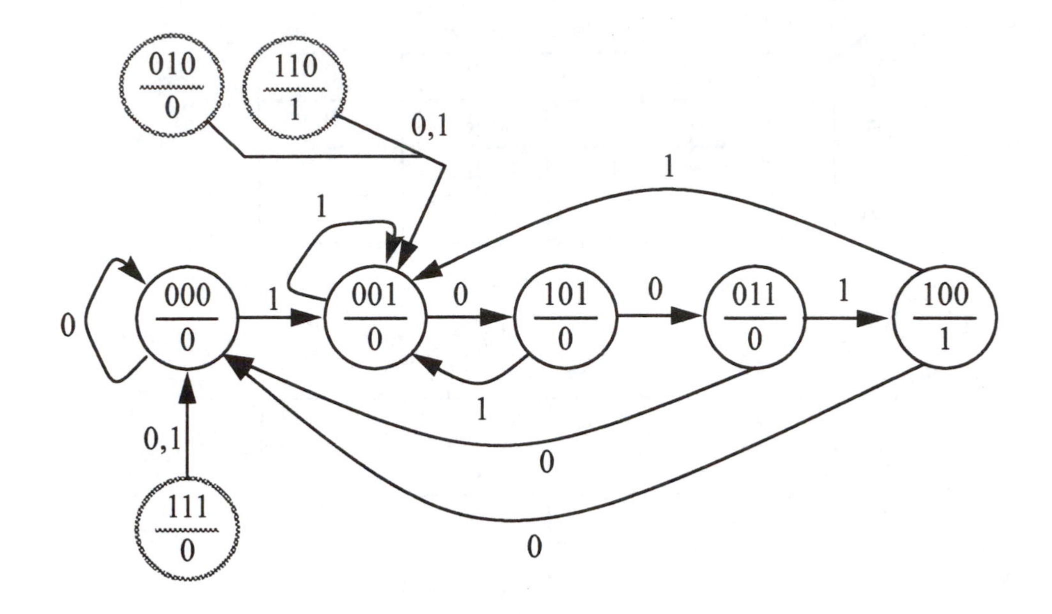

c.

$$X = 1001\ldots$$
$$Z = 0001\ldots$$

The Moore machine detects the nonresetting sequence 1001 which is the BCD encoding for 9.

Chapter 9. Sequential Design - Synthesis

9-1. Solution.

a. Moore Machine

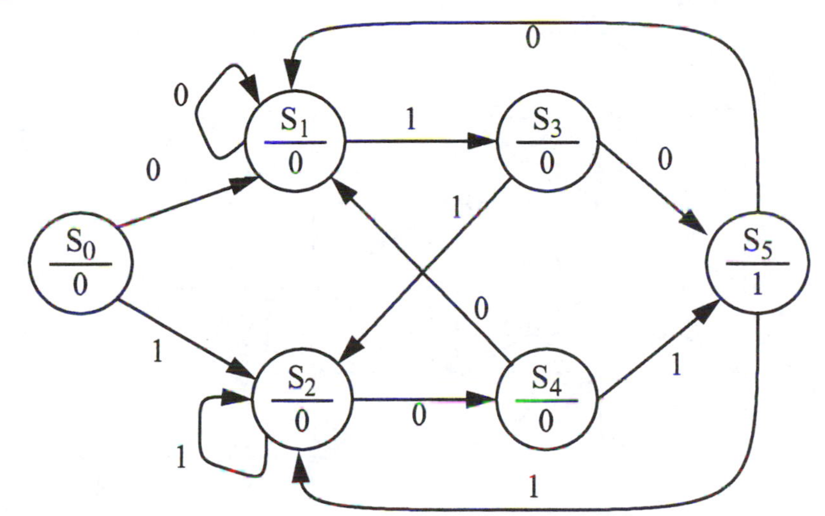

PS	NS		Z
	$X = 0$	$X = 1$	
S_0	S_1	S_2	0
S_1	S_1	S_3	0
S_2	S_4	S_2	0
S_3	S_5	S_2	0
S_4	S_1	S_5	0
S_5	S_1	S_2	1

Implication Table

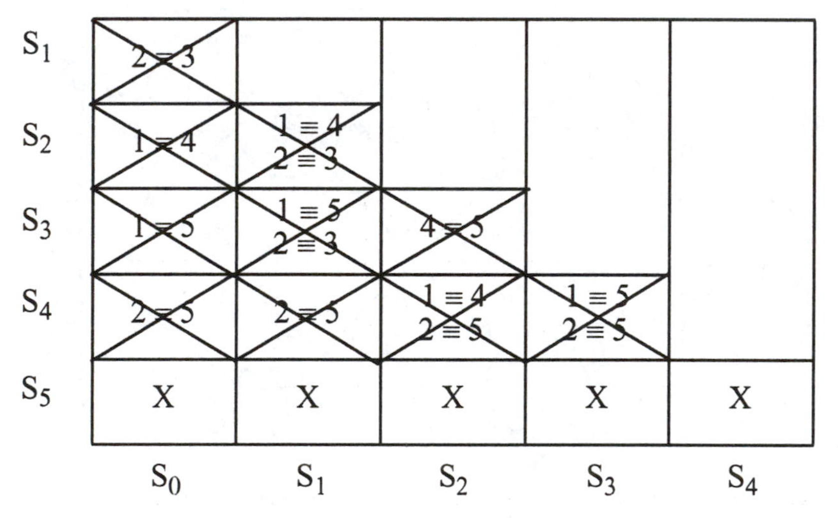

b. Mealy Machine

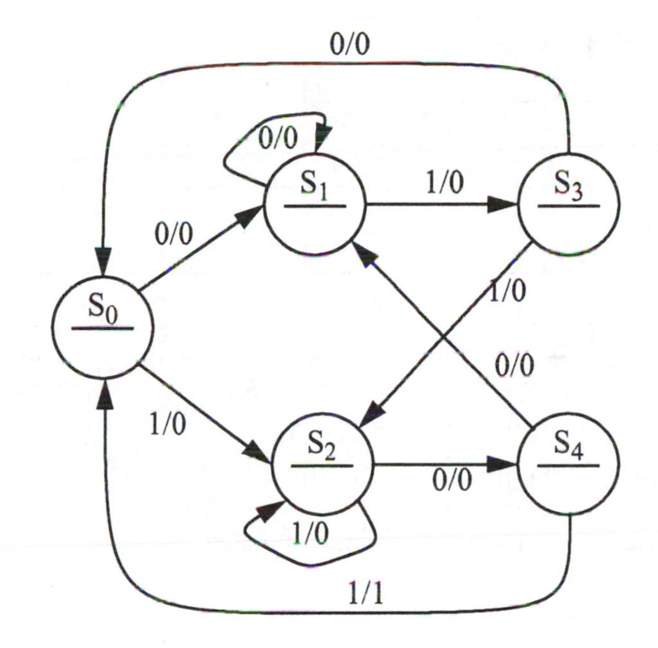

PS	NS		Z	
	$X = 0$	$X = 1$	$X = 0$	$X = 1$
S_0	S_1	S_2	0	0
S_1	S_1	S_3	0	0
S_2	S_4	S_2	0	0
S_3	S_0	S_2	1	0
S_4	S_1	S_0	0	1

Implication Table

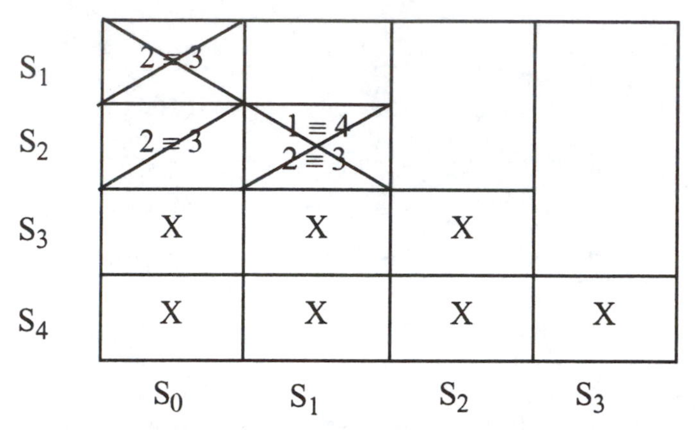

c. Both Moore and Mealy machines require 3 bistable memory devices.

Moore Machine: 6 used states and 2 unused states.

Mealy Machine: 5 used states and 3 unused states.

9-2. Solution.

a. Moore Machine

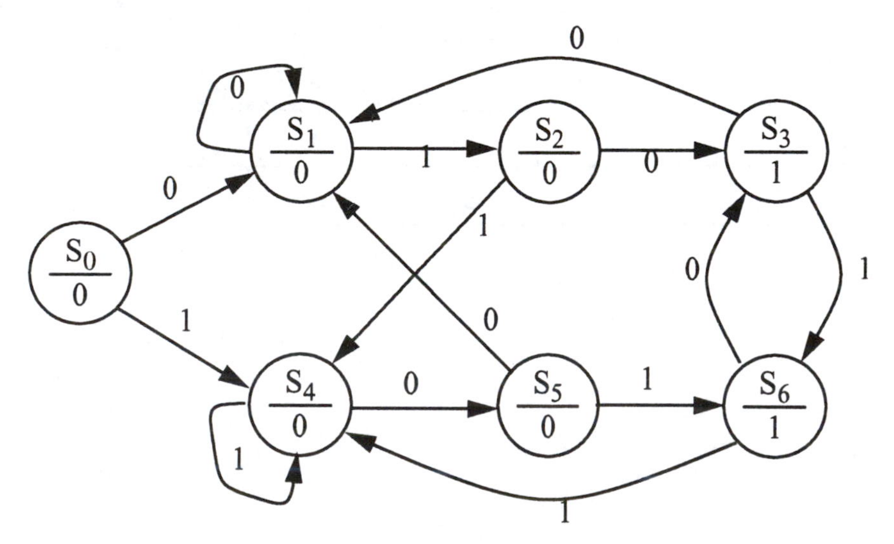

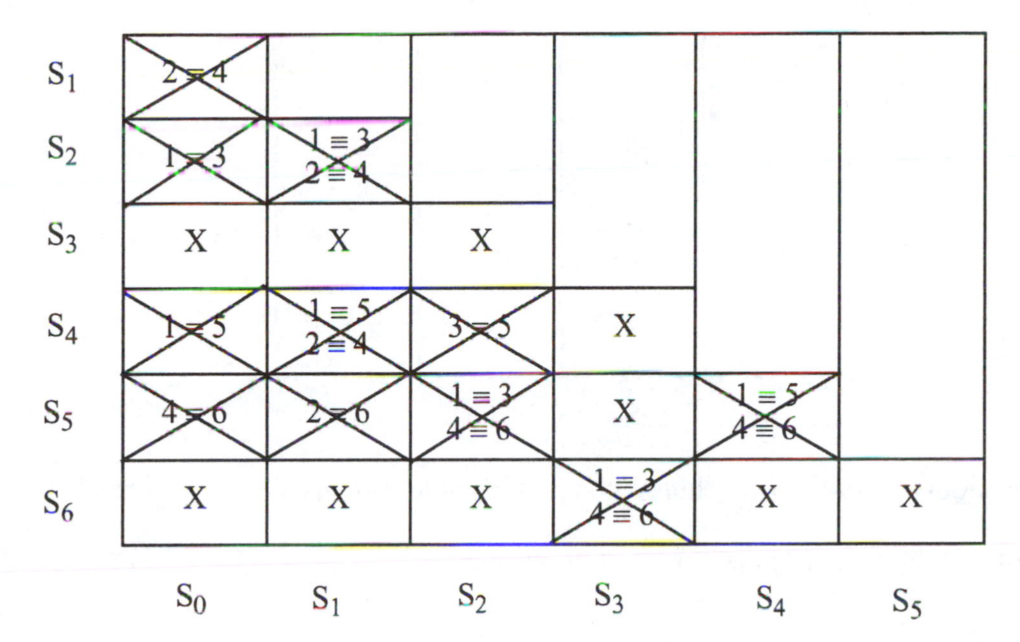

	S_0	S_1	S_2	S_3	S_4	S_5
S_1	$2 \equiv 4$					
S_2	$1 \equiv 3$	$1 \equiv 3$ $2 \equiv 4$				
S_3	X	X	X			
S_4	$1 \equiv 5$	$1 \equiv 5$ $2 \equiv 4$	$3 \equiv 5$	X		
S_5	$4 \equiv 6$	$2 \equiv 6$	$1 \equiv 3$ $4 \equiv 6$	X	$1 \equiv 5$ $4 \equiv 6$	
S_6	X	X	X	$1 \equiv 3$ $4 \equiv 6$	X	X

b. Mealy Machine

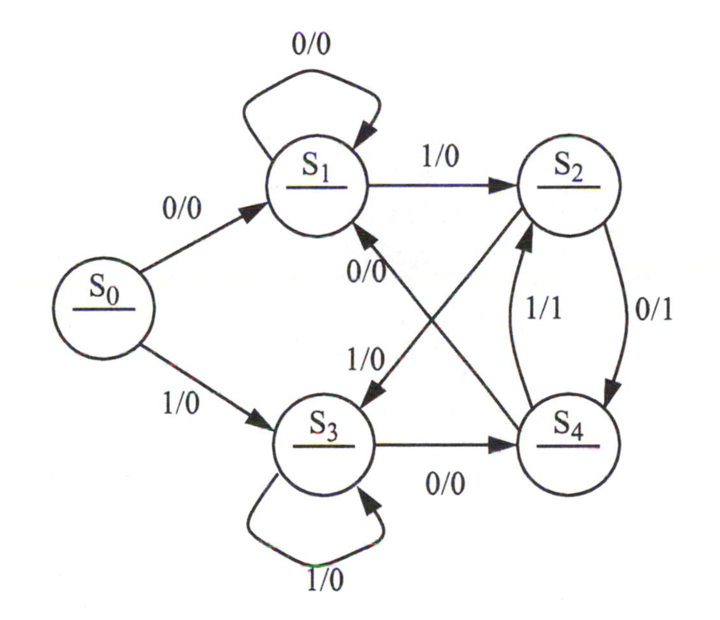

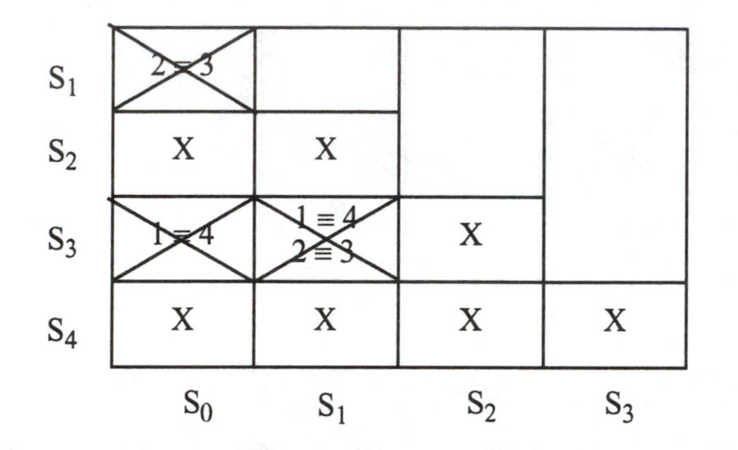

c. Both Moore and Mealy machines require 3 bistable memory devices.

Moore Machine: 7 used states and 1 unused states.

Mealy Machine: 5 used states and 3 unused states.

9-3. Solution.

a. Moore Machine

PS	NS		Z
	$X = 0$	$X = 1$	
S_0	S_1	S_2	0
S_1	S_1	S_3	0
S_2	S_4	S_2	0
S_3	S_5	S_2	0
S_4	S_1	S_5	0
S_5	S_1	S_2	1

Armstrong-Humphrey Rule #1: S_0,S_1,S_4,S_5 S_0,S_2,S_3,S_5

Armstrong-Humphrey Rule #2: S_1,S_2 S_1,S_3 S_2,S_4 S_2,S_5 S_1,S_5

Armstrong-Humphrey Rule #3: S_0,S_1,S_2,S_3,S_4

Q_A \ Q_BQ_C	00	01	11	10
0	S_0	S_1	S_5	S_4
1	S_3	S_2		

$Q_A Q_B Q_C$	$Q_A^+ Q_B^+ Q_C^+$		Z
	$X = 0$	$X = 1$	
000	001	101	0
001	001	100	0
101	010	101	0
100	011	101	0
010	001	011	0
011	001	101	1

$Q_A^+ = D_A$

Q_AQ_B \ Q_CX	00	01	11	10
00	0	1	1	0
01	0	0	1	0
11	-	-	-	-
10	0	1	1	0

$Q_B^+ = D_B$

Q_AQ_B \ Q_CX	00	01	11	10
00	0	0	0	0
01	0	1	0	0
11	-	-	-	-
10	1	0	0	1

$Q_C^+ = D_C$

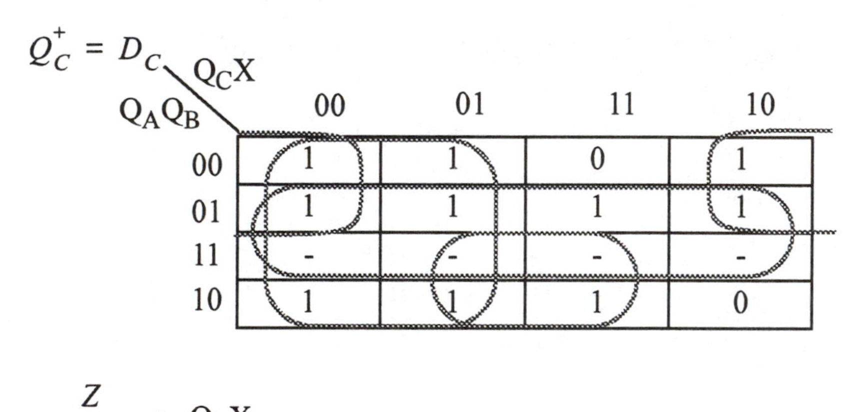

Q_AQ_B \ Q_CX	00	01	11	10
00	1	1	0	1
01	1	1	1	1
11	-	-	-	-
10	1	1	1	0

Z

Q_AQ_B \ Q_CX	00	01	11	10
00	0	0	1	0
01	0	0	-	-

$$D_A = Q_A^+ = (Q_C \bullet X) + (\overline{Q}_B \bullet X)$$

$$D_B = Q_B^+ = (Q_A \bullet \overline{X}) + (Q_B \bullet \overline{Q}_C \bullet X)$$

$$D_C = Q_C^+ = Q_B + \overline{Q}_C + (\overline{Q}_A \bullet \overline{X}) + (Q_A \bullet X)$$

$$Z = Q_B \bullet Q_C$$

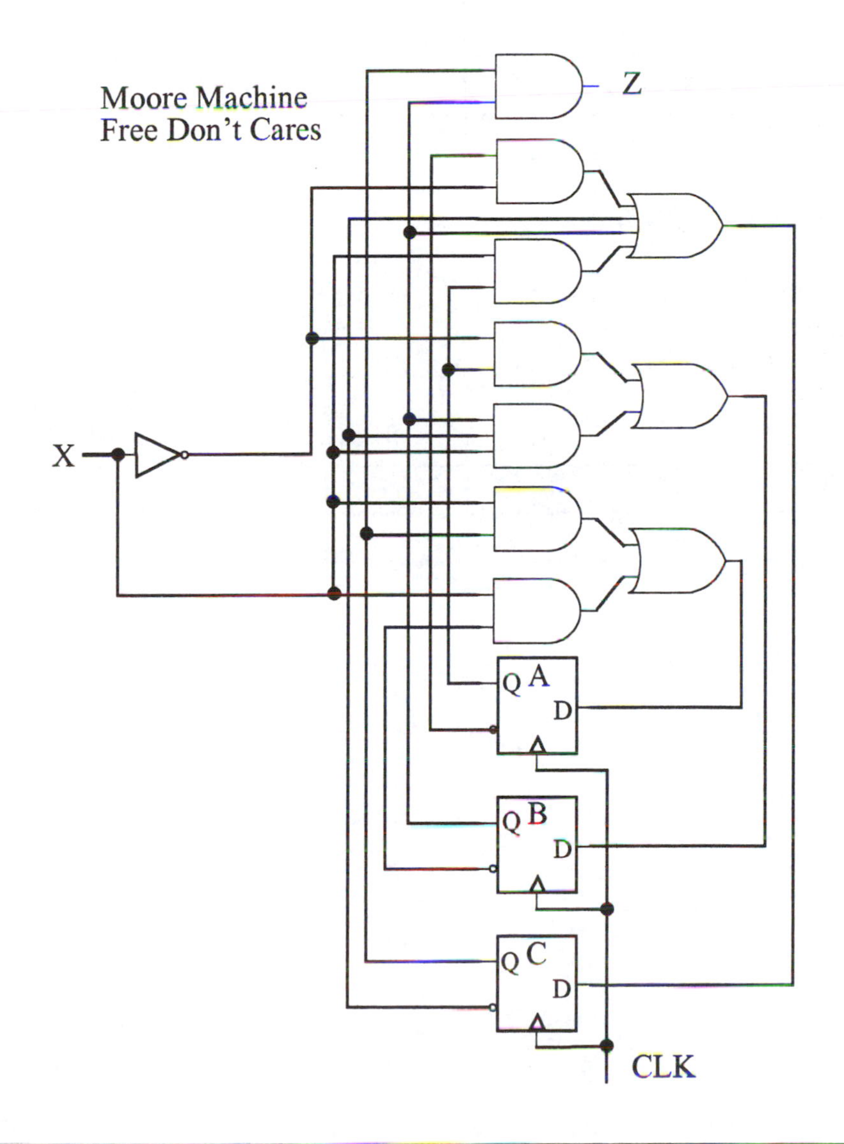

b.

PS	NS		Z
	$X = 0$	$X = 1$	
S_0	S_1	S_2	0
S_1	S_1	S_3	0
S_2	S_4	S_2	0
S_3	S_5	S_2	0
S_4	S_1	S_5	0
S_5	S_1	S_2	1
S_6	S_0	S_0	0
S_7	S_0	S_0	0

Armstrong-Humphrey Rule #1: S_0,S_1,S_4,S_5 S_6,S_7 S_0,S_2,S_3,S_5

Armstrong-Humphrey Rule #2: S_1,S_2 S_1,S_3 S_2,S_4 S_2,S_5 S_1,S_5

Armstrong-Humphrey Rule #3: S_0,S_1,S_2,S_3,S_6,S_7

Q_A \ Q_BQ_C	00	01	11	10
0	S_0	S_1	S_5	S_4
1	S_3	S_2		

$Q_A Q_B Q_C$	$Q_A^+ Q_B^+ Q_C^+$		Z
	$X = 0$	$X = 1$	
000	001	101	0
001	001	100	0
101	010	101	0
100	011	101	0
010	001	011	0
011	001	101	1
110	000	000	0
111	000	000	0

$Q_A^+ = D_A$

$Q_A Q_B$ \ $Q_C X$	00	01	11	10
00	0	1	1	0
01	0	0	1	0
11	0	0	0	0
10	0	1	1	0

$Q_B^+ = D_B$

$Q_A Q_B$ \ $Q_C X$	00	01	11	10
00	0	0	0	0
01	0	1	0	0
11	0	0	0	0
10	1	0	0	1

$Q_C^+ = D_C$

Q_AQ_B \ Q_CX	00	01	11	10
00	1	1	0	1
01	1	1	1	1
11	0	0	0	0
10	1	1	1	0

Z

Q_AQ_B \ Q_CX	00	01	11	10
00	0	0	1	0
01	0	0	0	0

$$D_A = Q_A^+ = (\overline{Q}_B \bullet X) + (\overline{Q}_A \bullet Q_C \bullet X)$$

$$D_B = Q_B^+ = (Q_A \bullet \overline{Q}_B \bullet \overline{X}) + (\overline{Q}_A \bullet Q_B \bullet \overline{Q}_C \bullet X)$$

$$D_C = Q_C^+ = (\overline{Q}_A + \overline{Q}_B) \bullet (\overline{Q}_A + \overline{Q}_C + X) \bullet (Q_A + Q_B + \overline{Q}_C + \overline{X})$$

$$Z = \overline{Q}_A \bullet Q_B \bullet Q_C$$

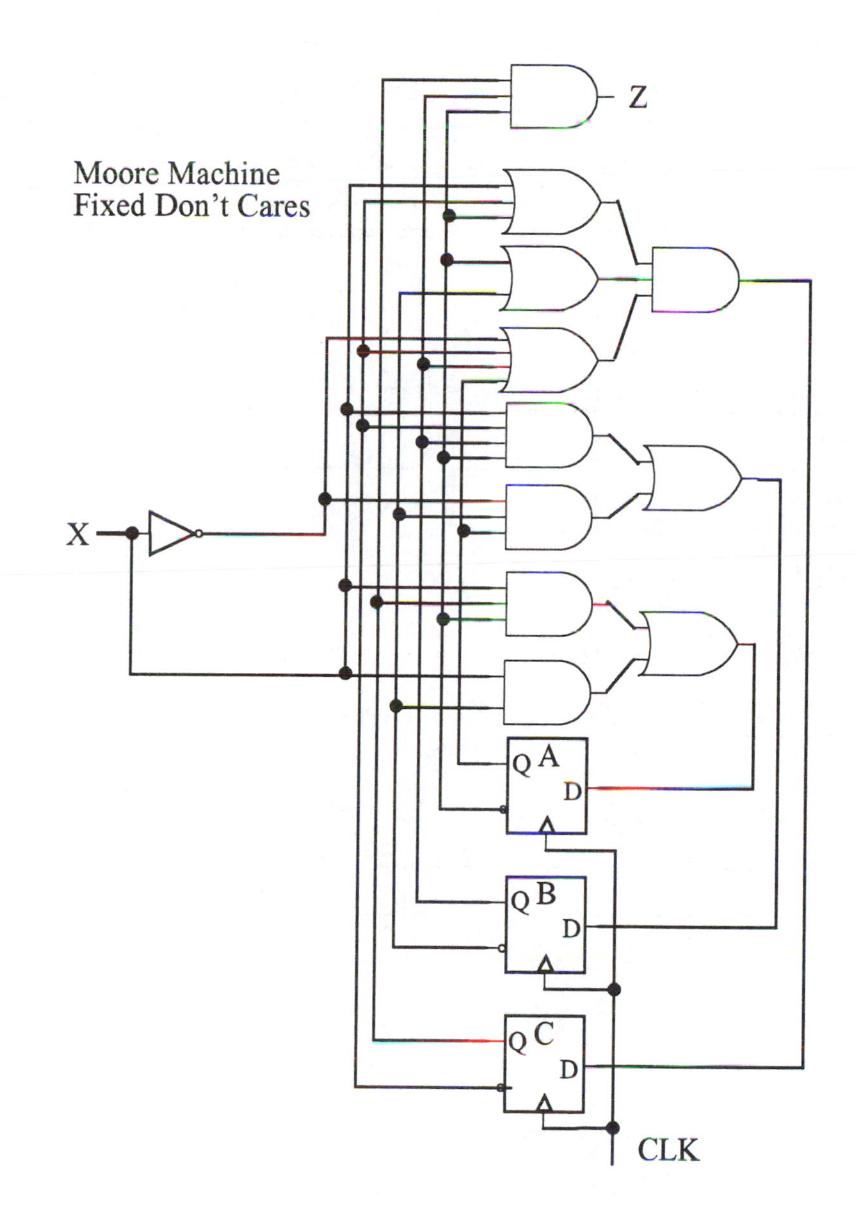
Z
Moore Machine
Fixed Don't Cares
X
Q A
D
Q B
D
Q C
D
CLK

9-4. Solution.

a.

PS	NS		Z	
	$X = 0$	$X = 1$	$X = 0$	$X = 1$
S_0	S_1	S_2	0	0
S_1	S_1	S_3	0	0
S_2	S_4	S_2	0	0
S_3	S_0	S_2	1	0
S_4	S_1	S_0	0	1

Armstrong-Humphrey Rule #1: S_0, S_1, S_4 S_0, S_2, S_3

Armstrong-Humphrey Rule #2: S_1, S_2 S_1, S_3 S_2, S_4 S_0, S_2 S_0, S_1

Armstrong-Humphrey Rule #3: S_0, S_1, S_2, S_4 S_0, S_1, S_2, S_3

Q_A \ Q_BQ_C	00	01	11	10
0	S_0	S_1	S_3	S_2
1		S_4		

$Q_A Q_B Q_C$	$Q_A^+ Q_B^+ Q_C^+$		Z	
	$X = 0$	$X = 1$	$X = 0$	$X = 1$
000	001	010	0	0
001	001	011	0	0
010	101	010	0	0
011	000	010	1	0
101	001	000	0	1

$Q_A^+ = D_A$

Q_AQ_B \ Q_CX	00	01	11	10
00	0	0	0	0
01	1	0	0	0
11	-	-	-	-
10	-	-	0	0

$Q_B^+ = D_B$

Q_AQ_B \ Q_CX	00	01	11	10
00	0	1	1	0
01	0	1	1	0
11	-	-	-	-
10	-	-	0	0

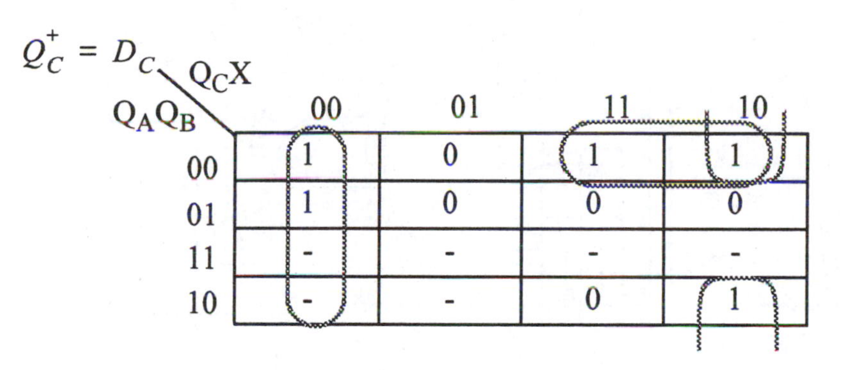

$Q_C^+ = D_C$

Q_AQ_B \ Q_CX	00	01	11	10
00	1	0	1	1
01	1	0	0	0
11	-	-	-	-
10	-	-	0	1

Z

Q_AQ_B \ Q_CX	00	01	11	10
00	0	0	0	0
01	0	0	0	1
11	-	-	-	-
10	-	-	1	0

$D_A = (Q_A \bullet \overline{Q}_C \bullet \overline{X})$

$D_B = \overline{Q}_A \bullet X$

$D_C = (\overline{Q}_C \bullet \overline{X}) + (\overline{Q}_A \bullet \overline{Q}_B \bullet Q_C) + (\overline{Q}_B \bullet Q_C \bullet \overline{X})$

$Z = (Q_A \bullet X) + (Q_B \bullet Q_C \bullet \overline{X})$

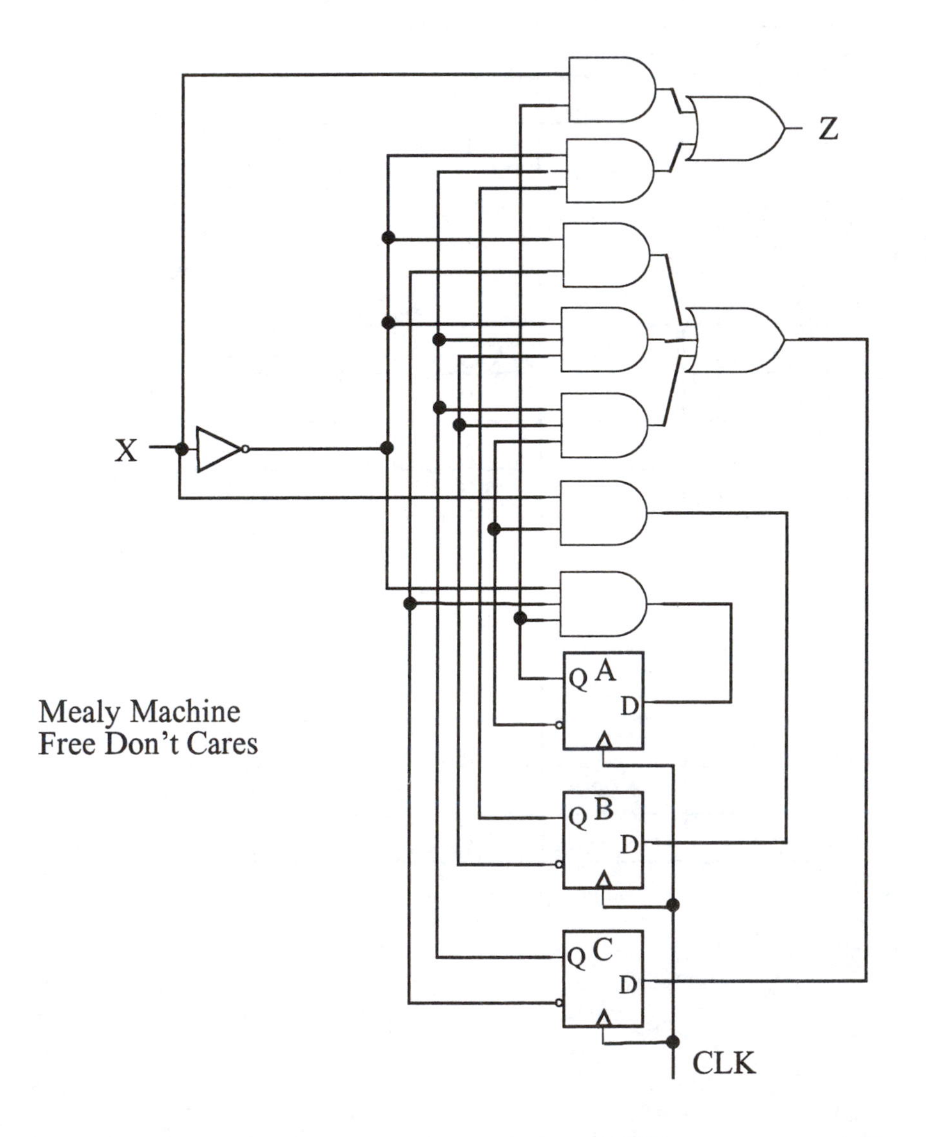

b.

PS	NS		Z	
	$X = 0$	$X = 1$	$X = 0$	$X = 1$
S_0	S_1	S_2	0	0
S_1	S_1	S_3	0	0
S_2	S_4	S_2	0	0
S_3	S_0	S_2	1	0
S_4	S_1	S_0	0	1
S_5	S_0	S_0	0	0
S_6	S_0	S_0	0	0
S_7	S_0	S_0	0	0

Armstrong-Humphrey Rule #1: S_0,S_1,S_4 S_0,S_2,S_3 S_3,S_5,S_6,S_7 S_4,S_5,S_6,S_7

Armstrong-Humphrey Rule #2: S_1,S_2 S_1,S_3 S_2,S_4 S_0,S_2 S_0,S_1

Armstrong-Humphrey Rule #3: $S_0,S_1,S_2,S_4,S_5,S_6,S_7$ $S_0,S_1,S_2,S_3,S_5,S_6,S_7$

$Q_A \backslash Q_BQ_C$	00	01	11	10
0	S_0	S_1	S_3	S_2
1	S_5	S_4	S_7	S_6

$Q_A Q_B Q_C$	$Q_A^+ Q_B^+ Q_C^+$		Z	
	$X = 0$	$X = 1$	$X = 0$	$X = 1$
000	001	010	0	0
001	001	011	0	0
010	101	010	0	0
011	000	010	1	0
101	001	000	0	1
100	000	000	0	0
110	000	000	0	0

$Q_A^+ = D_A$

$Q_A Q_B \backslash Q_C X$	00	01	11	10
00	0	0	0	0
01	1	0	0	0
11	0	0	0	0
10	0	0	0	0

$Q_B^+ = D_B$

$Q_A Q_B \backslash Q_C X$	00	01	11	10
00	0	1	1	0
01	0	1	1	0
11	0	0	0	0
10	0	0	0	0

$Q_C^+ = D_C$

Q_AQ_B \ Q_CX	00	01	11	10
00	1	0	1	1
01	1	0	0	0
11	0	0	0	0
10	0	0	0	1

Z

Q_AQ_B \ Q_CX	00	01	11	10
00	0	0	0	0
01	0	0	0	1
11	0	0	0	0
10	0	0	1	0

$$D_A = Q_A^+ = (\overline{Q}_A \bullet Q_B \bullet \overline{Q}_C \bullet \overline{X})$$

$$D_B = Q_B^+ = (\overline{Q}_A \bullet X)$$

$$D_C = (\overline{Q}_A \bullet \overline{Q}_C \bullet \overline{X}) + (\overline{Q}_A \bullet \overline{Q}_B \bullet Q_C) + (\overline{Q}_B \bullet Q_C \bullet \overline{X})$$

$$Z = (\overline{Q}_A \bullet Q_B \bullet Q_C \bullet \overline{X}) + (Q_A \bullet \overline{Q}_B \bullet Q_C \bullet X)$$

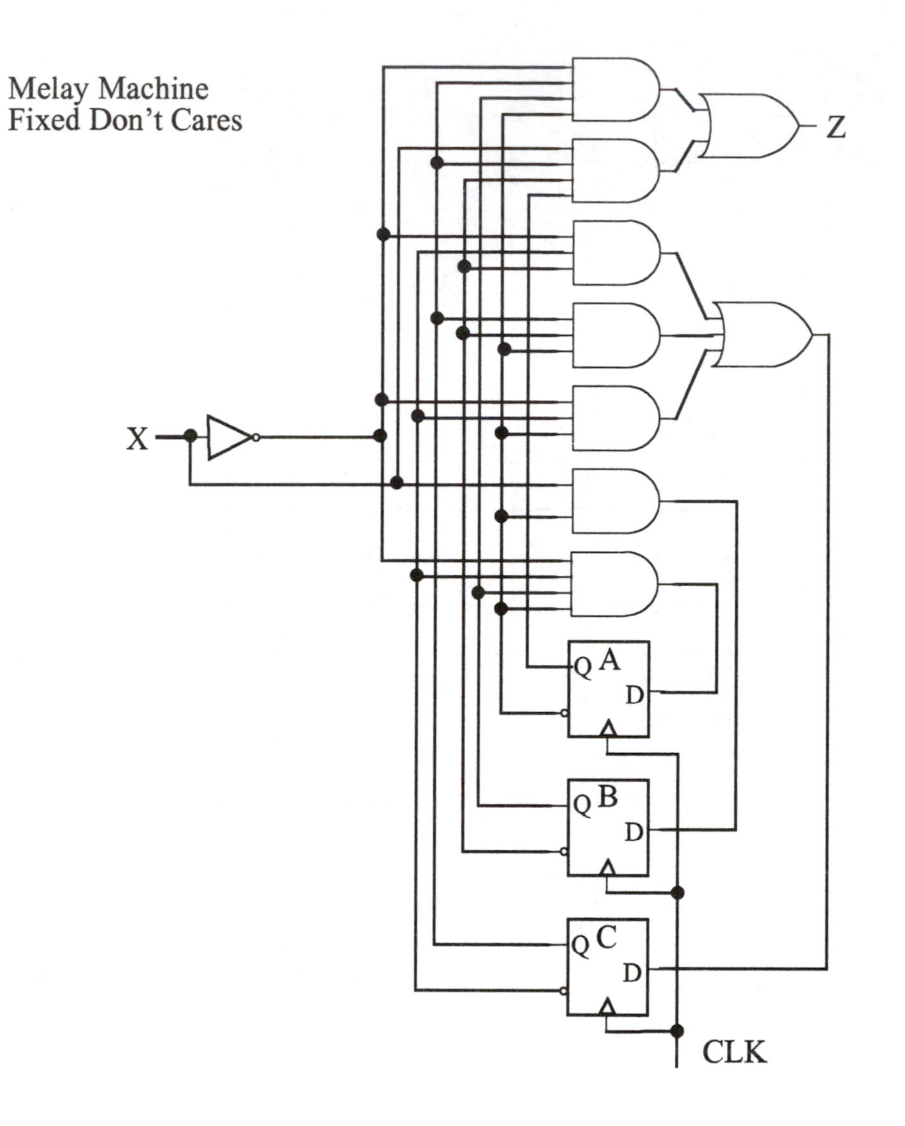
Melay Machine
Fixed Don't Cares
Z
X
Q
A
D
Q
B
D
Q
C
D
CLK

9-5. Solution.

PS	NS		Z	
	$X = 0$	$X = 1$	$X = 0$	$X = 1$
a	a	c	0	1
b	a	c	0	1
c	d	c	1	0
d	b	a	0	1

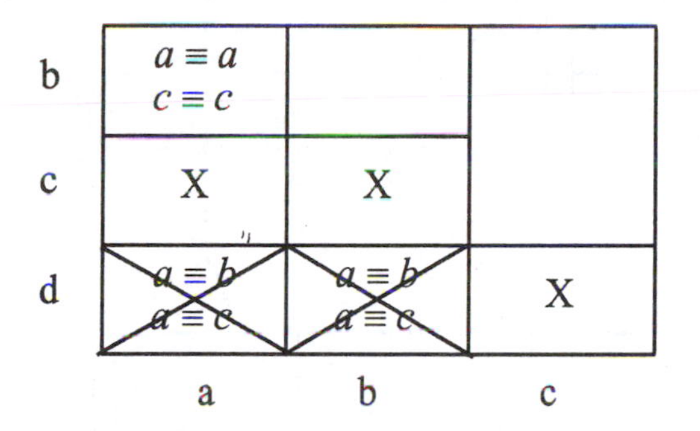

Yes. State a is equivalent to state b

9-6. Solution.

PS	NS		Z
	$X = 0$	$X = 1$	
a	e	c	0
b	h	d	0
c	d	g	0
d	h	e	1
e	f	d	1
f	b	e	0
g	e	a	0
h	f	d	0

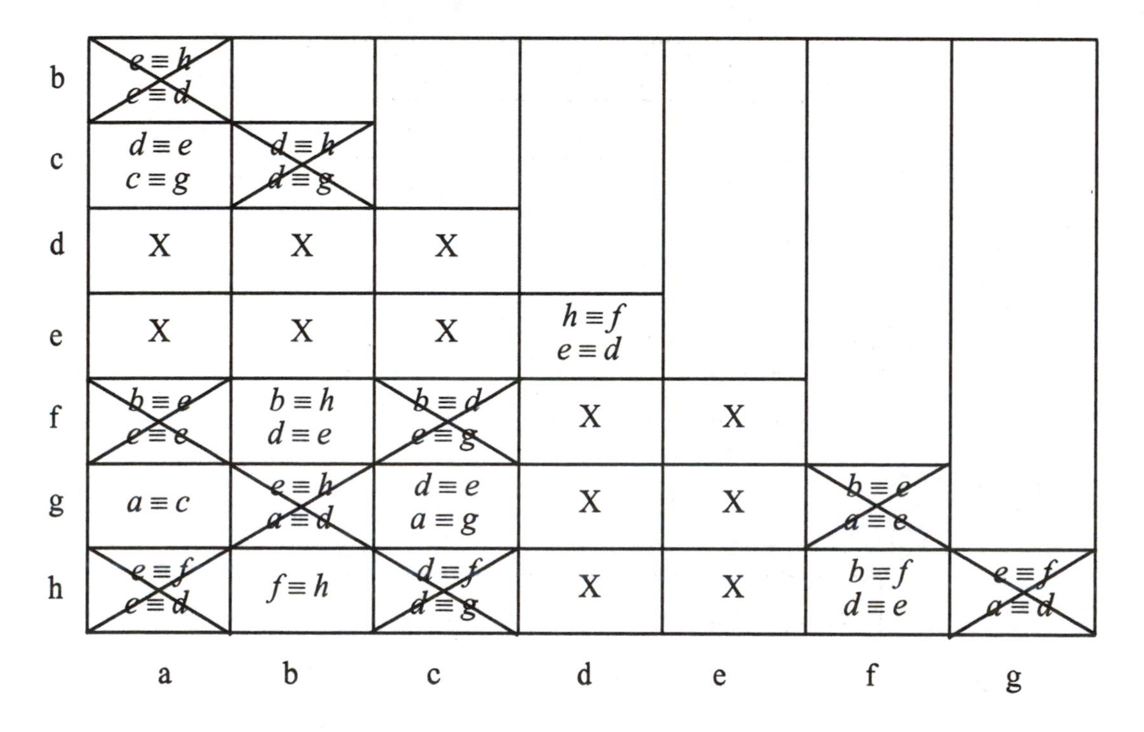

Yes. Equivalent states are $d \equiv e$, $a \equiv c \equiv g$, and $b \equiv h \equiv f$

9-7. Solution.

a.

Present State	Next State	Output $Z_2Z_1Z_0$
S_0	S_1	001
S_1	S_2	011
S_2	S_3	101
S_3	S_0	111

S_0=00, S_1=01, S_2=11, S_3=10

$Q_A Q_B$	$Q_A^+ Q_B^+$	$Z_2Z_1Z_0$
00	01	001
01	11	011
11	10	101
10	00	111

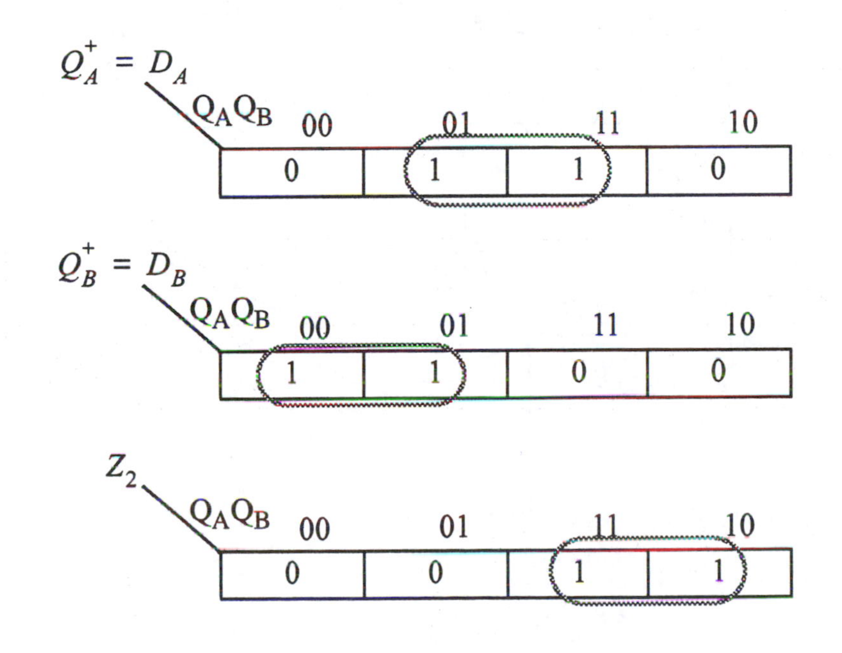

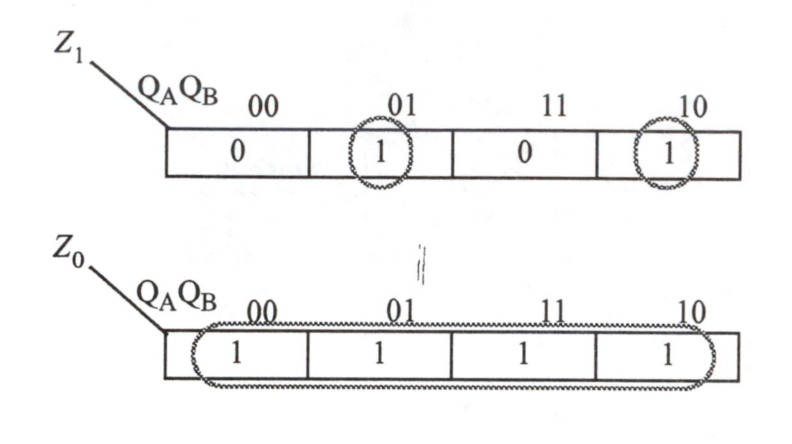

$D_A = Q_B$

$D_B = \overline{Q}_A$

$Z_2 = Q_A$

$Z_1 = Q_A \oplus Q_B$

$Z_0 = 1$

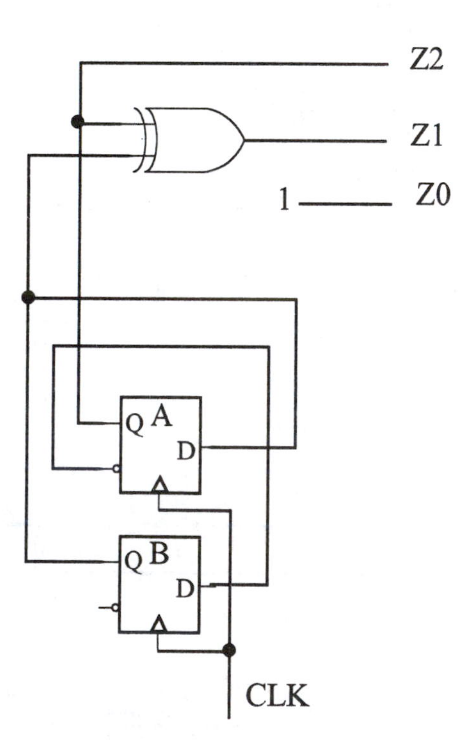

b.

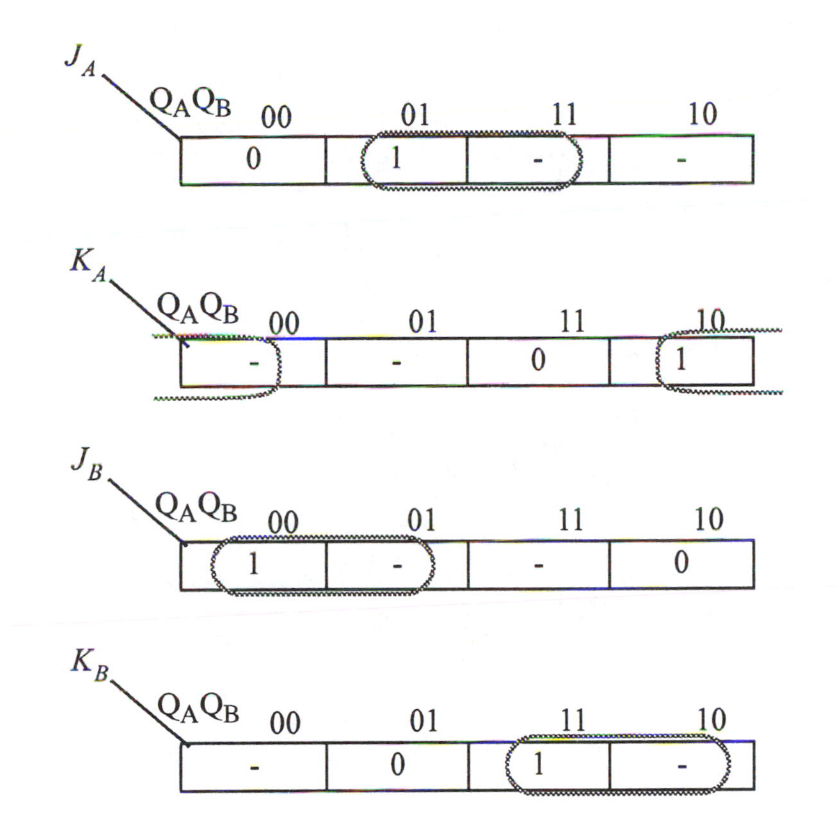

$$J_A = Q_B$$

$$K_A = \overline{Q}_B$$

$$J_B = \overline{Q}_A$$

$$K_B = Q_A$$

$$Z_2 = Q_A$$

$$Z_1 = Q_A \oplus Q_B$$

$$Z_0 = 1$$

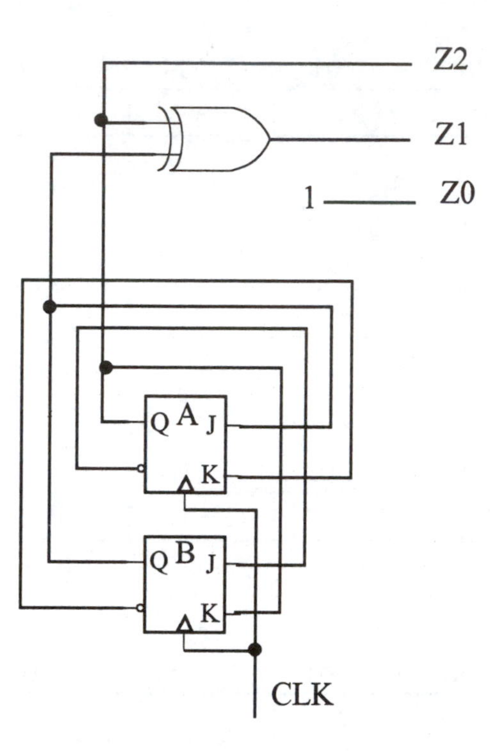

9-8. Solution.

a.

Present State	Next State (EN_BAR UP_DWN)				Output
	00	01	10	11	Z_1Z_0
S_0	S_1	S_3	S_0	S_0	00
S_1	S_2	S_0	S_1	S_1	01
S_2	S_3	S_1	S_2	S_2	10
S_3	S_0	S_2	S_3	S_3	11

Armstrong-Humphrey Rule #1:

Armstrong-Humphrey Rule #2: S_0,S_1,S_3 S_0,S_1,S_2 S_1,S_2,S_3 S_0,S_2,S_3

Armstrong-Humphrey Rule #3: S_0,S_1 S_2,S_3 S_0,S_2 S_1,S_3

Q_AQ_B	00	01	11	10
	S_0	S_1	S_3	S_2

State Assignment: S_0=00, S_1=01, S_2=10, S_3=11

Present State $Q_A Q_B$	**Next State (EN_BAR UP_DWN)** $Q_A^+ Q_B^+$				**Output** Z_1Z_0
	00	**01**	**10**	**11**	
00	01	11	00	00	00
01	10	00	01	01	01
10	11	01	10	10	10
11	00	10	11	11	11

Q_A^+

Q_AQ_B \ EN_BAR UP_DWN	00	01	11	10
00	0	1	0	0
01	1	0	0	0
11	0	1	1	1
10	1	0	1	1

Q_B^+

Q_AQ_B \ EN_BAR UP_DWN	00	01	11	10
00	1	1	0	0
01	0	0	1	1
11	0	0	1	1
10	1	1	0	0

T_A — Q_AQ_B \ EN_BAR UP_DWN

Q_AQ_B	00	01	11	10
00	0	1	0	0
01	1	0	0	0
11	1	0	0	0
10	0	1	0	0

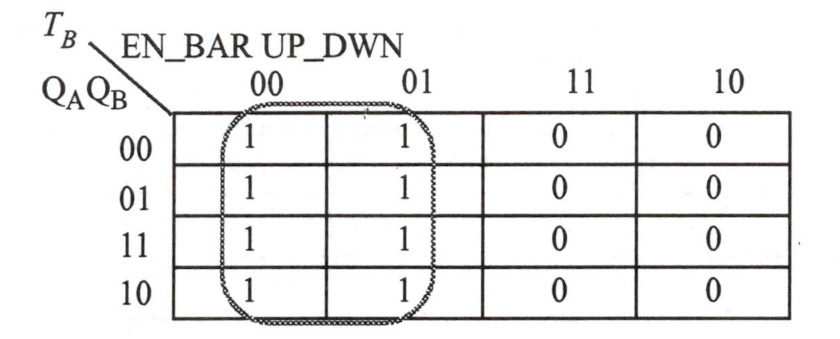

T_B — Q_AQ_B \ EN_BAR UP_DWN

Q_AQ_B	00	01	11	10
00	1	1	0	0
01	1	1	0	0
11	1	1	0	0
10	1	1	0	0

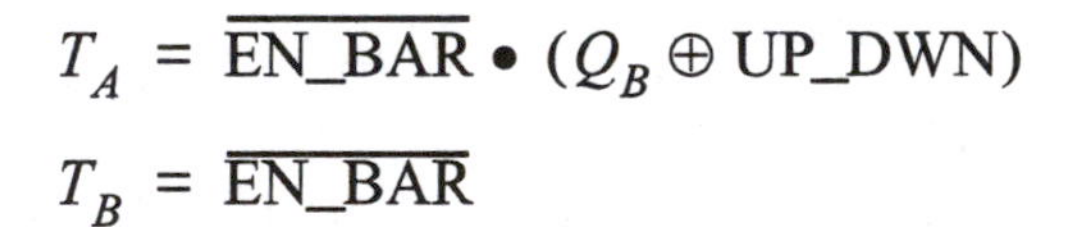

$$T_A = \overline{\text{EN\_BAR}} \bullet (Q_B \oplus \text{UP\_DWN})$$

$$T_B = \overline{\text{EN\_BAR}}$$

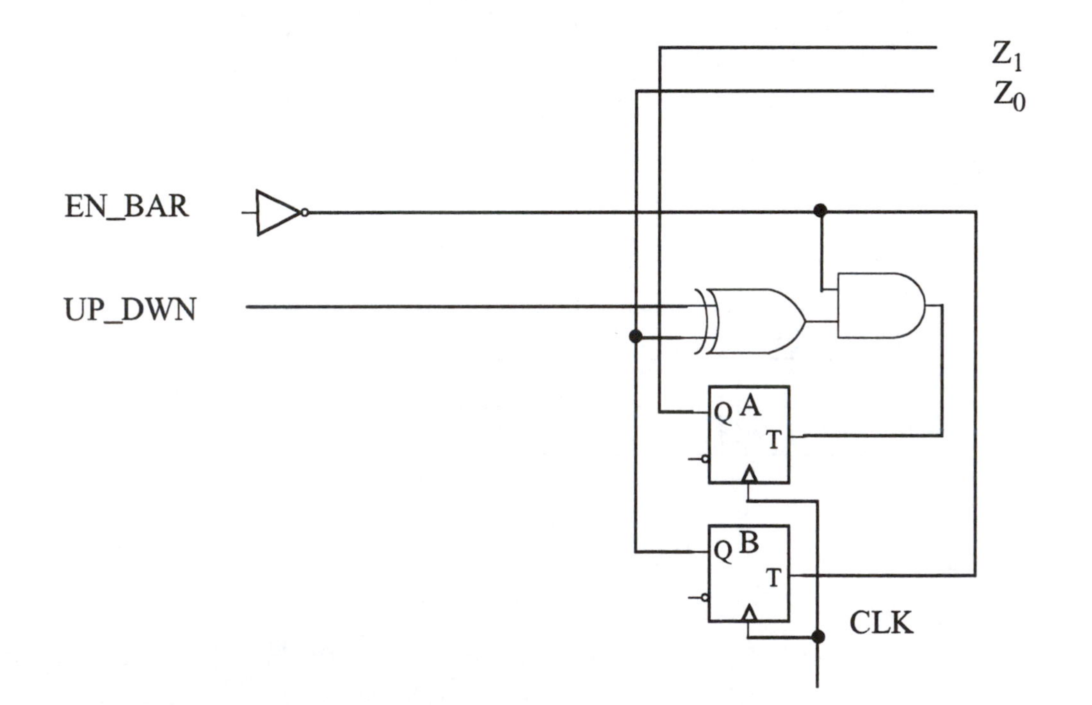

b.

S_A

Q_AQ_B \ EN_BAR UP_DWN	00	01	11	10
00	0	1	0	0
01	1	0	0	0
11	0	-	-	-
10	-	0	-	-

R_A

Q_AQ_B \ EN_BAR UP_DWN	00	01	11	10
00	-	0	-	-
01	0	-	-	-
11	1	0	0	0
10	0	1	0	0

S_B

Q_AQ_B \ EN_BAR UP_DWN	00	01	11	10
00	1	1	0	0
01	0	0	-	-
11	0	0	-	-
10	1	1	0	0

R_B

Q_AQ_B \ EN_BAR UP_DWN	00	01	11	10
00	0	0	-	-
01	1	1	0	0
11	1	1	0	0
10	0	0	-	-

$$S_A = \overline{Q}_A \bullet \overline{\text{EN\_BAR}} \bullet (Q_B \oplus \text{UP\_DWN})$$

$$R_A = Q_A \bullet \overline{\text{EN\_BAR}} \bullet (Q_B \oplus \text{UP\_DWN})$$

$$S_B = \overline{Q}_B \bullet \overline{\text{EN\_BAR}}$$

$$R_B = Q_B \bullet \overline{\text{EN\_BAR}}$$

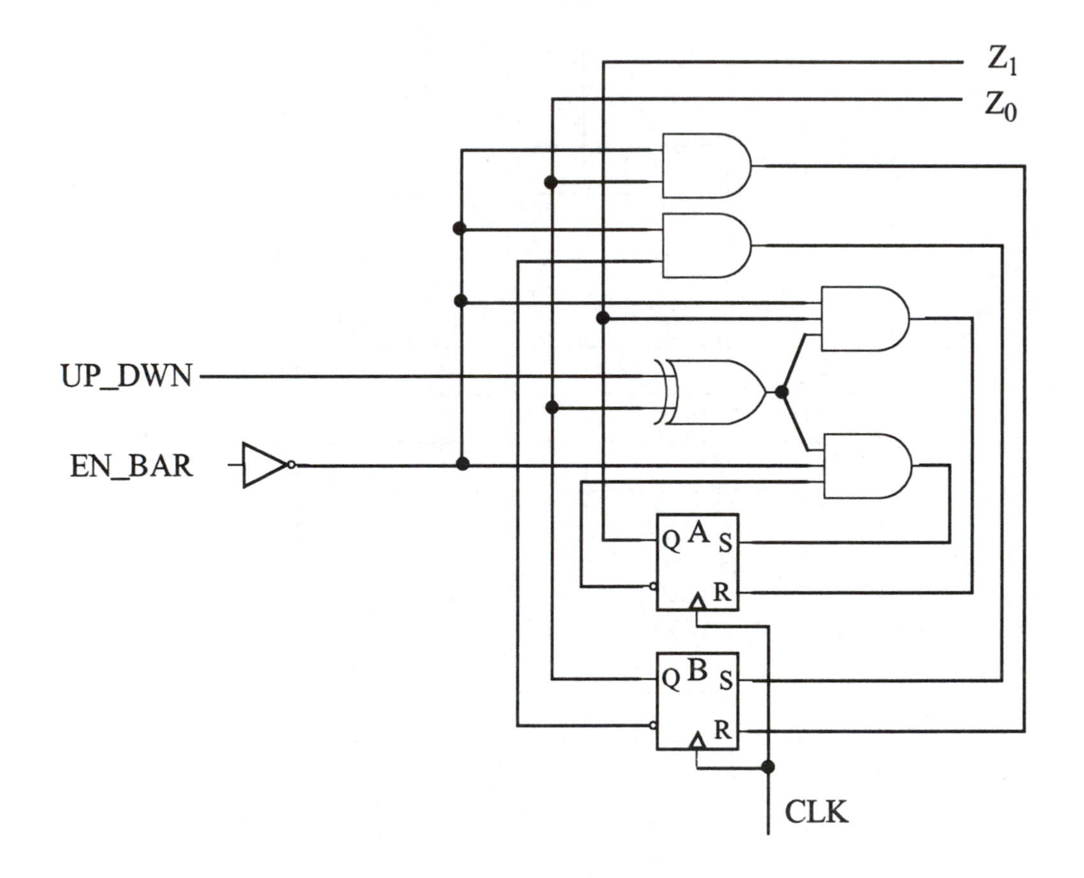

9-9. Solution.

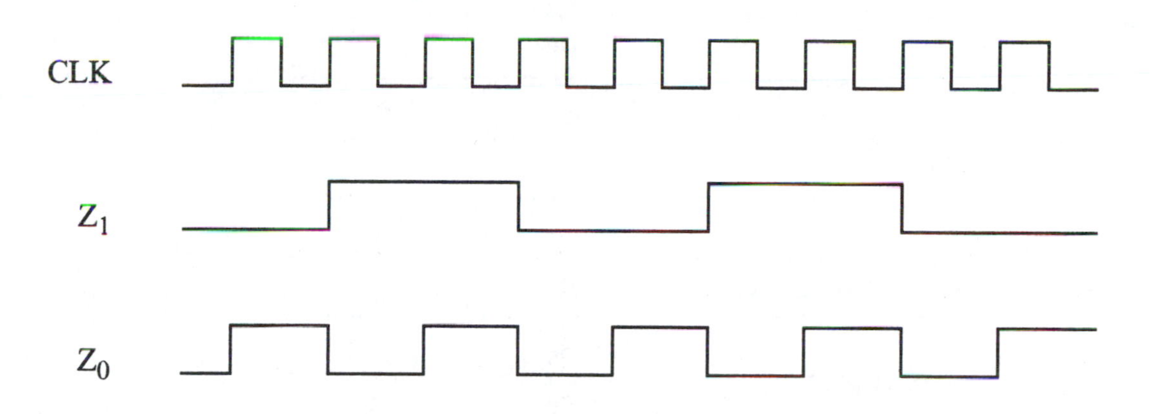

a. The frequency of Z0 is 1/2 the frequency of CLK, i.e., divide-by-two

b. The frequency of Z1 is 1/4 the frequency of CLK, i.e., divide-by-four

9-10. Solution.

Present State	Next State	Output DIV6
S_0	S_1	0
S_1	S_2	0
S_2	S_3	0
S_3	S_4	1
S_4	S_5	1
S_5	S_0	1

State Assignment: S_0=000, S_1=001, S_2=010, S_3=011, S_4=100, S_5=101

Present State	Next State	Output DIV10
S_0	S_1	0
S_1	S_2	0
S_2	S_3	0
S_3	S_4	0
S_4	S_5	0
S_5	S_6	1
S_6	S_7	1
S_7	S_8	1
S_8	S_9	1
S_9	S_0	1

State Assignment: S_0=0000, S_1=0001, S_2=0010, S_3=0011, S_4=0100, S_5=0101, S_6=0110, S_7=0111, S_8=1000, S_9=1001

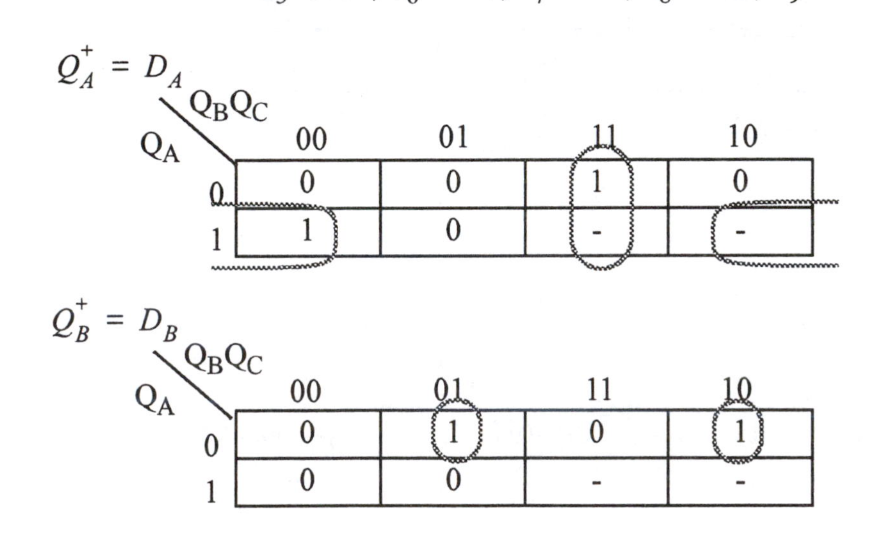

$Q_C^+ = D_C$

Q_A \ Q_BQ_C	00	01	11	10
0	1	0	0	1
1	1	0	-	-

$Q_A^+ = D_A$

Q_AQ_B \ Q_CQ_D	00	01	11	10
00	0	0	0	0
01	0	0	1	0
11	-	-	-	-
10	1	0	-	-

$Q_B^+ = D_B$

Q_AQ_B \ Q_CQ_D	00	01	11	10
00	0	0	1	0
01	1	1	0	1
11	-	-	-	-
10	0	0	-	-

$Q_C^+ = D_C$

Q_AQ_B \ Q_CQ_D	00	01	11	10
00	0	1	0	1
01	0	1	0	1
11	-	-	-	-
10	0	0	-	-

$Q_D^+ = D_D$

Q_AQ_B \ Q_CQ_D	00	01	11	10
00	1	0	0	1
01	1	0	0	1
11	-	-	-	-
10	1	0	-	-

$$D_A = (Q_A \bullet \overline{Q}_B) + (Q_B \bullet Q_C)$$

$$D_B = \overline{Q}_A \bullet (Q_B \oplus Q_C)$$

$$D_C = \overline{Q}_C$$

$$D_A = (Q_A \bullet \overline{Q}_C \bullet \overline{Q}_D) + (Q_B \bullet Q_C \bullet Q_D)$$

$$D_B = (Q_B \bullet \overline{Q}_C) + (Q_B \bullet \overline{Q}_D) + (\overline{Q}_B \bullet Q_C \bullet Q_D)$$

$$D_C = \overline{Q}_A \bullet (Q_C \oplus Q_D)$$

$$D_D = \overline{Q}_D$$

DIV6

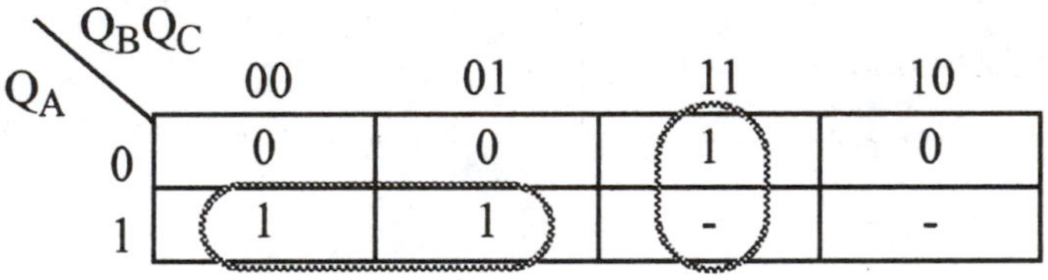

Q_A \ Q_BQ_C	00	01	11	10
0	0	0	1	0
1	1	1	-	-

DIV10

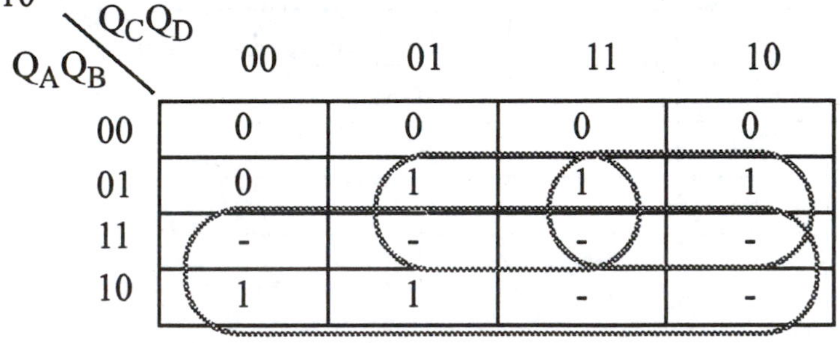

Q_AQ_B \ Q_CQ_D	00	01	11	10
00	0	0	0	0
01	0	1	1	1
11	-	-	-	-
10	1	1	-	-

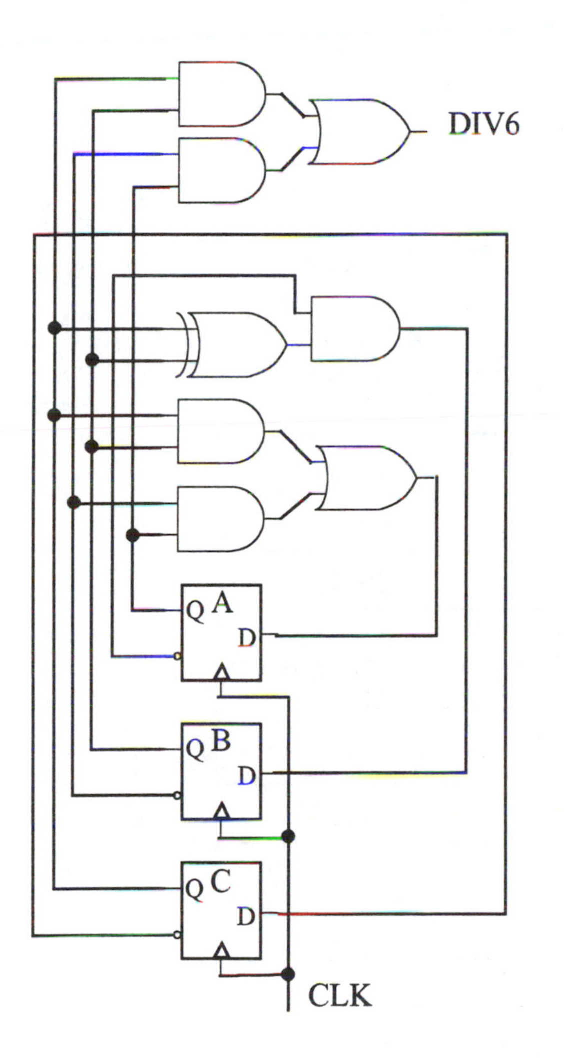
DIV6
Q A
D
Q B
D
Q C
D
CLK

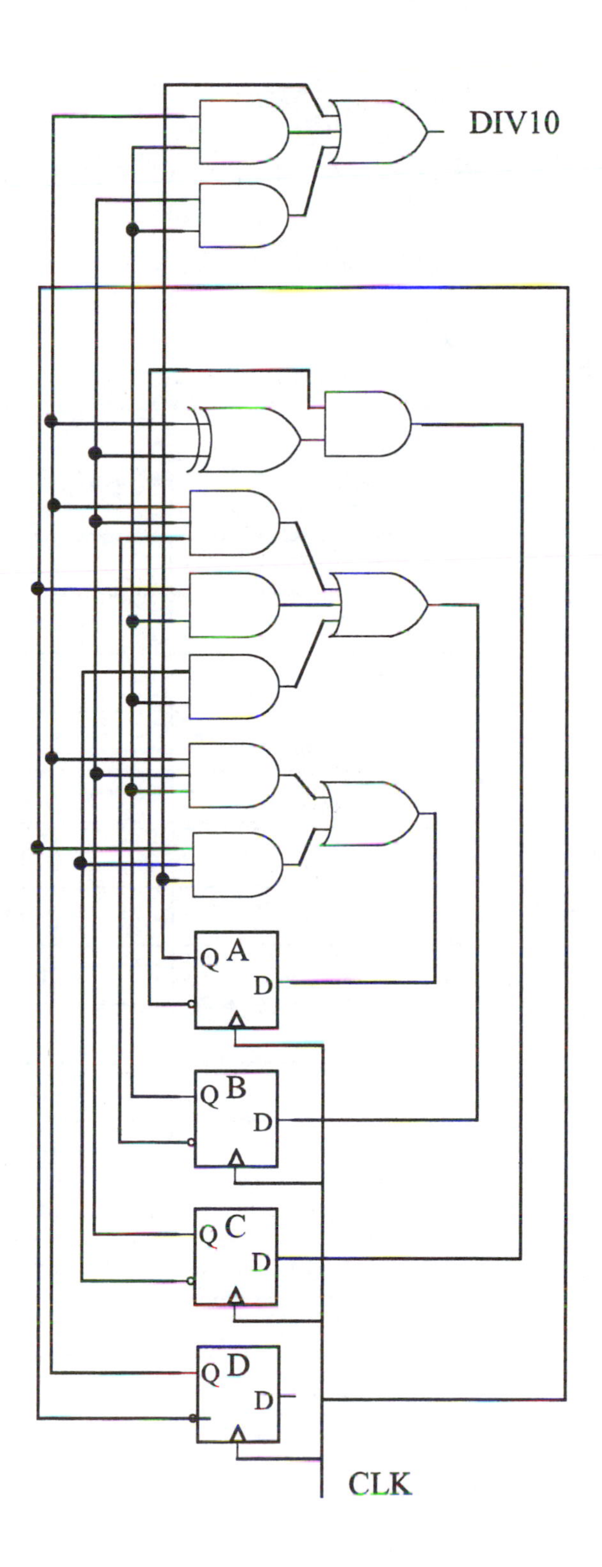
DIV10
Q A
D
Q B
D
Q C
D
Q D
D
CLK

9-11. Solution.

$$D_A = (Q_A \bullet Q_B) + (Q_A \bullet \overline{Q}_D) + (Q_A \bullet Q_C) + (Q_B \bullet Q_C \bullet Q_D) = Q_A^+$$

$$D_B = (Q_B \bullet \overline{Q}_C) + (Q_B \bullet \overline{Q}_D) + (\overline{Q}_B \bullet Q_C \bullet Q_D) = Q_B^+$$

$$D_C = (Q_C \bullet \overline{Q}_D) + (\overline{Q}_A \bullet \overline{Q}_C \bullet Q_D) + (Q_A \bullet Q_B \bullet Q_D) = Q_C^+$$

$$D_D = \overline{Q}_D$$

$Q_A Q_B Q_C Q_D$	$Q_A^+ Q_B^+ Q_C^+ Q_D^+$	$Z_3 Z_2 Z_1 Z_0$
0000	0001	0000
0001	0010	0001
0010	0011	0010
0011	0100	0011
0100	0101	0100
0101	0110	0101
0110	0111	0110
0111	1000	0111
1000	1001	1000
1001	0000	1001
1010	1011	1010
1011	1100	1011
1100	1101	1100
1101	1110	1101
1110	1111	1110
1111	1010	1111

Q_A^+

Q_AQ_B \ Q_CQ_D	00	01	11	10
00	0	0	0	0
01	0	0	1	0
11	1	1	1	1
10	1	0	1	1

Q_B^+

Q_AQ_B \ Q_CQ_D	00	01	11	10
00	0	0	1	0
01	1	1	0	1
11	1	1	0	1
10	0	0	1	0

Q_C^+

Q_AQ_B \ Q_CQ_D	00	01	11	10
00	0	1	0	1
01	0	1	0	1
11	0	1	1	1
10	0	0	0	1

Q_D^+

Q_AQ_B \ Q_CQ_D	00	01	11	10
00	1	0	0	1
01	1	0	0	1
11	1	0	0	1
10	1	0	0	1

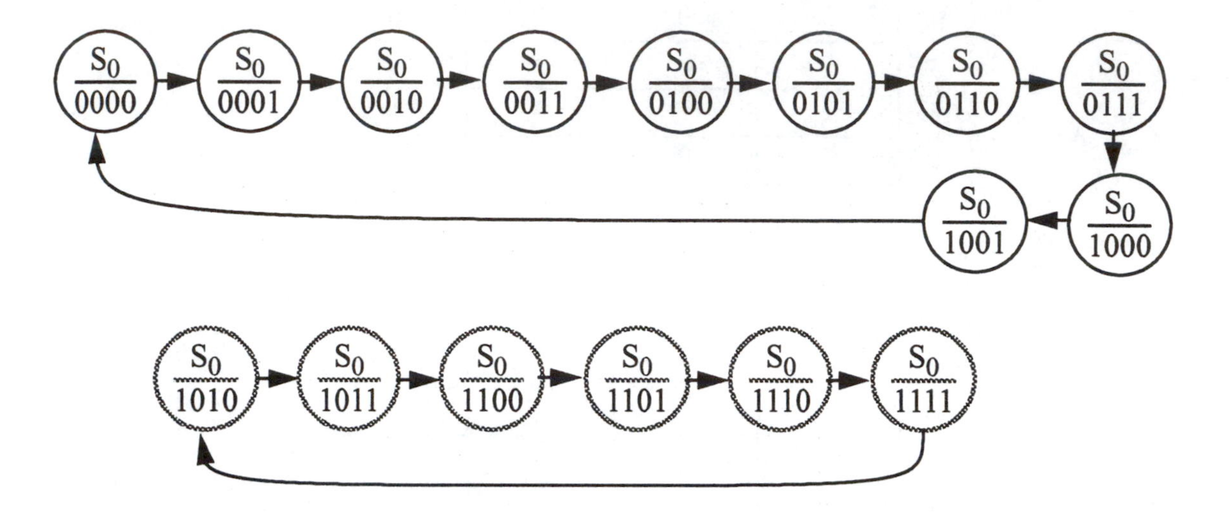

a. BCD Counter

b. The unused states form a loop, so once the counter enters the loop of unused states, it will stay in the loop and never return to the BCD counting loop.

9-12. Solution.

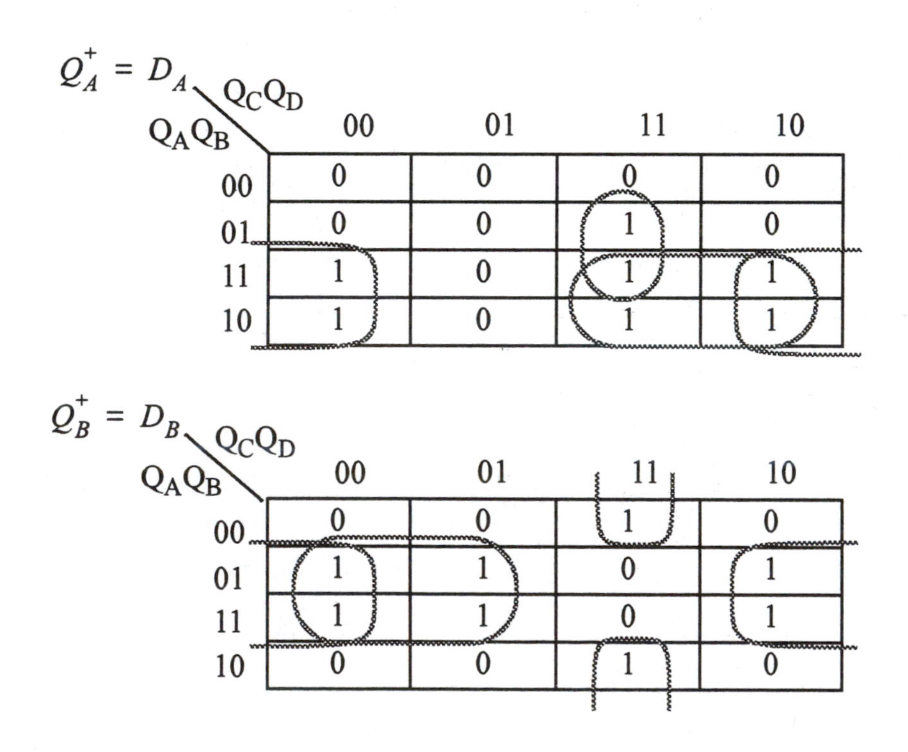

$Q_C^+ = D_C$

Q_AQ_B \ Q_CQ_D	00	01	11	10
00	0	1	0	1
01	0	1	0	1
11	0	0	0	1
10	0	0	0	1

$Q_D^+ = D_D$

Q_AQ_B \ Q_CQ_D	00	01	11	10
00	1	0	0	1
01	1	0	0	1
11	1	0	0	1
10	1	0	0	1

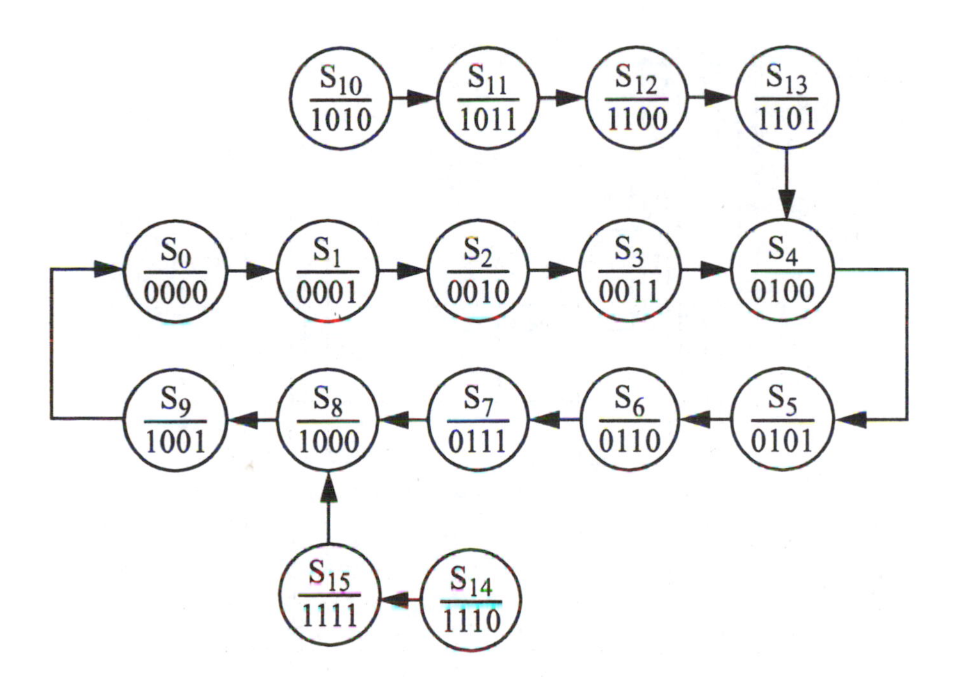

$$D_A = (Q_A \bullet \overline{Q}_D) + (Q_A \bullet Q_C) + (Q_B \bullet Q_C \bullet Q_D) = Q_A^+$$
$$D_B = (Q_B \bullet \overline{Q}_C) + (Q_B \bullet \overline{Q}_D) + (\overline{Q}_B \bullet Q_C \bullet Q_D) = Q_B^+$$
$$D_C = (Q_C \bullet \overline{Q}_D) + (\overline{Q}_A \bullet \overline{Q}_C \bullet Q_D) = Q_C^+$$
$$D_D = \overline{Q}_D$$

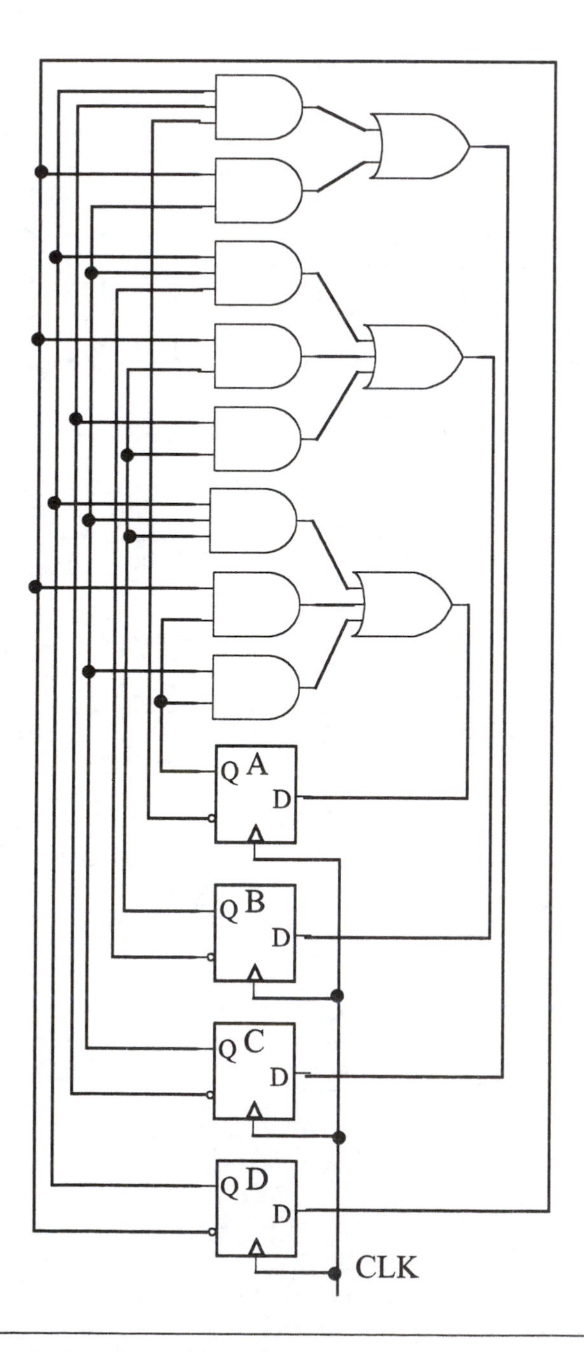

9-13. Solution.

Present State	Next State	a	b	c	d	e	f	g
S_0	S_1	1	1	1	1	1	1	0
S_1	S_2	0	1	1	0	0	0	0
S_2	S_3	1	1	0	1	1	0	1
S_3	S_4	1	1	1	1	0	0	1
S_4	S_5	0	1	1	0	0	1	1
S_5	S_6	1	0	1	1	0	1	1
S_6	S_7	1	0	1	1	1	1	1
S_7	S_8	1	1	1	0	0	0	0
S_8	S_9	1	1	1	1	1	1	1
S_9	S_0	1	1	1	0	0	1	1

Use self-correcting state assignment and next state logic of BCD counter designed in Problem 9-12.

$Q_A^+ = D_A$

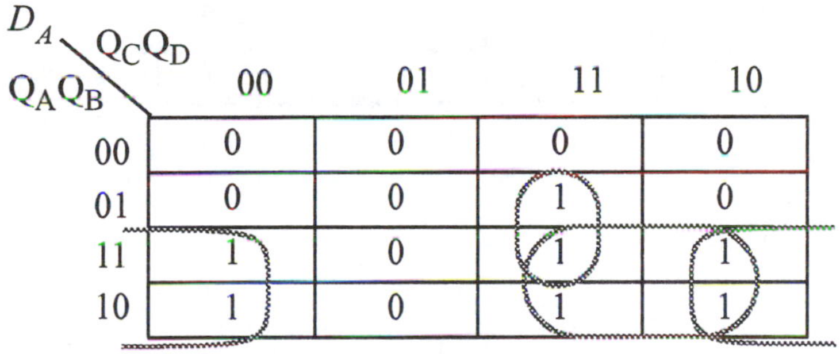

Q_AQ_B \ Q_CQ_D	00	01	11	10
00	0	0	0	0
01	0	0	1	0
11	1	0	1	1
10	1	0	1	1

$Q_B^+ = D_B$

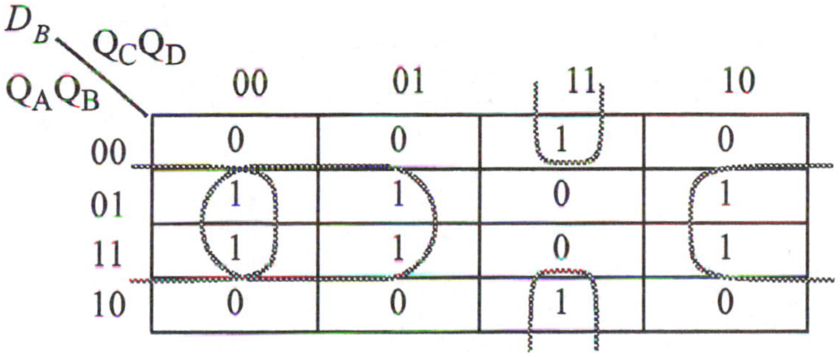

Q_AQ_B \ Q_CQ_D	00	01	11	10
00	0	0	1	0
01	1	1	0	1
11	1	1	0	1
10	0	0	1	0

$Q_C^+ = D_C$

Q_AQ_B \ Q_CQ_D	00	01	11	10
00	0	1	0	1
01	0	1	0	1
11	0	0	0	1
10	0	0	0	1

$Q_D^+ = D_D$

Q_AQ_B \ Q_CQ_D	00	01	11	10
00	1	0	0	1
01	1	0	0	1
11	1	0	0	1
10	1	0	0	1

a

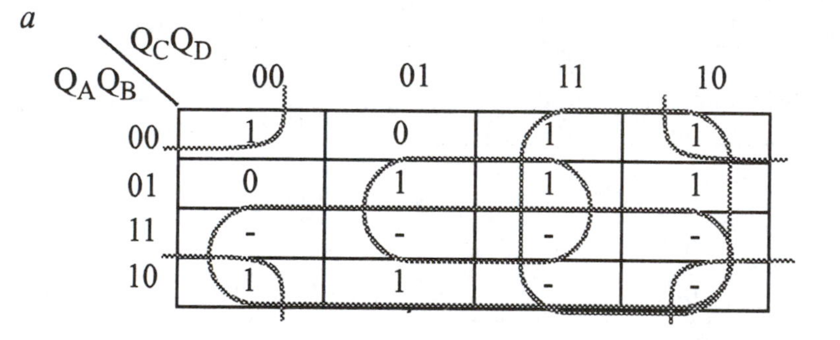

Q_AQ_B \ Q_CQ_D	00	01	11	10
00	1	0	1	1
01	0	1	1	1
11	-	-	-	-
10	1	1	-	-

b

Q_AQ_B \ Q_CQ_D	00	01	11	10
00	1	1	1	1
01	1	0	1	0
11	-	-	-	-
10	1	1	-	-

c

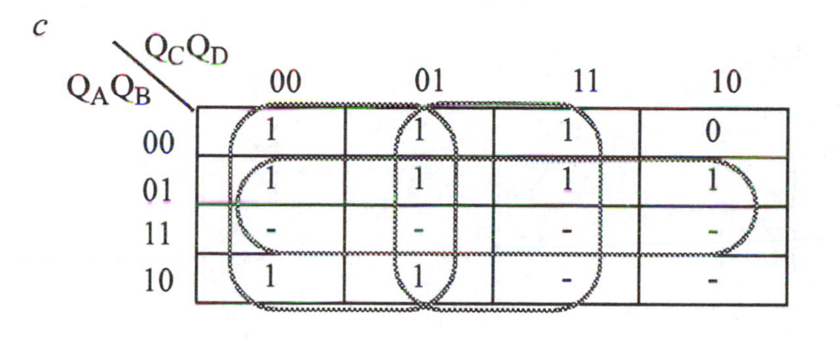

d

Q_AQ_B \ Q_CQ_D	00	01	11	10
00	1	0	1	1
01	0	1	0	1
11	-	-	-	-
10	1	0	-	-

e

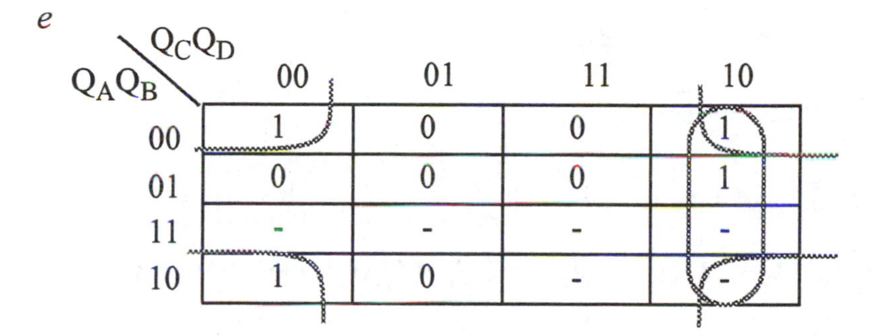

f

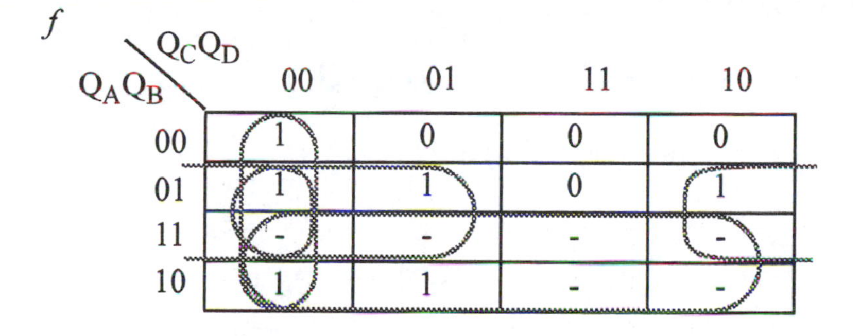

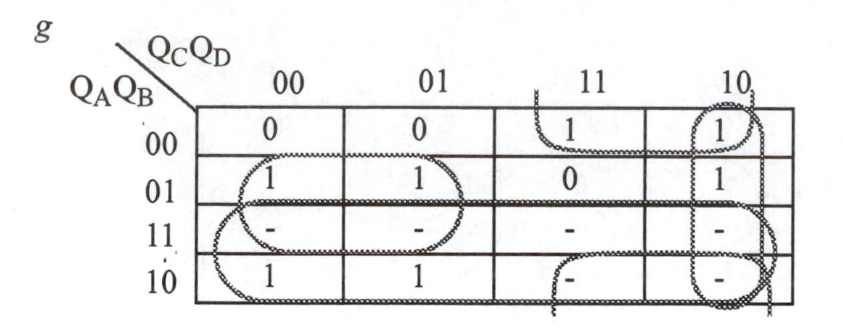

$$D_A = (Q_A \bullet \overline{Q}_D) + (Q_A \bullet Q_C) + (Q_B \bullet Q_C \bullet Q_D) = Q_A^+$$

$$D_B = (Q_B \bullet \overline{Q}_C) + (Q_B \bullet \overline{Q}_D) + (\overline{Q}_B \bullet Q_C \bullet Q_D) = Q_B^+$$

$$D_C = (Q_C \bullet \overline{Q}_D) + (\overline{Q}_A \bullet \overline{Q}_C \bullet Q_D) = Q_C^+$$

$$D_D = \overline{Q}_D = Q_D^+$$

$$a = Q_A + Q_C + \overline{(Q_B \oplus Q_D)}$$

$$b = \overline{Q}_B + \overline{(Q_C \oplus Q_D)}$$

$$c = \overline{Q}_C + Q_B + Q_D$$

$$d = (\overline{Q}_B \bullet \overline{Q}_D) + (\overline{Q}_B \bullet Q_C) + (Q_C \bullet \overline{Q}_D) + (Q_B \bullet \overline{Q}_C \bullet Q_D)$$

$$e = (\overline{Q}_B \bullet \overline{Q}_D) + (Q_C \bullet \overline{Q}_D)$$

$$f = Q_A + (\overline{Q}_C \bullet \overline{Q}_D) + (Q_B \bullet \overline{Q}_C) + (Q_B \bullet \overline{Q}_D)$$

$$g = Q_A + (Q_B \bullet \overline{Q}_C) + (Q_C \bullet \overline{Q}_D) + (\overline{Q}_B \bullet Q_C)$$

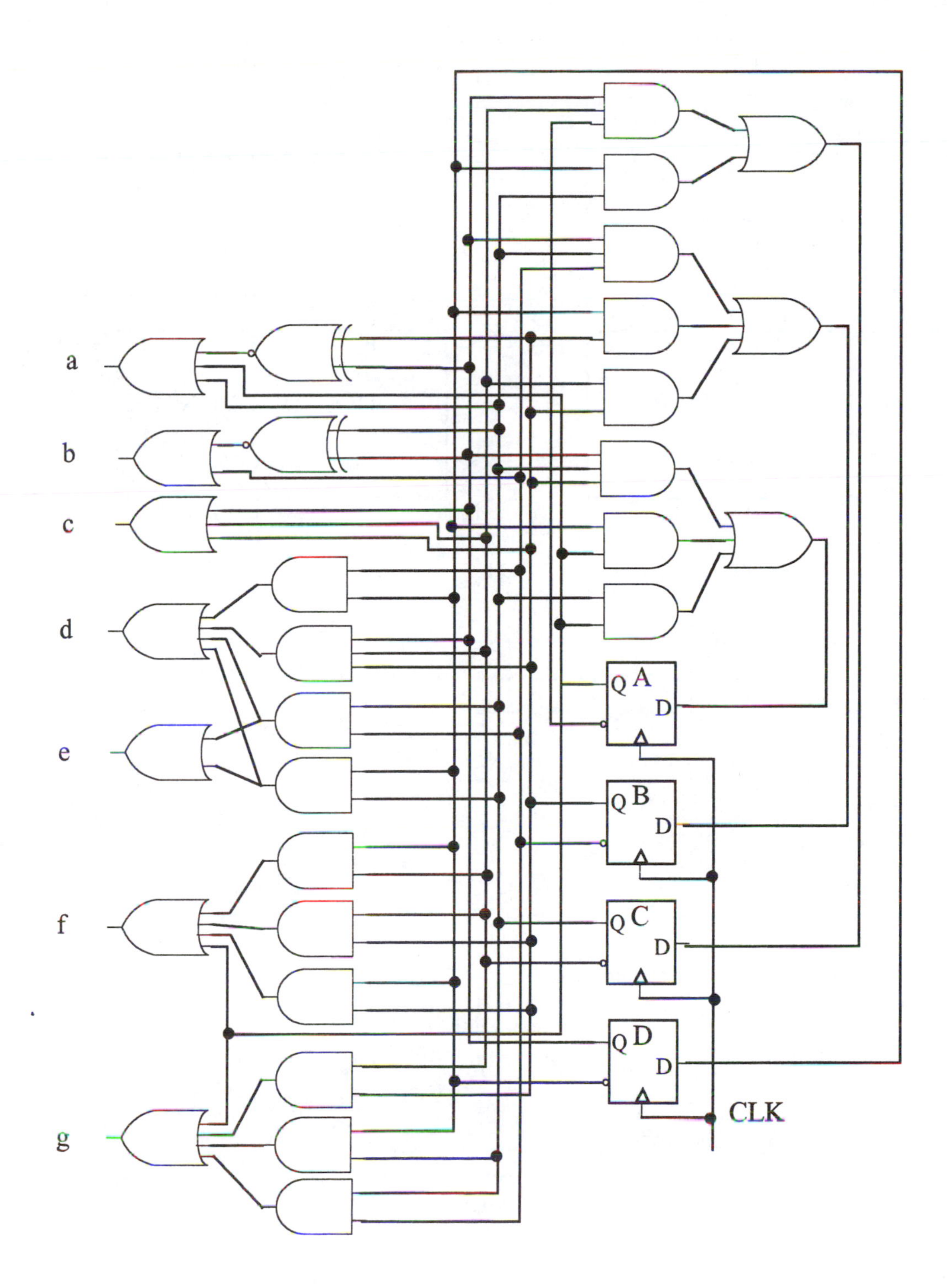
a
b
c
d
e
f
g
Q A
D
Q B
D
Q C
D
Q D
D
CLK

9-14. Solution.

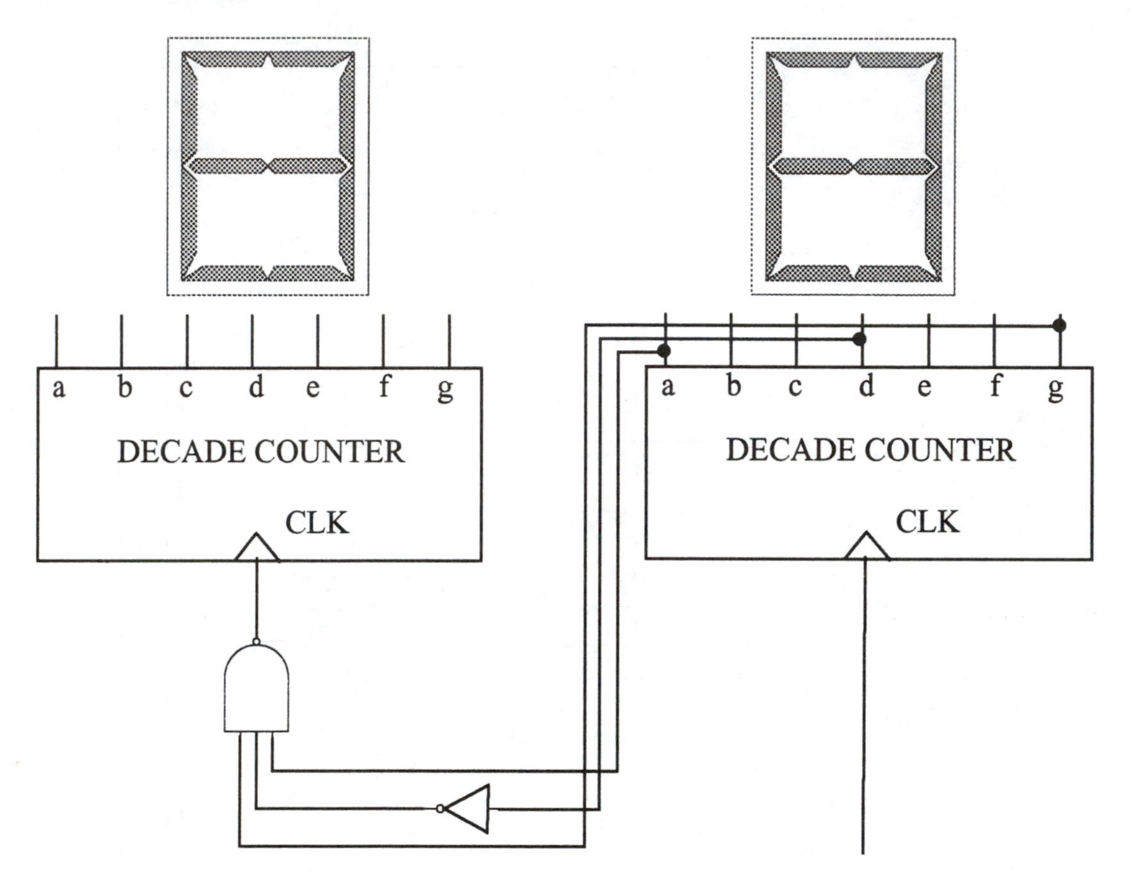

9-15. Solution.

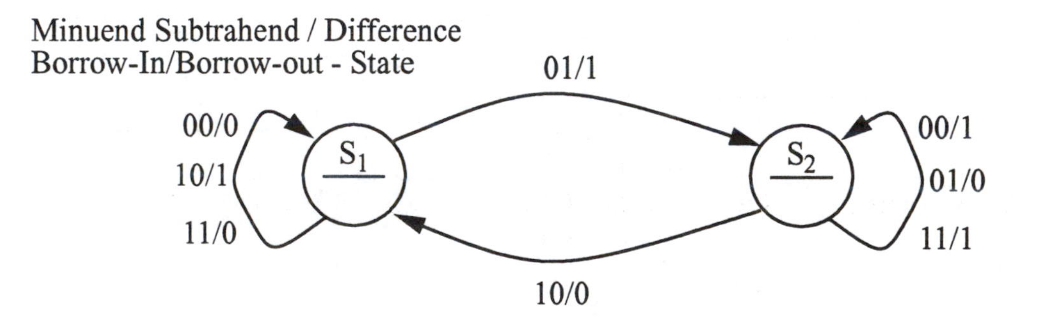

State Assignment: S1=0 and S2=1

Present State Q_A	Next State Q_A^+ Minuend (M) Subtrahend (S)				Difference Minuend (M) Subtrahend (S)			
	00	**01**	**10**	**11**	**00**	**01**	**10**	**11**
0	0	1	0	0	0	1	1	0
1	1	1	0	1	1	0	0	1

a. $Q_A^+ = D_A$

Q_A \ MS	00	01	11	10
0	0	1	0	0
1	1	1	1	0

DIFF

Q_A \ MS	00	01	11	10
0	0	1	0	1
1	1	0	1	0

$$D_A = (Q_A \bullet \overline{M}) + (Q_A \bullet S) + (\overline{M} \bullet S)$$

$$DIFF = Q_A \oplus M \oplus S$$

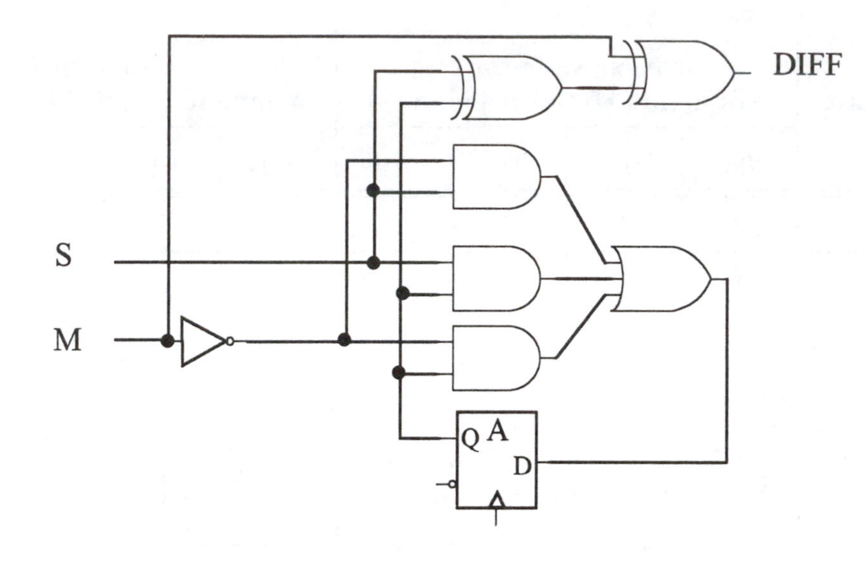

b.

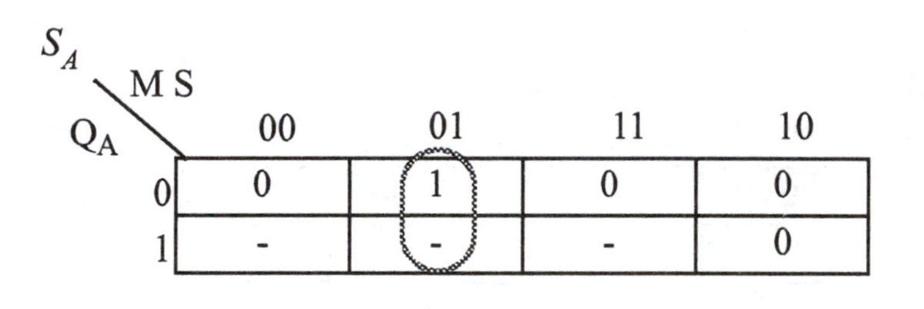

S_A

Q_A \ M S	00	01	11	10
0	0	1	0	0
1	-	-	-	0

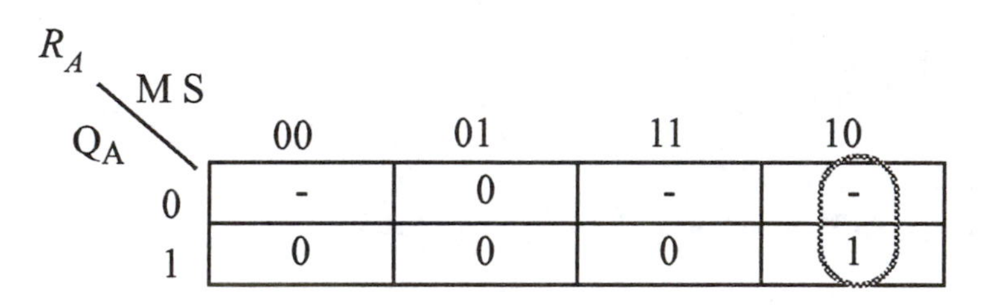

R_A

Q_A \ M S	00	01	11	10
0	-	0	-	-
1	0	0	0	1

$$S_A = (\overline{M} \bullet S)$$

$$R_A = (M \bullet \overline{S})$$

$$DIFF = Q_A \oplus M \oplus S$$

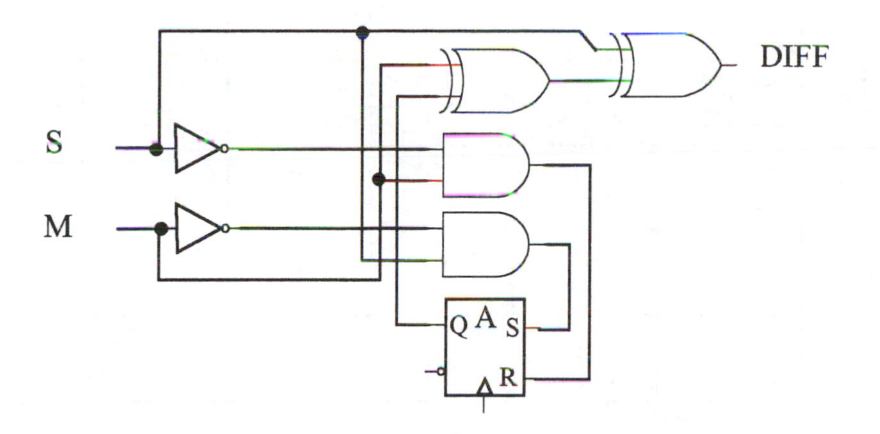

9-16. Solution.

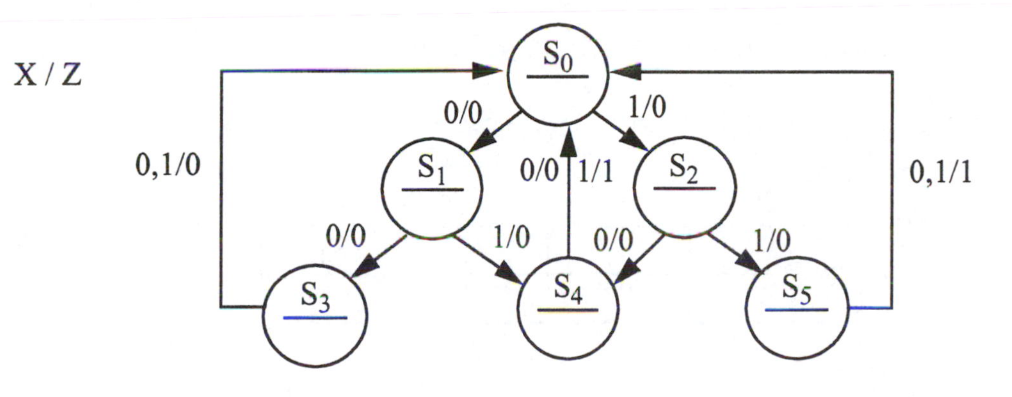

PS	NS		Z	
	$X = 0$	$X = 1$	$X = 0$	$X = 1$
S_0	S_1	S_2	0	0
S_1	S_3	S_4	0	0
S_2	S_4	S_5	0	0
S_3	S_0	S_0	0	0
S_4	S_0	S_0	0	1
S_5	S_0	S_0	1	1

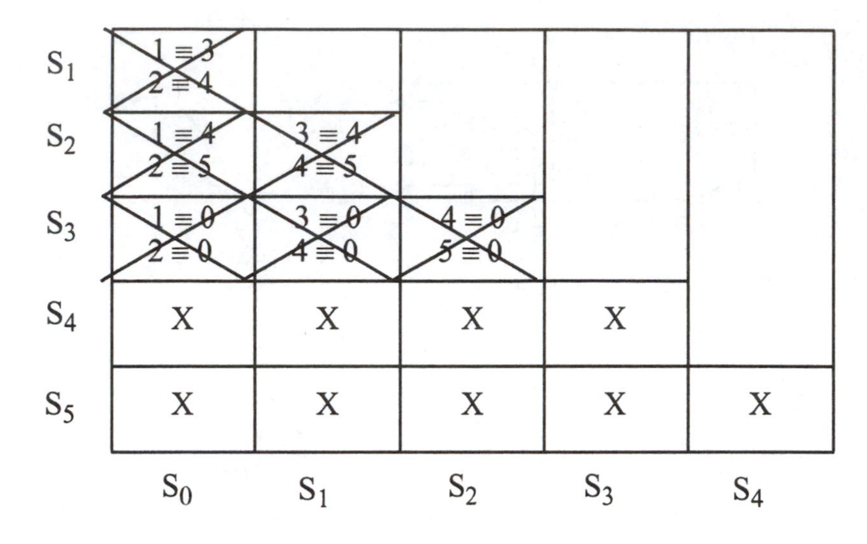

Armstrong-Humphrey Rule #1: S_3,S_4,S_5

Armstrong-Humphrey Rule #2: S_1,S_2 S_3,S_4 S_4,S_5

Armstrong-Humphrey Rule #3: S_0,S_1,S_2,S_3,S_4 S_0,S_1,S_2,S_3 S_4,S_5

Q_A \ Q_BQ_C	00	01	11	10
0	S_0	S_1	S_5	S_4
1	S_3	S_2		

Present State $Q_A Q_B Q_C$	Next State $Q_A^+ Q_B^+ Q_C^+$		Output Z	
	X=0	**X=1**	**X=0**	**X=1**
010	110	100	0	0
110	000	001	0	0
100	001	011	0	0
000	010	010	0	0
001	010	010	0	1
011	010	010	1	1

$Q_A^+ = D_A$

Q_AQ_B \ Q_CX	00	01	11	10
00	0	0	0	
01	1	1	0	0
11	0	0	-	-
10	0	0	-	-

$Q_B^+ = D_B$

Q_AQ_B \ Q_CX	00	01	11	10
00	1	1	1	1
01	1	0	1	1
11	0	0	-	-
10	0	1	-	-

$Q_C^+ = D_C$

Q_AQ_B \ Q_CX	00	01	11	10
00	0	0	0	0
01	0	0	0	0
11	0	1	-	-
10	1	1	-	-

Z

Q_AQ_B \ Q_CX	00	01	11	10
00	0	0	1	0
01	0	0	1	1
11	0	0	-	-
10	0	0	-	-

$$D_A = Q_A^+ = (\overline{Q}_A \bullet Q_B \bullet \overline{Q}_C)$$

$$D_B = Q_B^+ = Q_A + (\overline{Q}_A \bullet \overline{Q}_B) + (\overline{Q}_A \bullet X) + (\overline{Q}_B \bullet X)$$

$$D_C = Q_C^+ = (Q_A \bullet \overline{Q}_B) + (Q_A \bullet X)$$

$$Z = (Q_C \bullet X) + (Q_B \bullet Q_C)$$

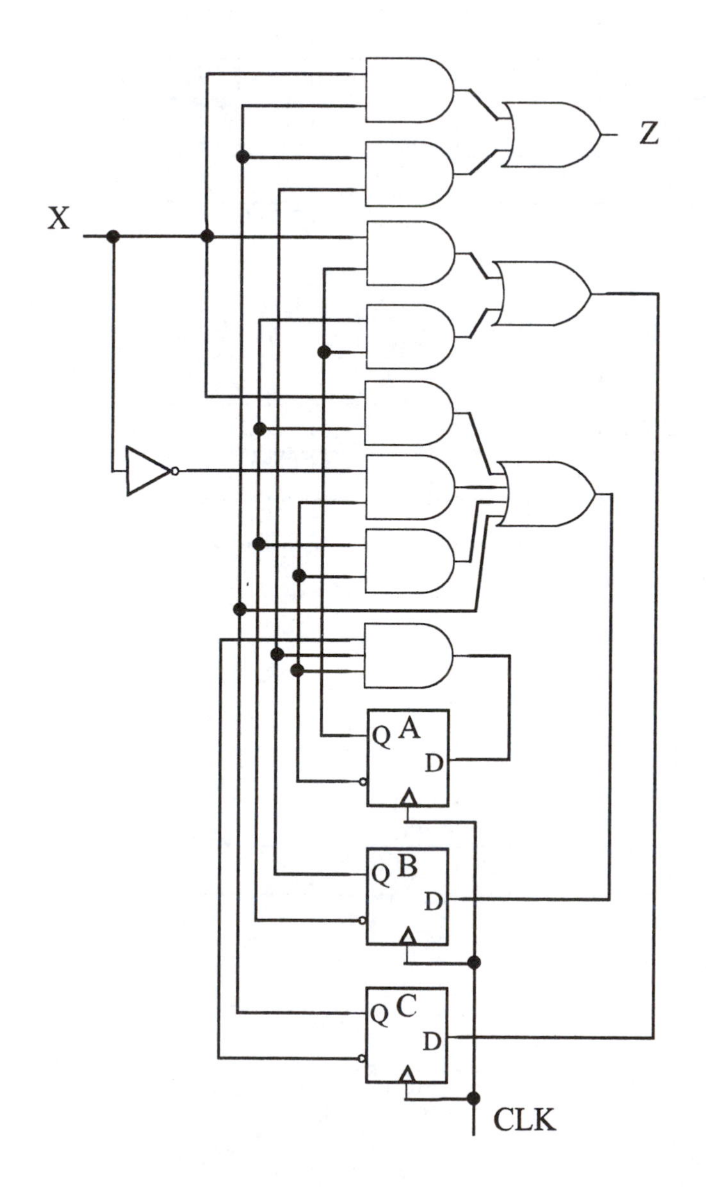

9-17. Solution.

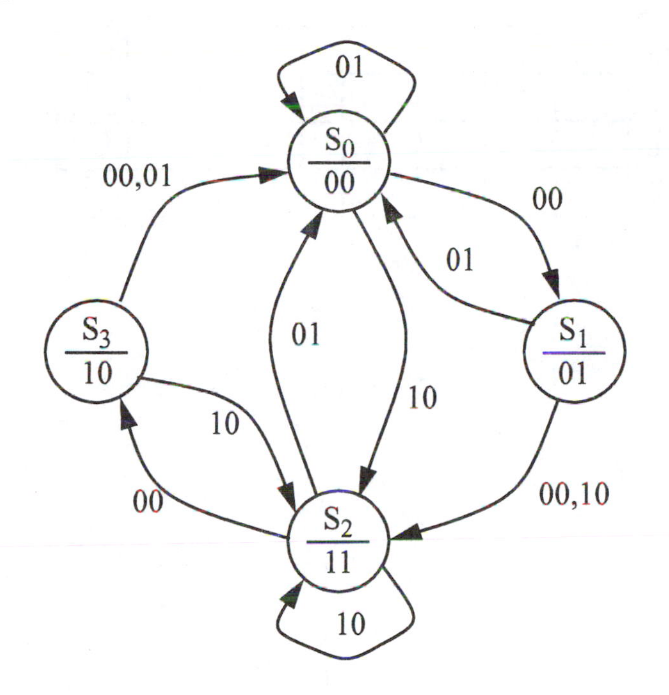

Present State $Q_A Q_B$	Next State $Q_A^+ Q_B^+$ SET (S) CLEAR (C)				Output $Z_1 Z_0$
	00	**01**	**10**	**11**	
00	01	00	11	-	00
01	11	00	11	-	01
10	00	00	11	-	10
11	10	00	11	-	11

$Q_A^+ = D_A$

Q_AQ_B \ SC	00	01	11	10
00	0	0	-	1
01	1	0	-	1
11	1	0	-	1
10	0	0	-	1

$Q_B^+ = D_B$

Q_AQ_B \ SC	00	01	11	10
00	1	0	-	1
01	1	0	-	1
11	0	0	-	1
10	0	0	-	1

$$D_A = (Q_B \bullet \overline{C}) + S$$

$$D_B = (\overline{Q}_A \bullet \overline{C}) + S$$

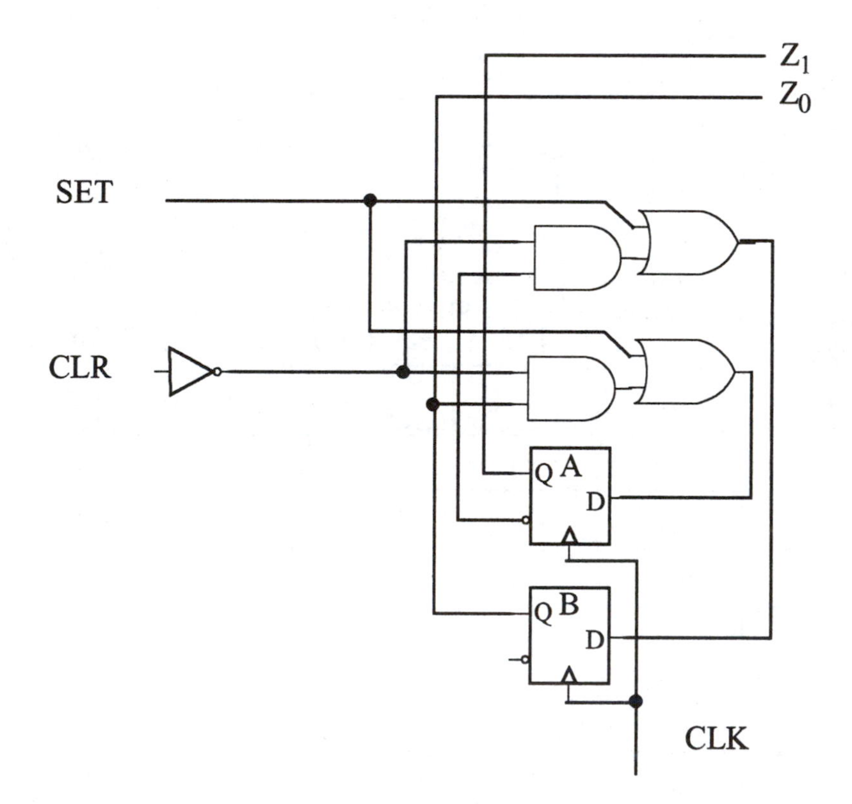

9-18. Solution.

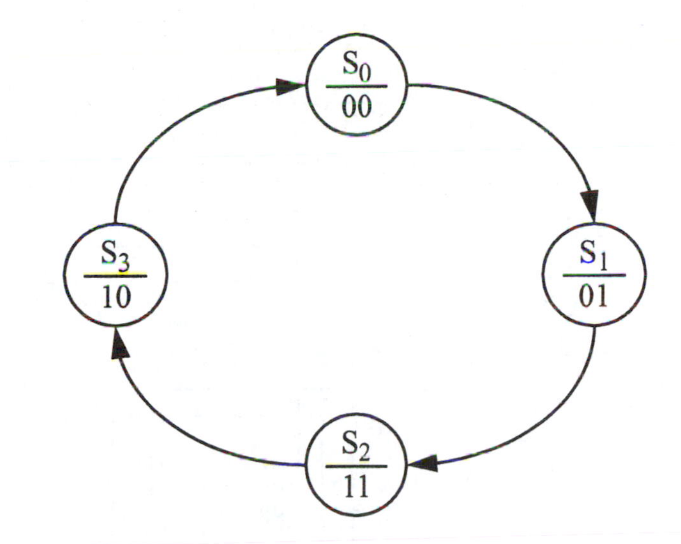

State Assignment: S_0=00, S_1=01, S_2=11, S_3=10

Present State	Next State	Output Z_1Z_0
00	01	00
01	11	01
10	00	10
11	10	11

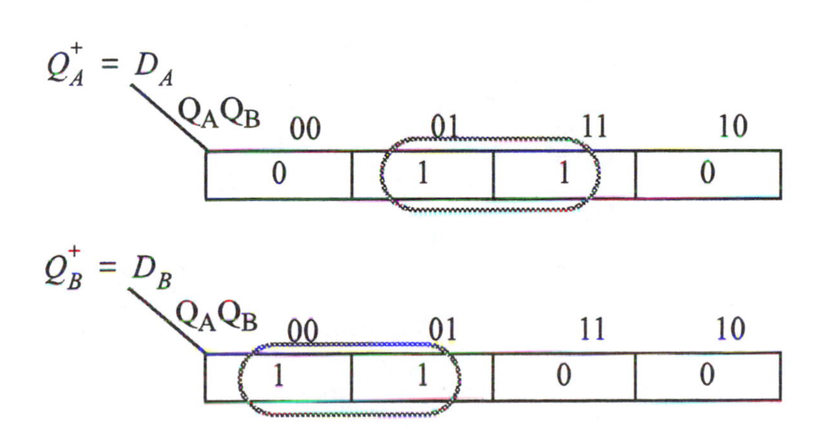

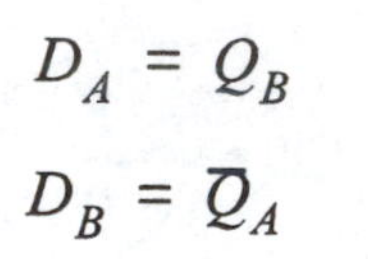

$$D_A = Q_B$$

$$D_B = \overline{Q}_A$$

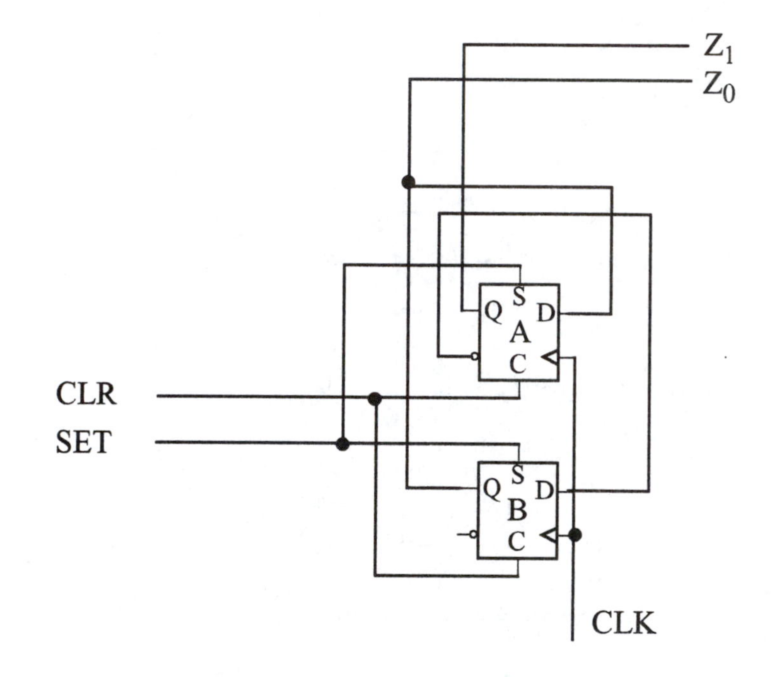

Chapter 10. Sequential Design - Implementation

10-1. Solution.

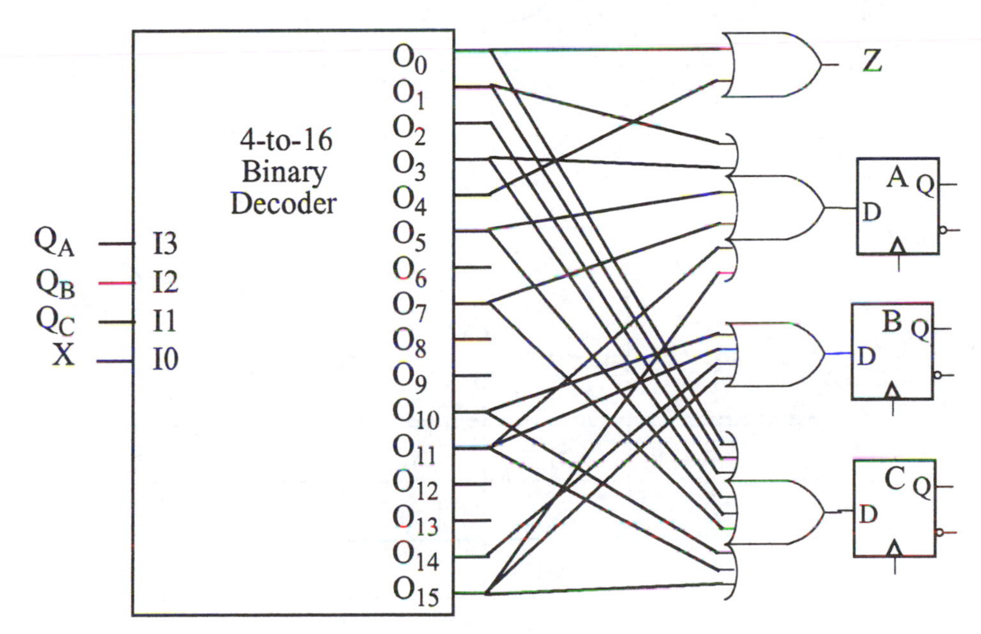

10-2. Solution.

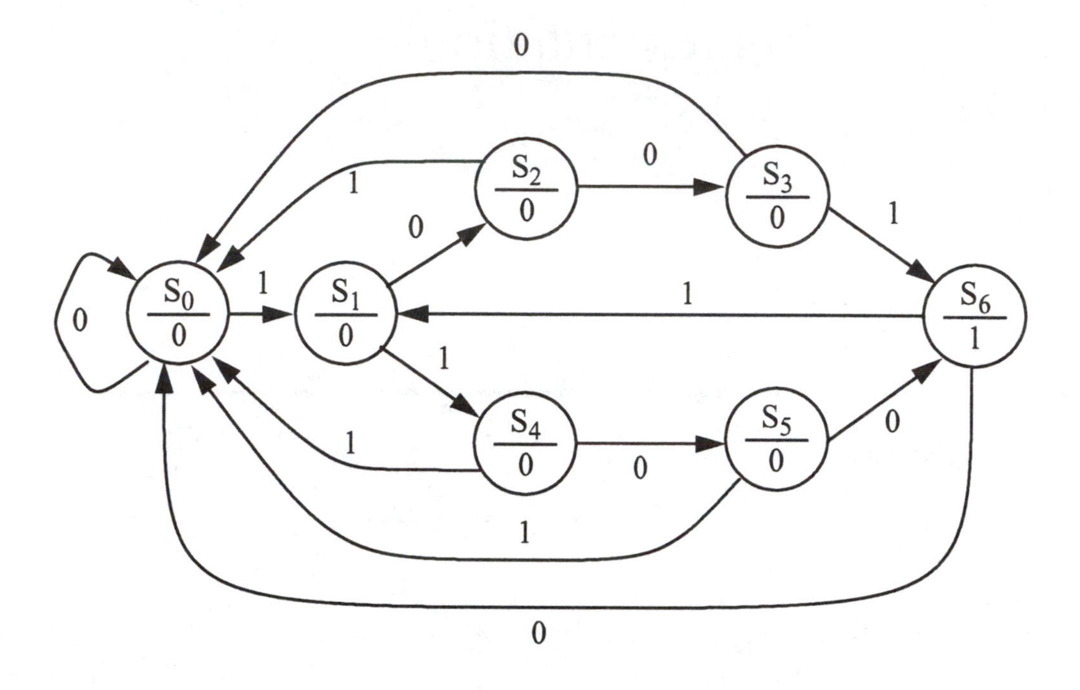

PS	NS		Z
	$X = 0$	$X = 1$	
S_0	S_0	S_1	0
S_1	S_2	S_4	0
S_2	S_3	S_0	0
S_3	S_0	S_6	0
S_4	S_5	S_0	0
S_5	S_6	S_0	1
S_6	S_0	S_1	1

	S_0	S_1	S_2	S_3	S_4	S_5
S_1	$0 \equiv 2$ $1 \equiv 4$					
S_2	$0 \equiv 3$ $1 \equiv 0$	$2 \equiv 3$ $0 \equiv 4$				
S_3	$1 \equiv 6$	$0 \equiv 2$ $4 \equiv 6$	$0 \equiv 3$ $0 \equiv 6$			
S_4	$0 \equiv 5$ $1 \equiv 0$	$2 \equiv 5$ $0 \equiv 4$	$3 \equiv 5$	$0 \equiv 5$ $0 \equiv 6$		
S_5	$0 \equiv 6$ $1 \equiv 0$	$2 \equiv 6$ $0 \equiv 4$	$3 \equiv 6$	$0 \equiv 6$	$5 \equiv 6$	
S_6	X	X	X	X	X	X

Armstrong-Humphrey Rule #1: S_0,S_3,S_6 S_2,S_4,S_5 S_0,S_6

Armstrong-Humphrey Rule #2: S_0,S_1 S_2,S_4 S_0,S_3 S_0,S_6 S_0,S_5

Armstrong-Humphrey Rule #3: S_0,S_1,S_2,S_3,S_4,S_5

Q_A \ Q_BQ_C	00	01	11	10
0	S_0	S_6	S_5	S_1
1	S_3	S_2	S_4	

$Q_A Q_B Q_C$	$Q_A^+ Q_B^+ Q_C^+$		Z
	$X = 0$	$X = 1$	
000	000	010	0
010	101	111	0
101	100	000	0
100	000	001	0
111	011	000	0
011	001	000	0
001	000	010	1

$Q_A^+ = D_A$

Q_AQ_B \ Q_CX	00	01	11	10
00	0	0	0	0
01	1	1	0	0
11	-	-	0	0
10	0	0	0	1

$Q_B^+ = D_B$

Q_AQ_B \ Q_CX	00	01	11	10
00	0	1	1	0
01	0	1	0	0
11	-	-	0	1
10	0	0	0	0

$Q_C^+ = D_C$

Q_AQ_B \ Q_CX	00	01	11	10
00	0	0	0	0
01	1	1	0	1
11	-	-	0	1
10	0	1	0	0

Z

Q_A \ Q_BQ_C	00	01	11	10
0	0	1	0	0
1	0	0	0	-

a.

$$D_A = (Q_B \bullet \overline{Q}_C) + (Q_A \bullet \overline{Q}_B \bullet Q_C \bullet \overline{X})$$

$$D_B = (\overline{Q}_A \bullet \overline{Q}_C \bullet X) + (Q_A \bullet Q_B \bullet \overline{X}) + (\overline{Q}_A \bullet \overline{Q}_B \bullet X)$$

$$D_C = (Q_B \bullet \overline{X}) + (Q_B \bullet \overline{Q}_C) + (Q_A \bullet \overline{Q}_C \bullet X)$$

$$Z = \overline{Q}_A \bullet \overline{Q}_B \bullet Q_C$$

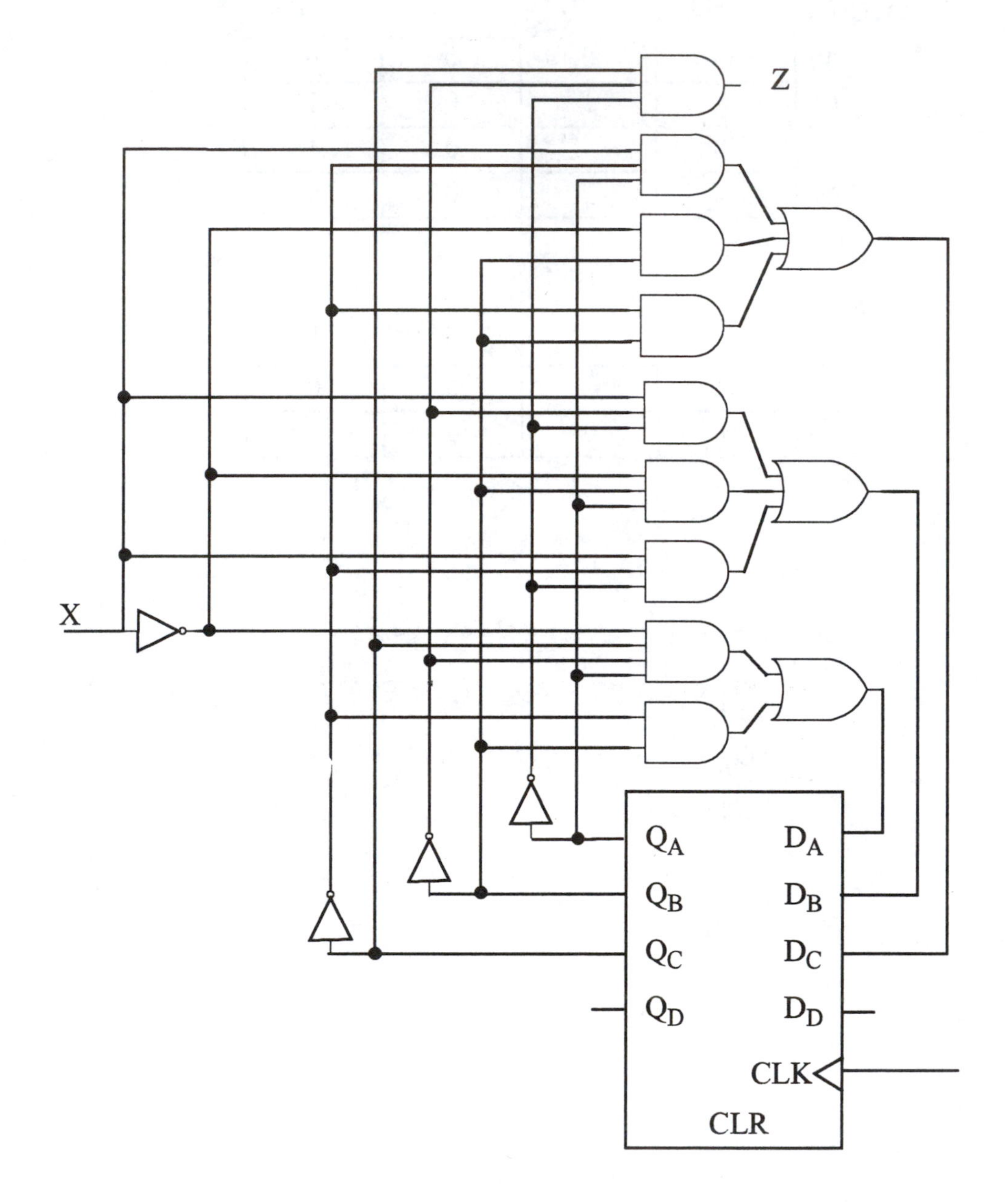

b.

$$D_A = (Q_B + Q_C) \bullet (\overline{Q}_C + \overline{X}) \bullet (\overline{Q}_B + \overline{Q}_C) \bullet (Q_A + \overline{Q}_C)$$

$$D_B = (Q_A + X) \bullet (\overline{Q}_A + Q_B) \bullet (\overline{Q}_A + \overline{X}) \bullet (Q_A + \overline{Q}_B + \overline{Q}_C)$$

$$D_C = (Q_A + Q_B) \bullet (Q_B + X) \bullet (\overline{Q}_C + \overline{X})$$

$$Z = \overline{Q_A + Q_B + \overline{Q}_C}$$

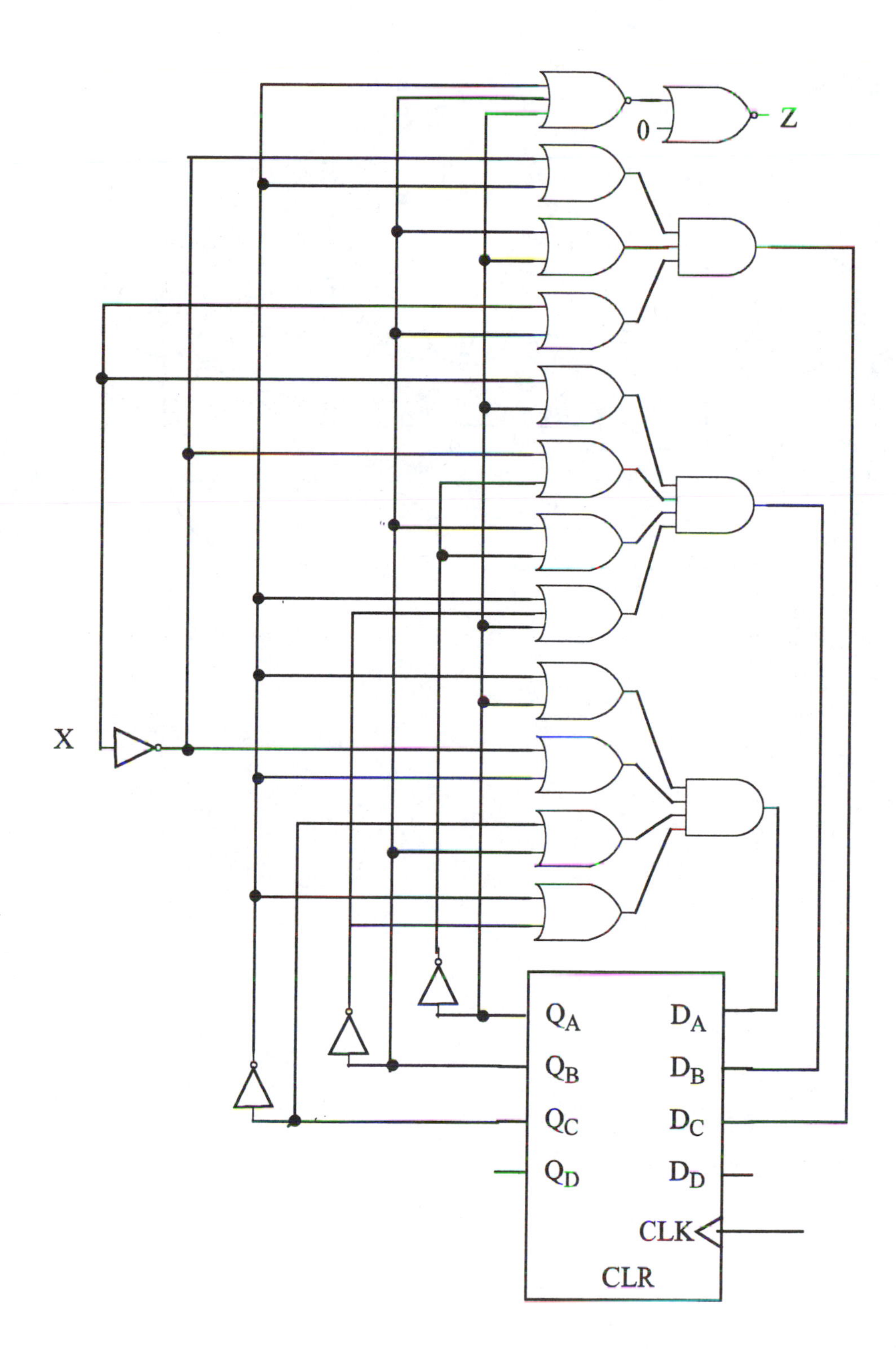
Z
0
X
QA
QB
QC
QD
DA
DB
DC
DD
CLK
CLR

c.

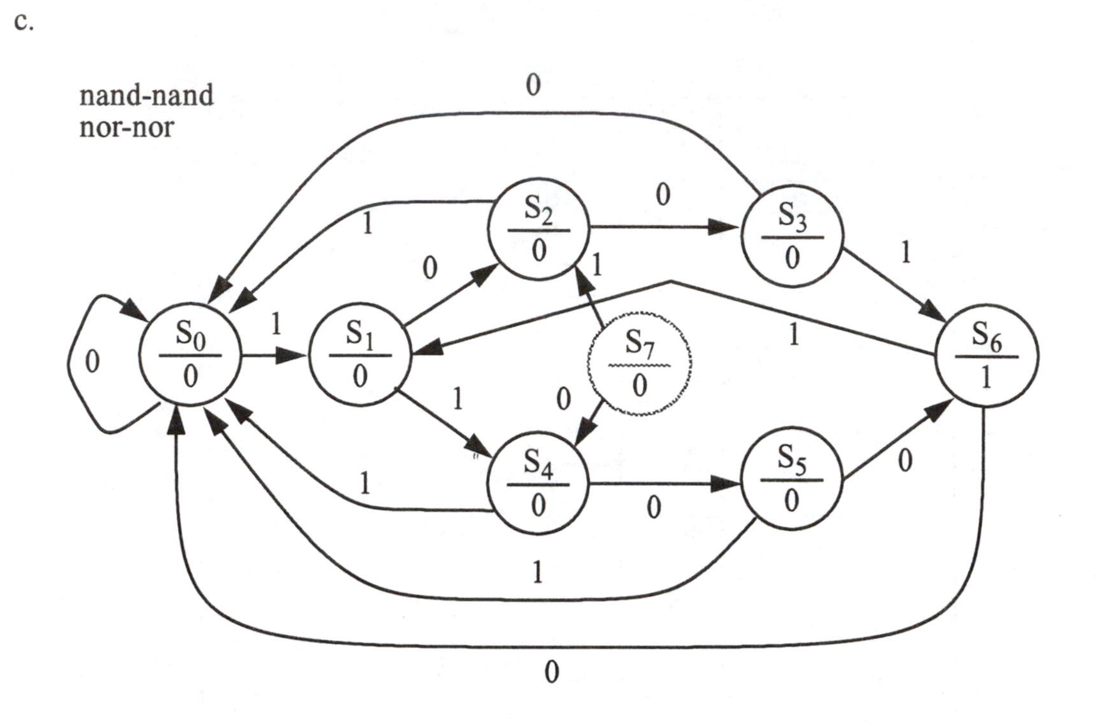

10-3. Solution.

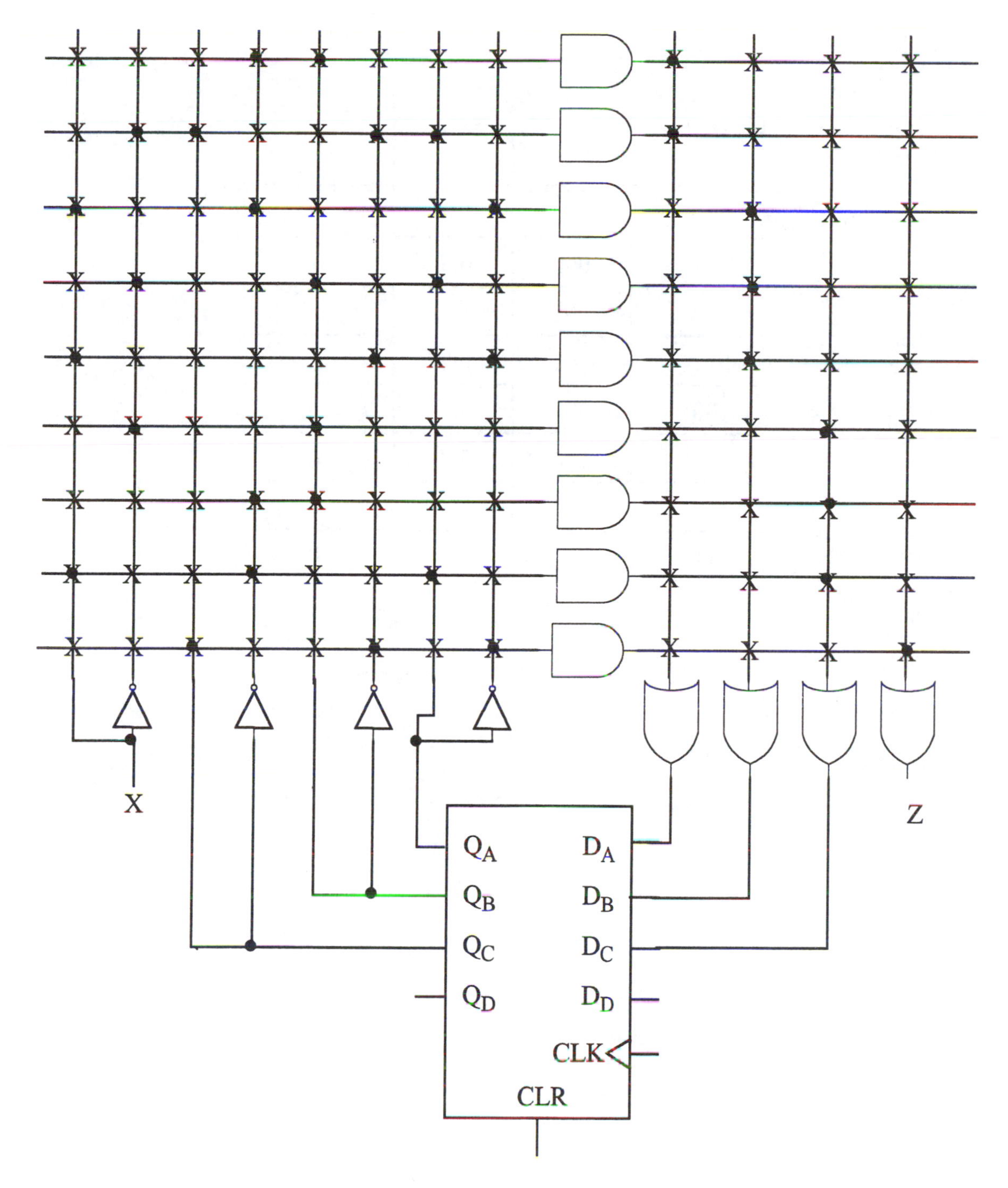

10-4. Solution.

$Q_A Q_B Q_C$	$Q_A^+ Q_B^+ Q_C^+$		Z
	$X = 0$	$X = 1$	
000	000	010	0
001	000	010	1
010	101	111	0
011	001	000	0
100	000	001	0
101	100	000	0
110	111	101	0
111	011	000	0

$Q_A\,Q_B\,Q_C\,X$	Q_A^+	Q_B^+	Q_C^+	Z
0000	0	0	0	0
0001	0	1	0	0
0010	0	1	0	0
0011	0	1	0	1
0100	1	0	1	0
0101	1	1	1	0
0110	0	0	1	0
0111	0	0	0	0
1000	0	0	0	0
1001	0	0	0	0
1001	0	0	1	0
1010	1	0	0	0
1011	0	0	0	0
1100	1	1	1	0
1101	1	0	1	0
1110	0	1	1	0
1111	0	0	0	0

a.

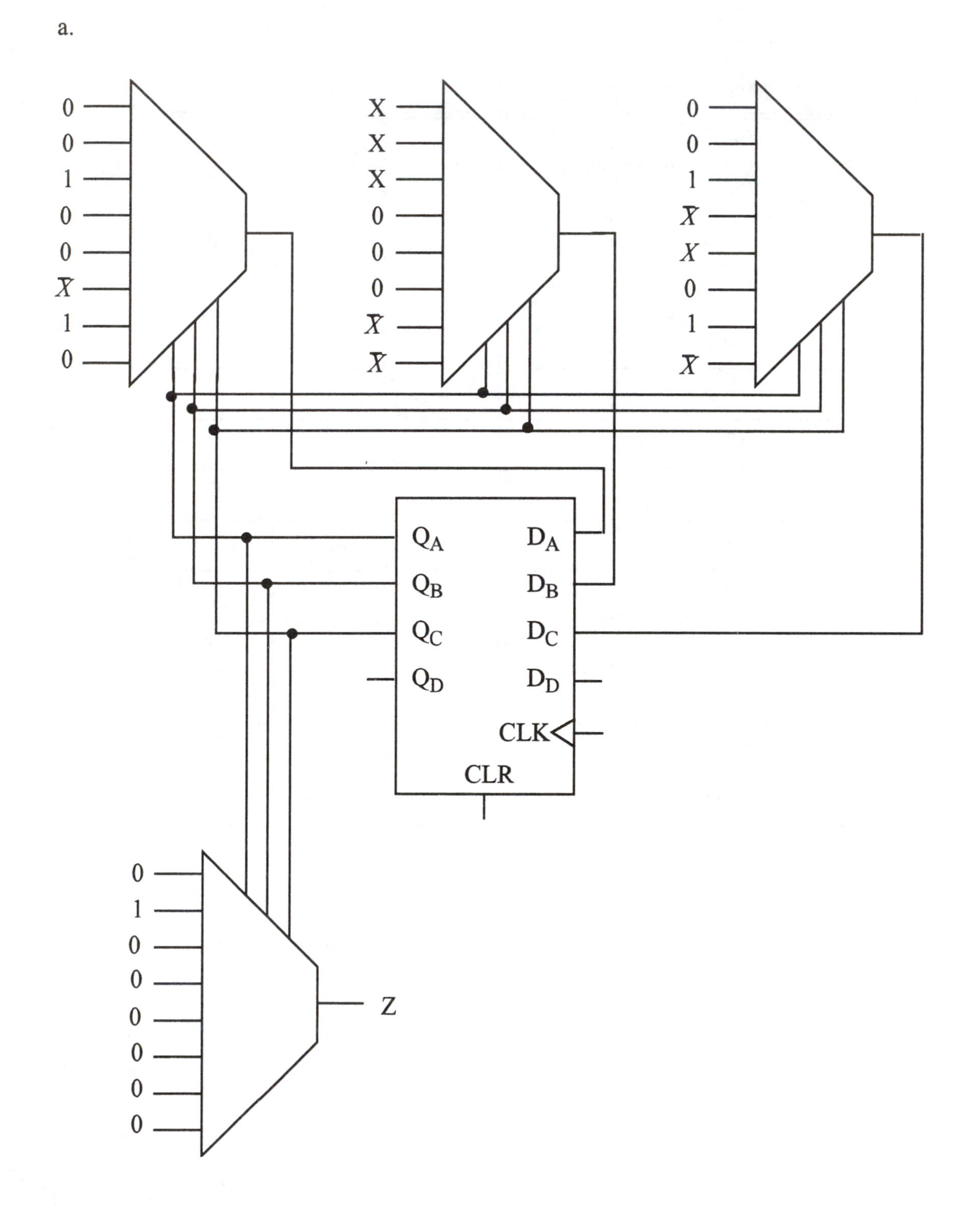

0
0
1
0
0
X̄
1
0
X
X
X
0
0
0
X̄
X̄
0
0
1
X̄
X
0
1
X̄
Q_A
Q_B
Q_C
Q_D
D_A
D_B
D_C
D_D
CLK
CLR
0
1
0
0
0
0
0
0
Z

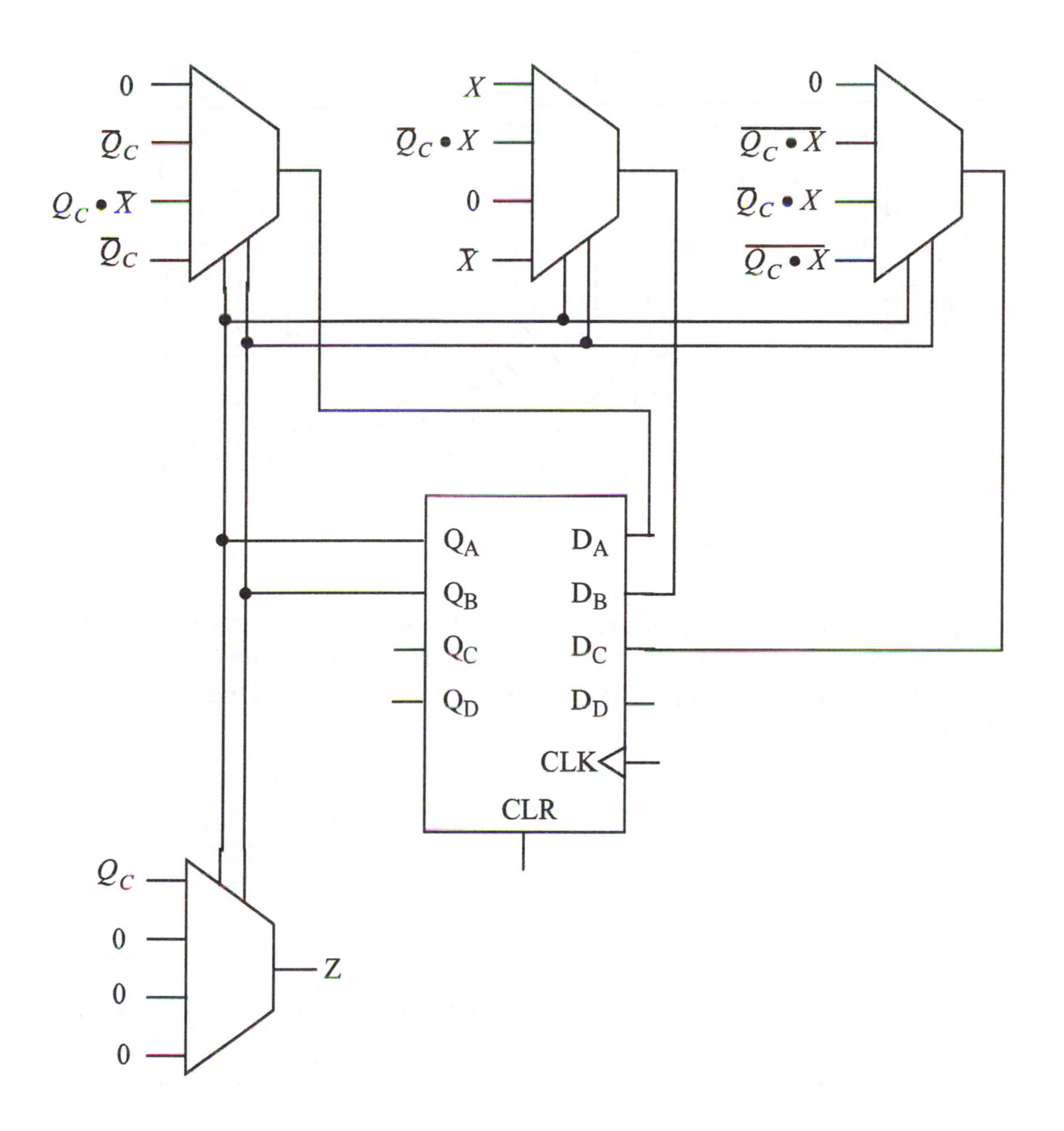
0
$\overline{Q}_C$
$Q_C \bullet \overline{X}$
$\overline{Q}_C$
X
$\overline{Q}_C \bullet X$
0
$\overline{X}$
0
$\overline{Q_C \bullet X}$
$\overline{Q}_C \bullet X$
$\overline{Q_C \bullet X}$
Q_A
Q_B
Q_C
Q_D
D_A
D_B
D_C
D_D
CLK
CLR
Q_C
0
0
0
Z

b.

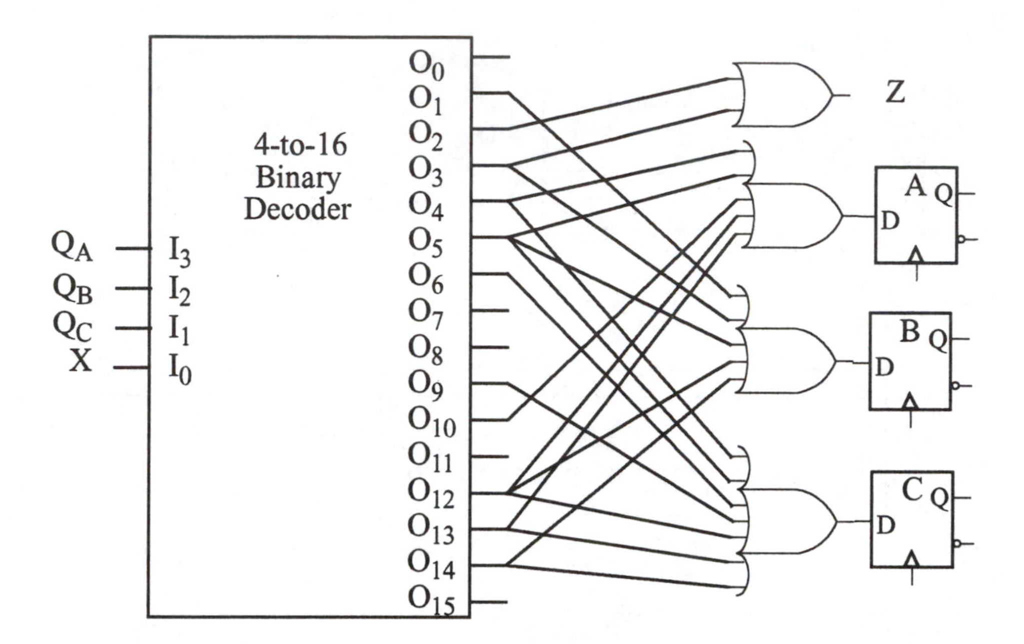

c. The binary decoder directly implements the entire truth table for the output and next state combinational logic, whereas the multiplexers allow some logic minimization in computing the expressions to drive data input. The binary decoder tends to yield high fan-in gates, whereas the multiplexers do not.

10-5. Solution.

$Q_A Q_B$	$Q_A^+ Q_B^+ Q_C^+$				UP DOWN			
	FR=00	**FR=01**	**FR=10**	**FR=11**	**FR=00**	**FR=01**	**FR=10**	**FR=11**
F_1	F_1	F_1	F_2	F_2	00	00	10	10
F_2	F_2	F_1	F_2	F_3	00	01	00	10
F_3	F_3	F_2	F_2	F_3	00	01	01	00

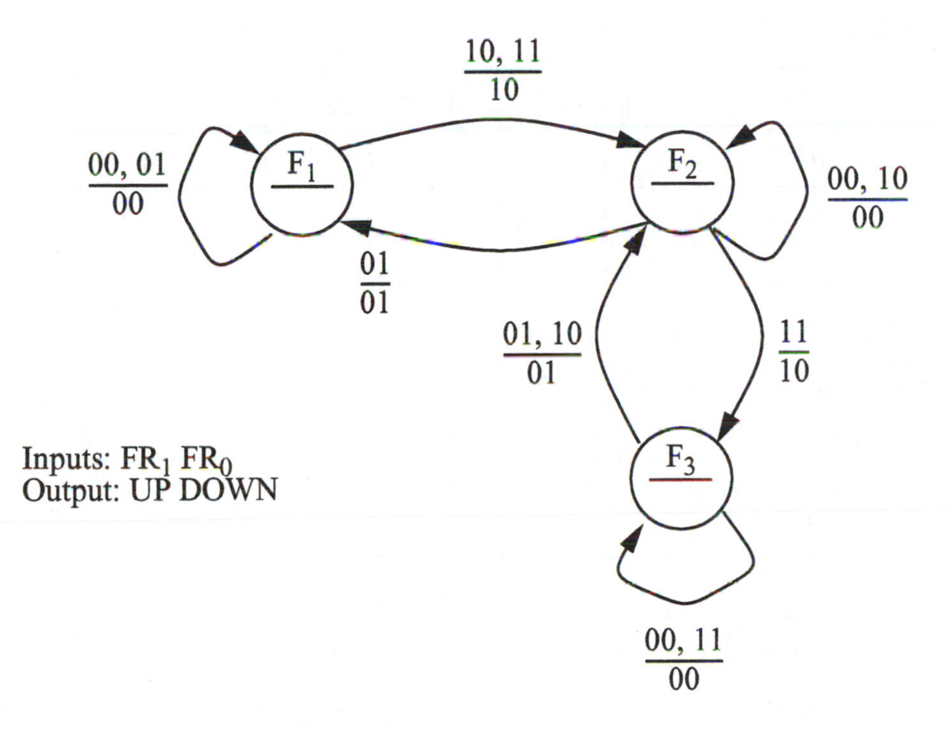

Armstrong-Humphrey Rule #1: F_1,F_2 F_1,F_2,F_3 F_2,F_3

Armstrong-Humphrey Rule #2: F_1,F_2 F_1,F_2,F_3 F_2,F_3

Armstrong-Humphrey Rule #3: F_1,F_2,F_3 F_2,F_3 F_1,F_2

Q_AQ_B 00	01	11	10
F_1		F_3	F_2

$Q_A Q_B$	$Q_A^+ Q_B^+$				*UP DOWN*			
	FR=00	**FR=01**	**FR=10**	**FR=11**	**FR=00**	**FR=01**	**FR=10**	**FR=11**
00	00	00	10	10	00	00	10	10
10	10	00	10	11	00	01	00	10
11	11	10	10	11	00	01	01	00

$Q_A^+ = D_A$

Q_AQ_B \ FR_1FR_0	00	01	11	10
00	0	0	1	1
01	-	-	-	-
11	1	1	1	1
10	1	0	1	1

$Q_B^+ = D_B$

Q_AQ_B \ FR_1FR_0	00	01	11	10
00	0	0	0	0
01	-	-	-	-
11	1	0	1	0
10	0	0	1	0

UP

Q_AQ_B \ FR_1FR_0	00	01	11	10
00	0	0	1	1
01	-	-	-	-
11	0	0	0	0
10	0	0	1	0

DOWN

Q_AQ_B \ FR_1FR_0	00	01	11	10
00	0	0	0	0
01	-	-	-	-
11	0	1	0	1
10	0	1	0	0

a.

$$D_A = FR_1 + Q_B + \left(Q_A \bullet \overline{FR_0}\right)$$

$$D_B = (Q_A \bullet FR_1 \bullet FR_0) + \left(Q_B \bullet \overline{FR_1} \bullet \overline{FR_0}\right)$$

$$UP = (\overline{Q}_A \bullet FR_1) + (\overline{Q}_B \bullet FR_1 \bullet FR_0)$$

$$DOWN = \left(Q_A \bullet \overline{FR_1} \bullet FR_0\right) + \left(Q_B \bullet FR_1 \bullet \overline{FR_0}\right)$$

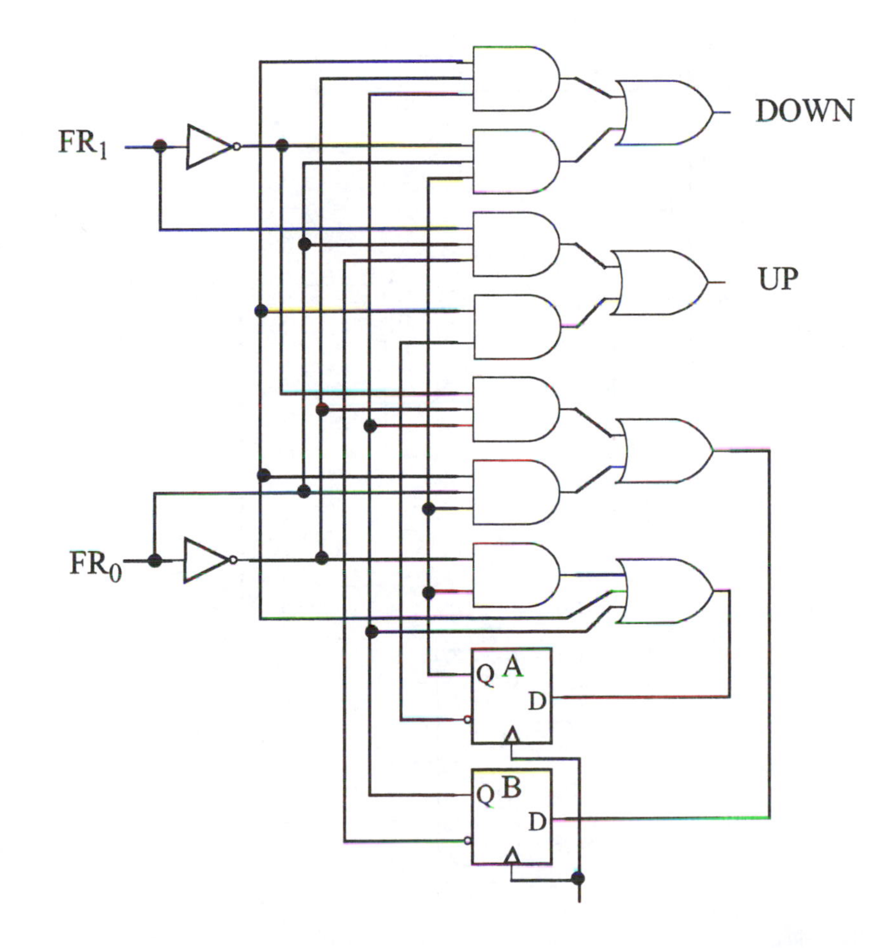

b.

$$D_A = (Q_A + FR_1) \bullet \left(Q_B + FR_1 + \overline{FR}_0 \right)$$
$$D_B = Q_A \bullet (Q_B + FR_0) \bullet \left(FR_1 + \overline{FR}_0 \right) \bullet \left(\overline{FR}_1 + FR_0 \right)$$
$$UP = FR_1 \bullet \overline{Q}_B \bullet (\overline{Q}_A + FR_0)$$
$$DOWN = Q_A \bullet (FR_1 + FR_0) \bullet (\overline{FR}_1 + \overline{FR}_0) \bullet (Q_B + FR_0)$$

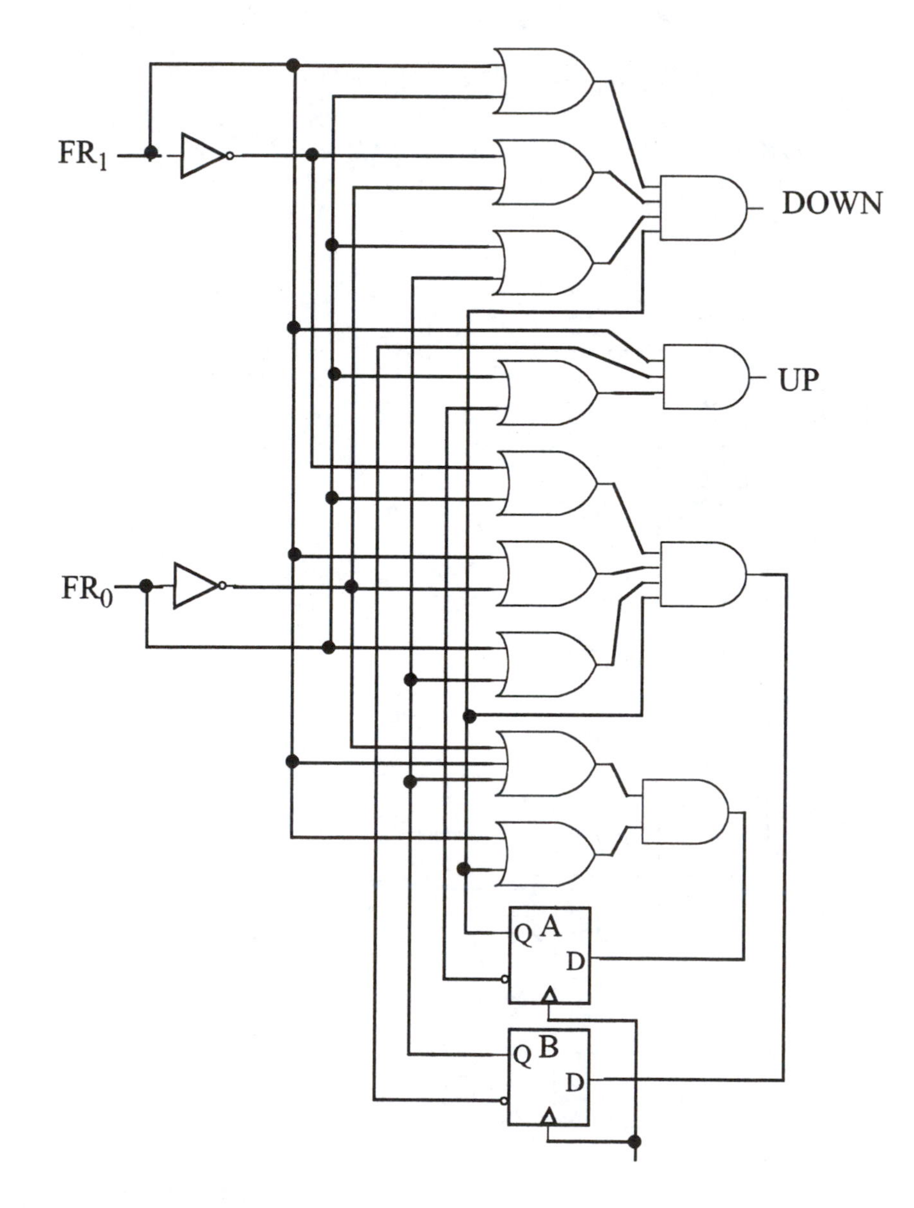

c.

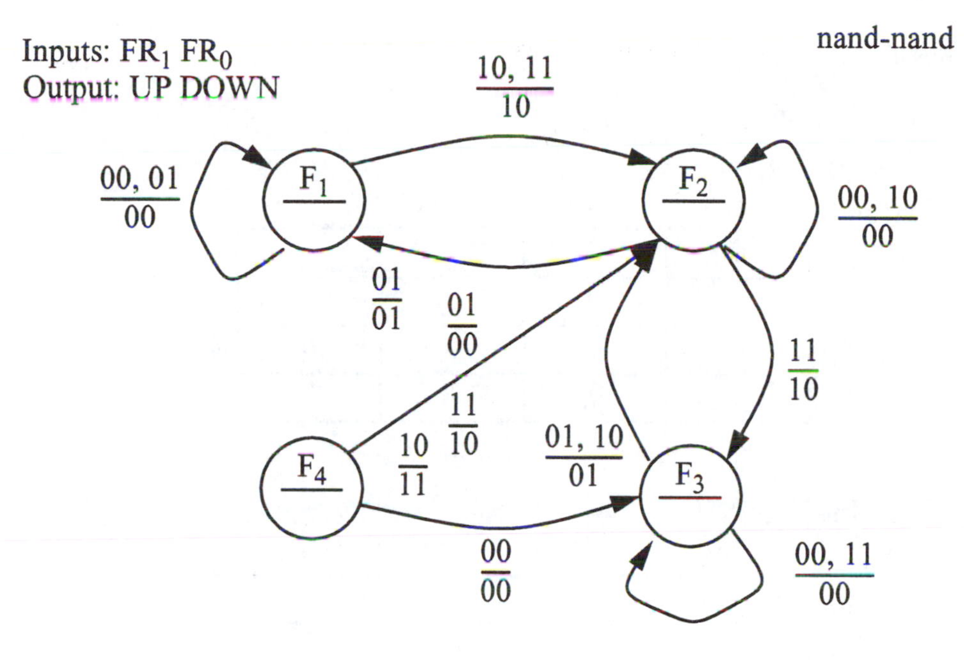
Inputs: FR1 FR0
Output: UP DOWN
nand-nand
10, 11
10
00, 01
00
F1
F2
00, 10
00
01
01
01
00
11
10
11
10
10
11
01, 10
01
F4
F3
00
00
00, 11
00

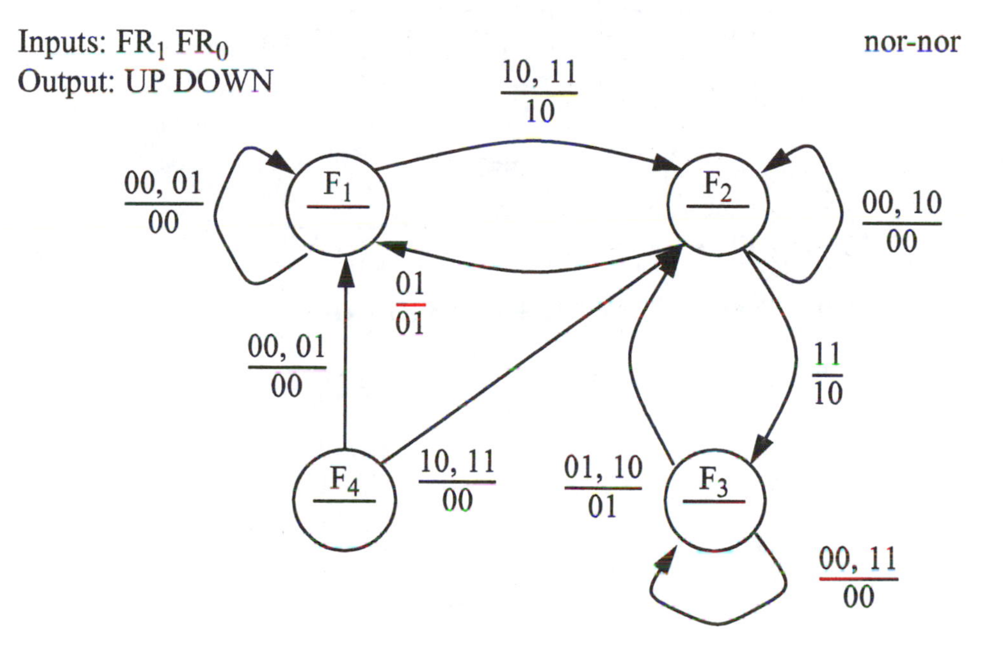
Inputs: FR1 FR0
Output: UP DOWN
nor-nor
10, 11
10
00, 01
00
F1
F2
00, 10
00
01
01
00, 01
00
11
10
F4
10, 11
00
01, 10
01
F3
00, 11
00

10-6. Solution.

a. nand-nand

UP

Q_AQ_B \ FR_1FR_0	00	01	11	10
00	0	0	1	1
01	0	0	0	0
11	0	0	0	0
10	0	0	1	0

DOWN

Q_AQ_B \ FR_1FR_0	00	01	11	10
00	0	0	0	0
01	0	0	0	0
11	0	1	0	1
10	0	1	0	0

$$D_A = FR_1 + Q_B + \left(Q_A \bullet \overline{FR}_0\right)$$

$$D_B = (Q_A \bullet FR_1 \bullet FR_0) + \left(Q_B \bullet \overline{FR}_1 \bullet \overline{FR}_0\right)$$

$$UP = (\overline{Q}_A \bullet \overline{Q}_B \bullet FR_1) + (\overline{Q}_B \bullet FR_1 \bullet FR_0)$$

$$DOWN = \left(Q_A \bullet \overline{FR}_1 \bullet FR_0\right) + \left(Q_A \bullet Q_B \bullet FR_1 \bullet \overline{FR}_0\right)$$

b. nor-nor

UP

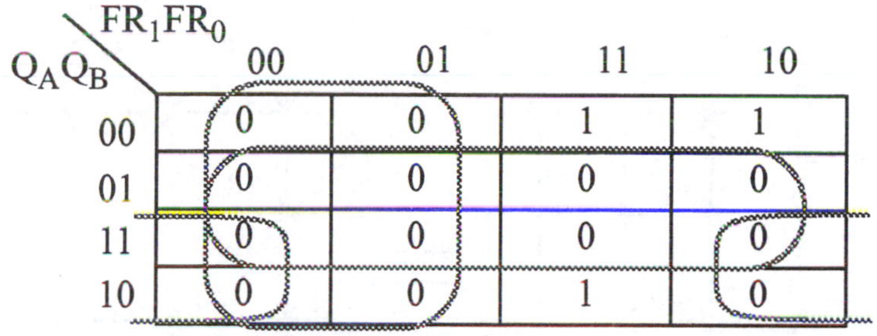

Q_AQ_B \ FR_1FR_0	00	01	11	10
00	0	0	1	1
01	0	0	0	0
11	0	0	0	0
10	0	0	1	0

DOWN

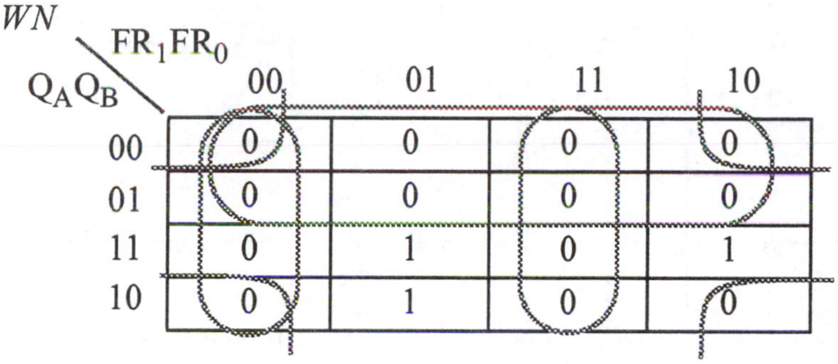

Q_AQ_B \ FR_1FR_0	00	01	11	10
00	0	0	0	0
01	0	0	0	0
11	0	1	0	1
10	0	1	0	0

$$D_A = (Q_A + FR_1) \bullet \left(Q_B + FR_1 + \overline{FR_0} \right)$$

$$D_B = Q_A \bullet (Q_B + FR_0) \bullet \left(FR_1 + \overline{FR_0} \right) \bullet \left(\overline{FR_1} + FR_0 \right)$$

$$UP = FR_1 \bullet \overline{Q}_B \bullet (\overline{Q}_A + FR_0)$$

$$DOWN = Q_A \bullet (FR_1 + FR_0) \bullet (\overline{FR_1} + \overline{FR_0}) \bullet (Q_B + FR_0)$$

10-7. Solution.

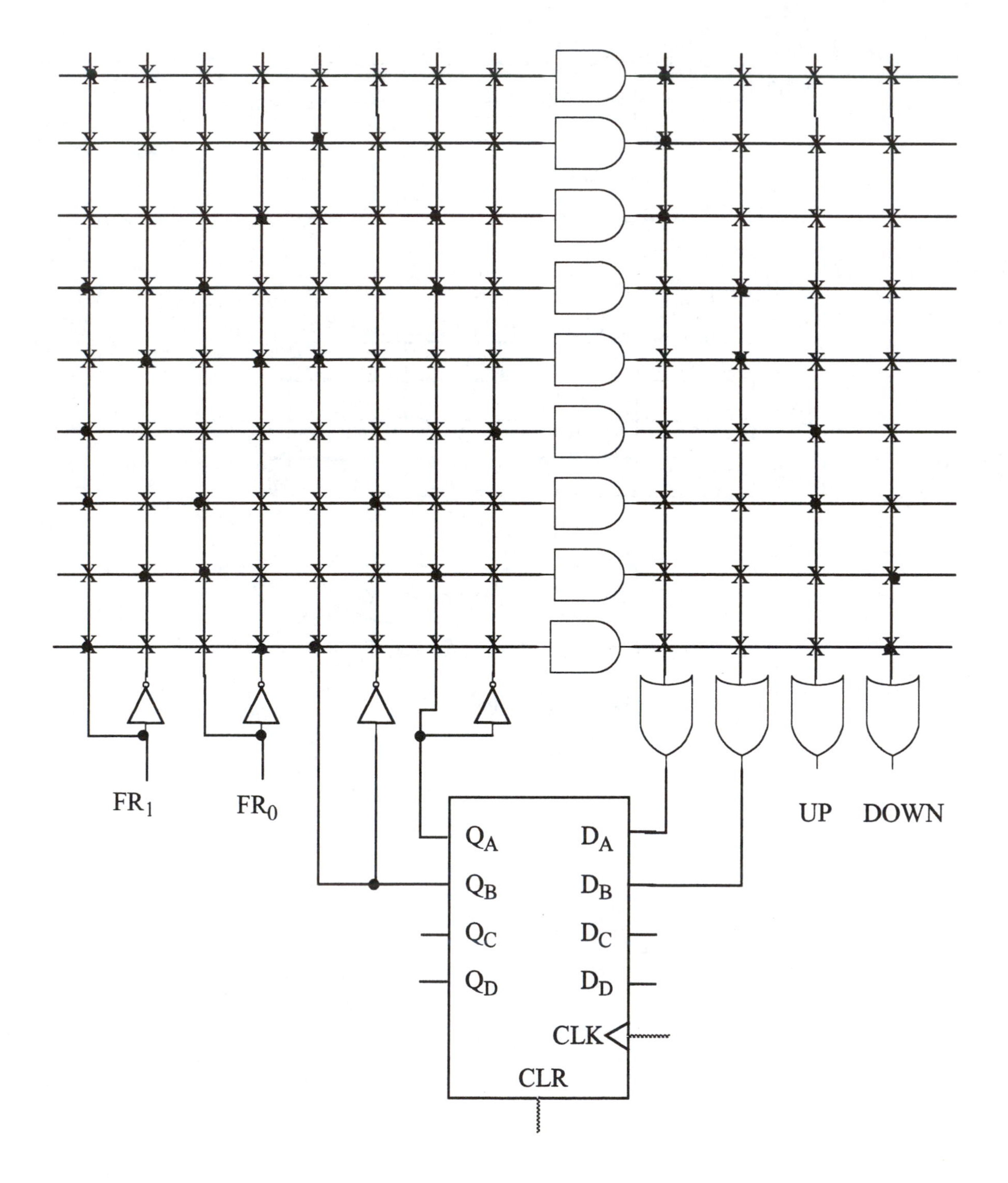

10-8. Solution.

$Q_A Q_B FR_1 FR_2$	Q_A^+	Q_B^+	UP	$DOWN$
0000	0	0	0	0
0001	0	0	0	0
0010	1	0	1	0
0011	1	0	1	1
0100	-	-	-	-
0101	-	-	-	-
0110	-	-	-	-
0111	-	-	-	-
1000	1	0	0	0
1001	0	0	0	1
1001	1	0	0	0
1010	1	1	1	0
1011	1	1	0	0
1100	1	0	0	1
1101	1	0	0	1
1110	1	1	0	0
1111	1	1	0	0

a.

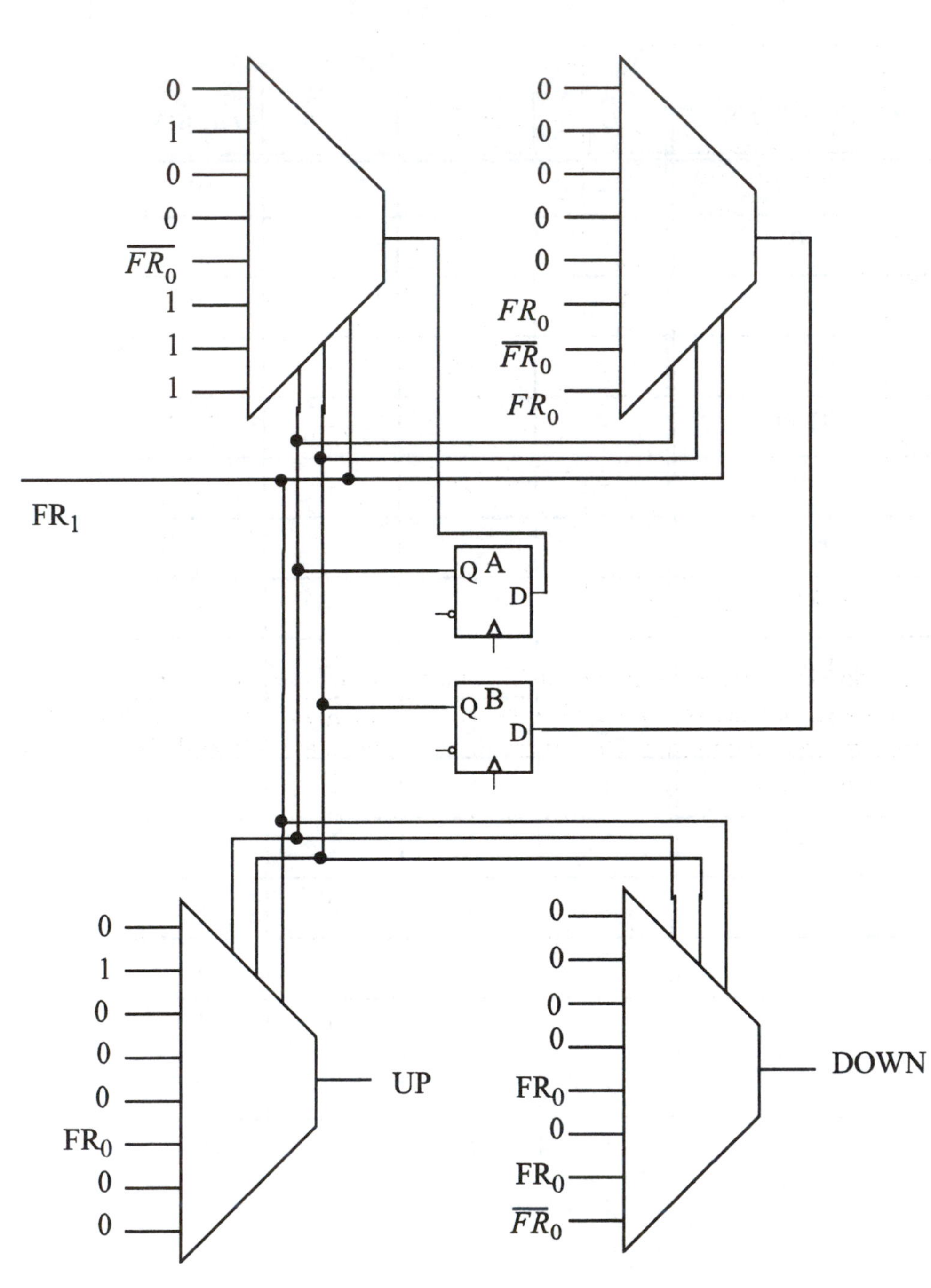

0
1
0
0
$\overline{FR_0}$
1
1
1
0
0
0
0
0
FR_0
$\overline{FR_0}$
FR_0
FR_1
Q A
D
Q B
D
0
1
0
0
0
FR_0
0
0
UP
0
0
0
0
FR_0
0
FR_0
$\overline{FR_0}$
DOWN

b.

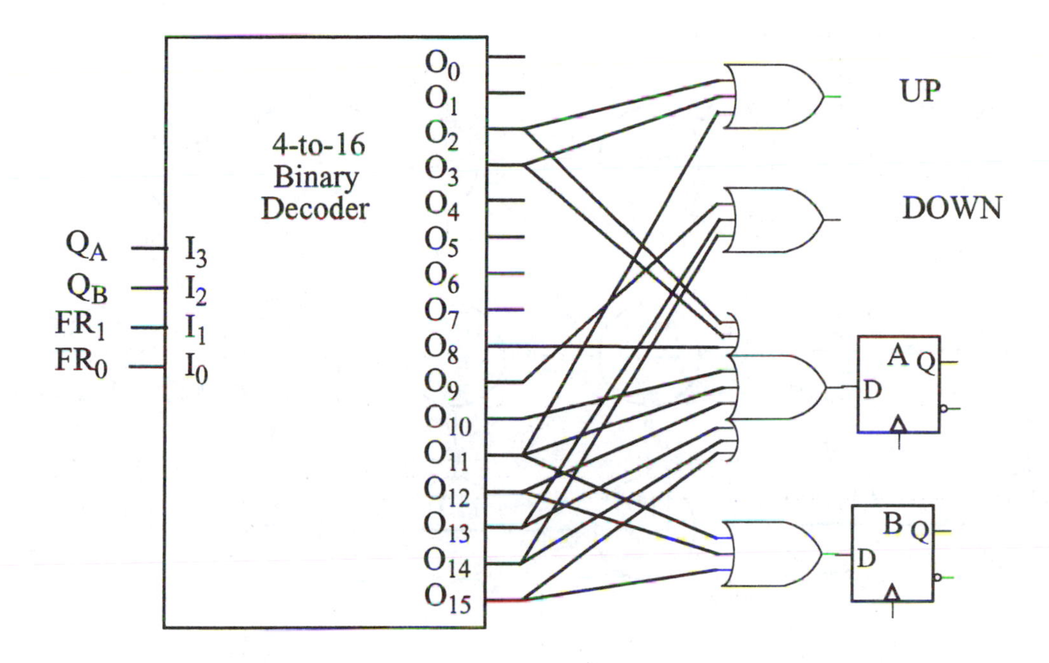

c. The binary decoder directly implements the entire truth table for the output and next state combinational logic, whereas the multiplexers allow some logic minimization in computing the expressions to drive data input. The binary decoder tends to yield high fan-in gates, whereas the multiplexers do not.

10-9. Solution.

Without self-correcting feedback logic.

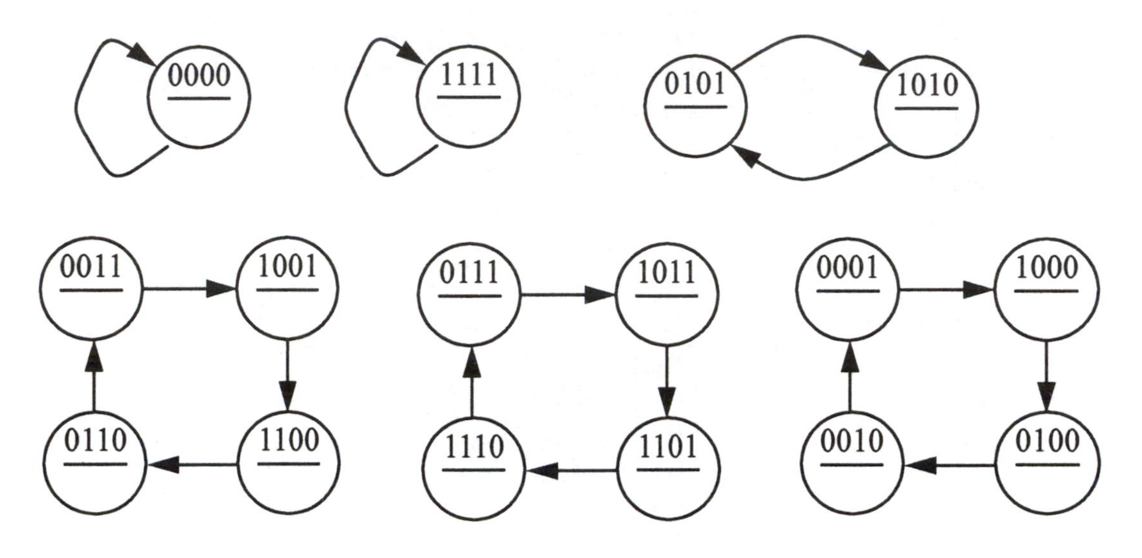

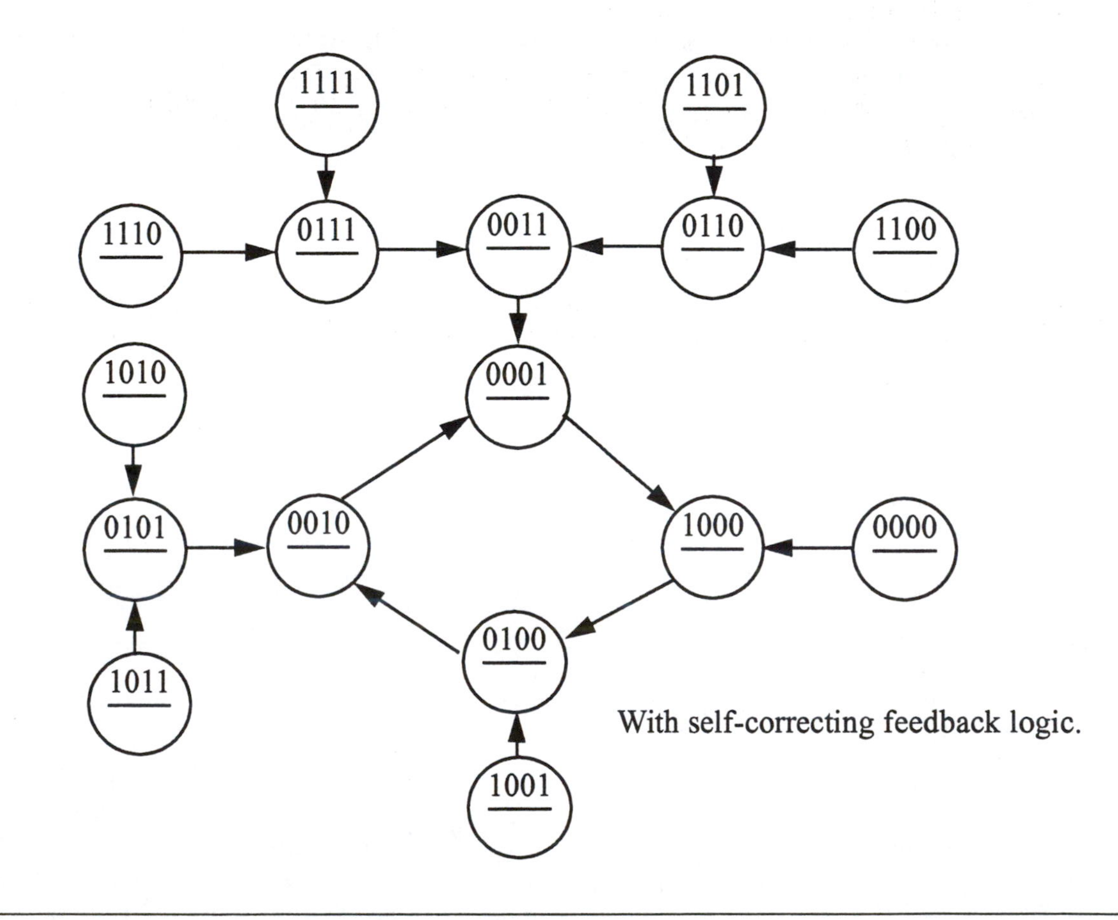

With self-correcting feedback logic.

10-10. Solution.

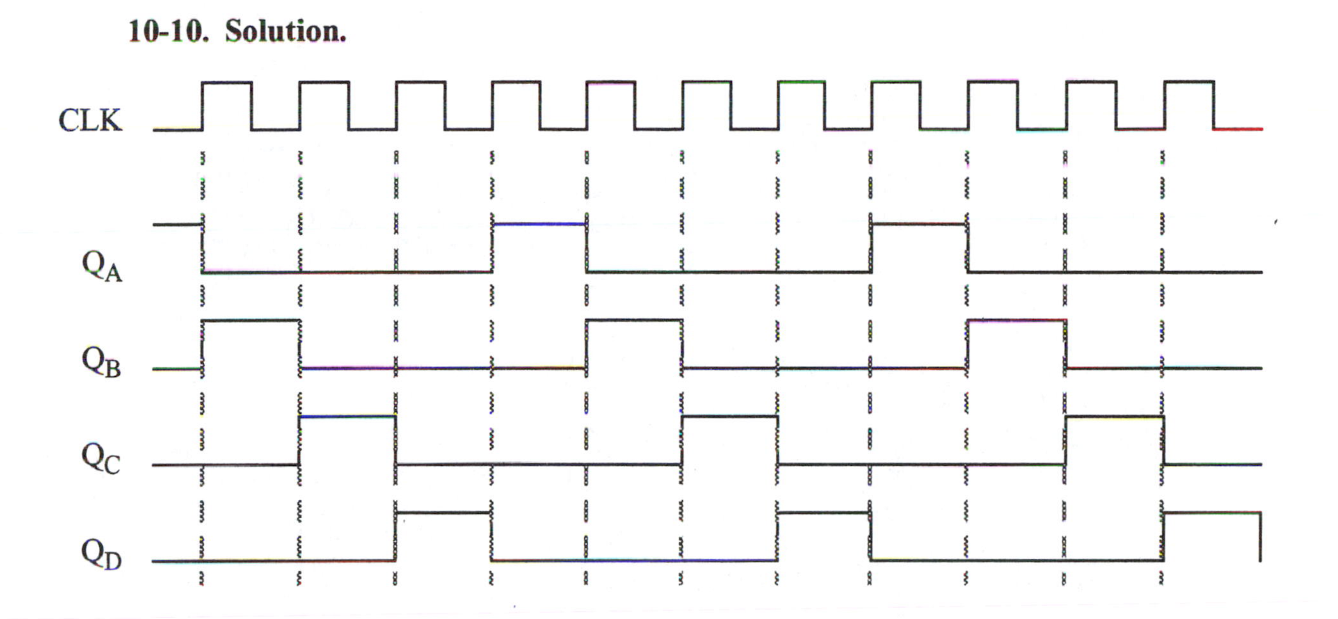

10-11. Solution.

Without self-correcting feedback logic.

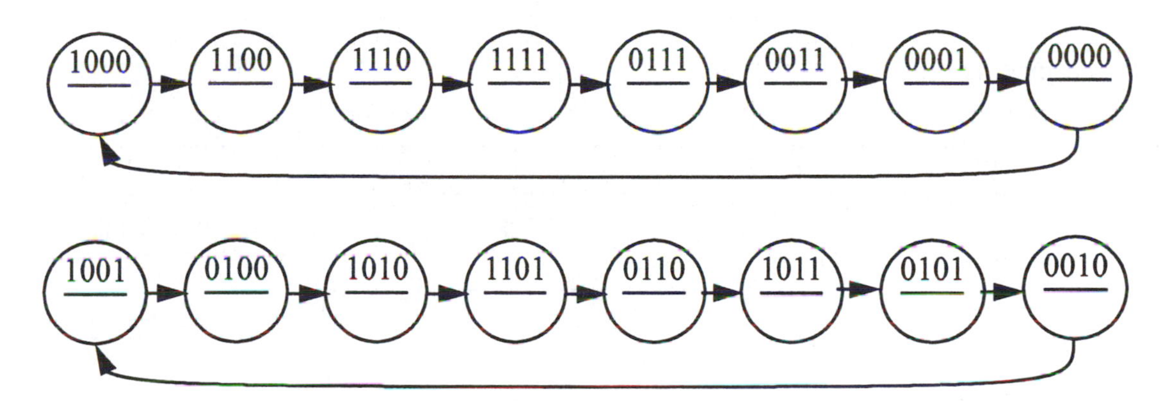

With self-correcting feedback logic.

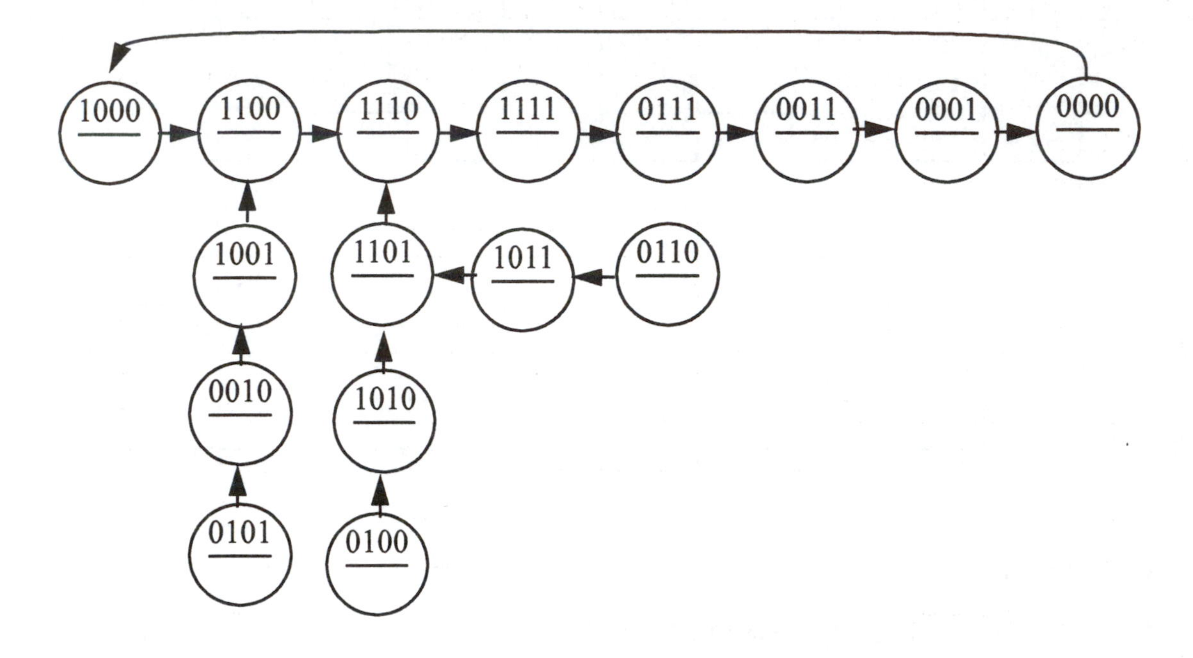

10-12. Solution.

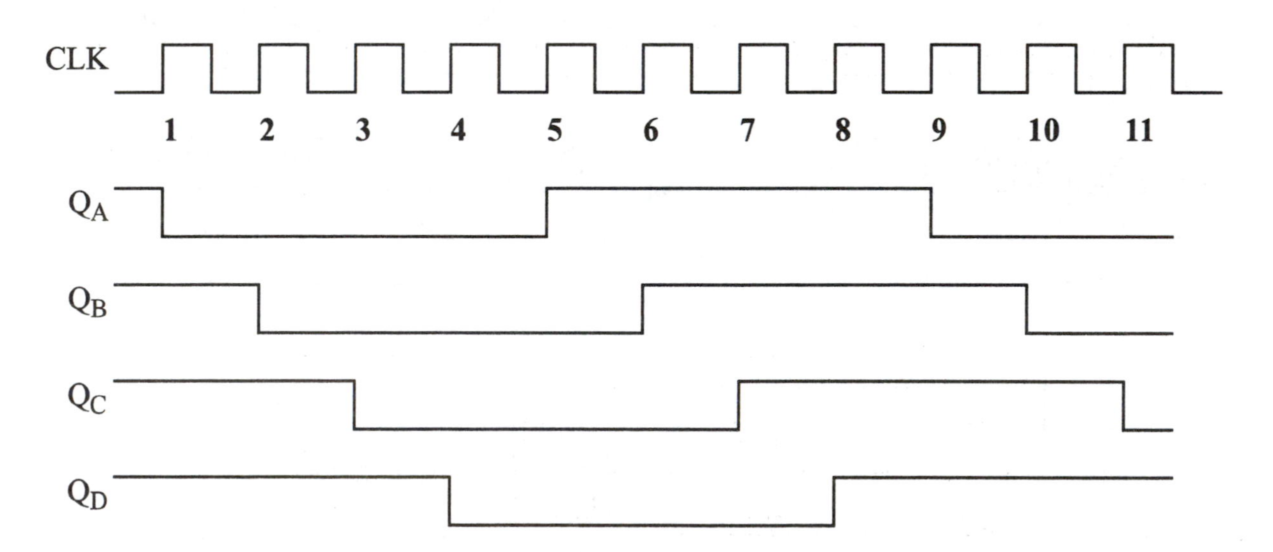

a. Q_A, Q_B, Q_C, and Q_D are all square waves having 1/8 the CLK frequency. Q_A, Q_B, Q_C, and Q_D are each shifted in time by one clock cycle relative to each other.

10-13. Solution.

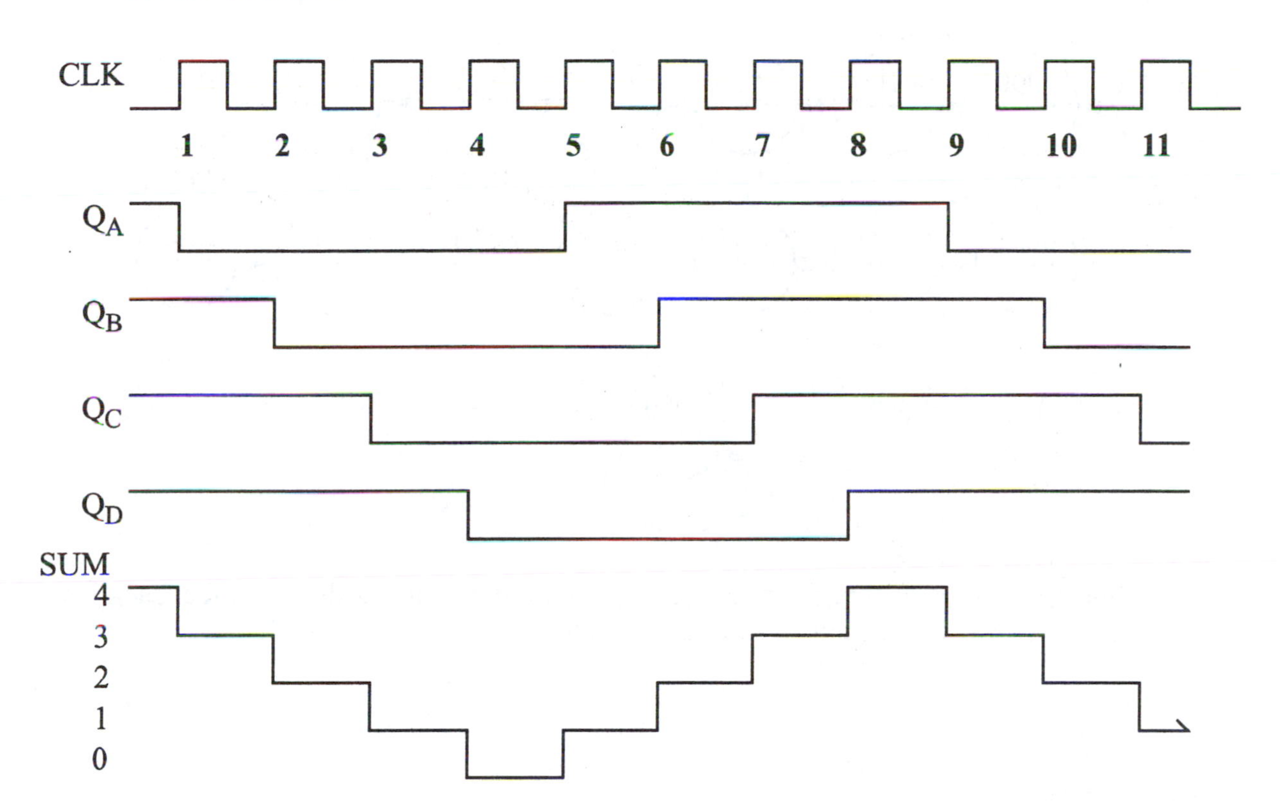

a. The new digital waveform is a stair-step approximation to a sinusoidal waveform having a period of 8 CLK cycles.

10-14. Solution.

a.

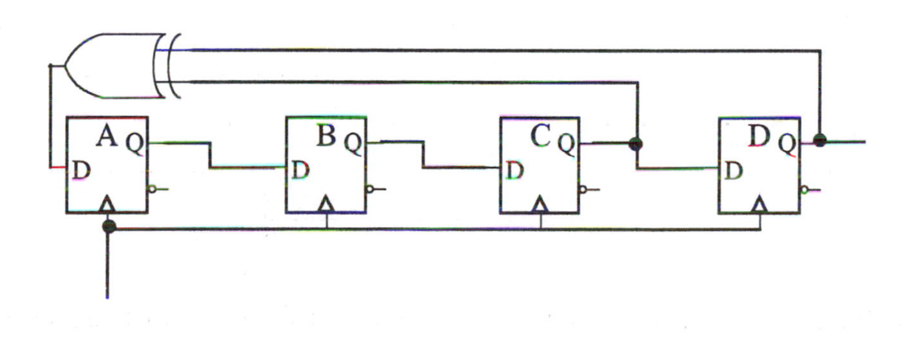

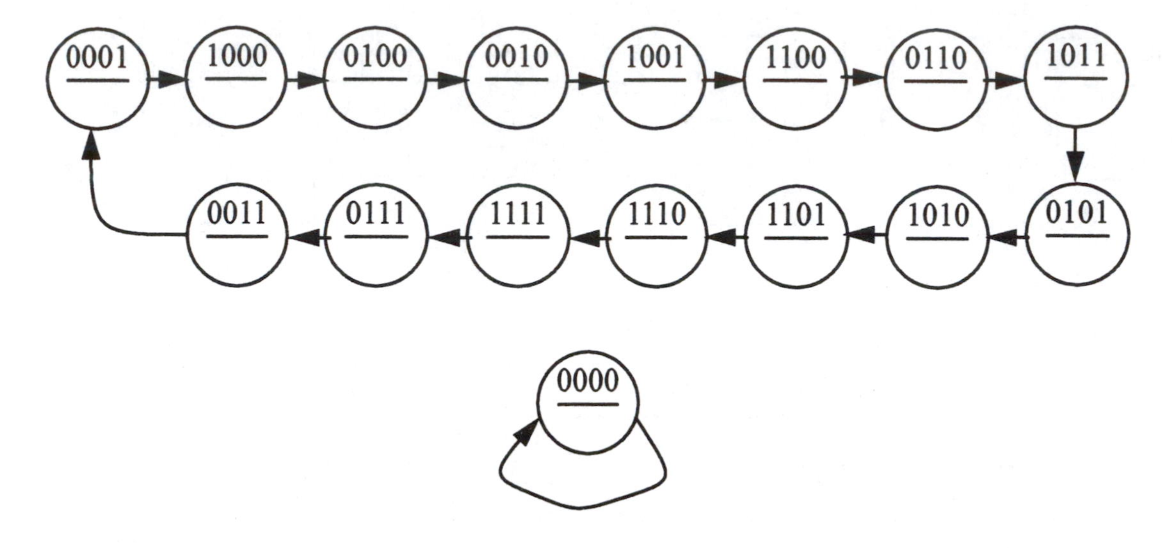

b. The linear feedback shift register always stays in the state 0000. Additional feedback logic must be combined with the exclusive or gate to address the unused state 0000.

10-15. Solution.

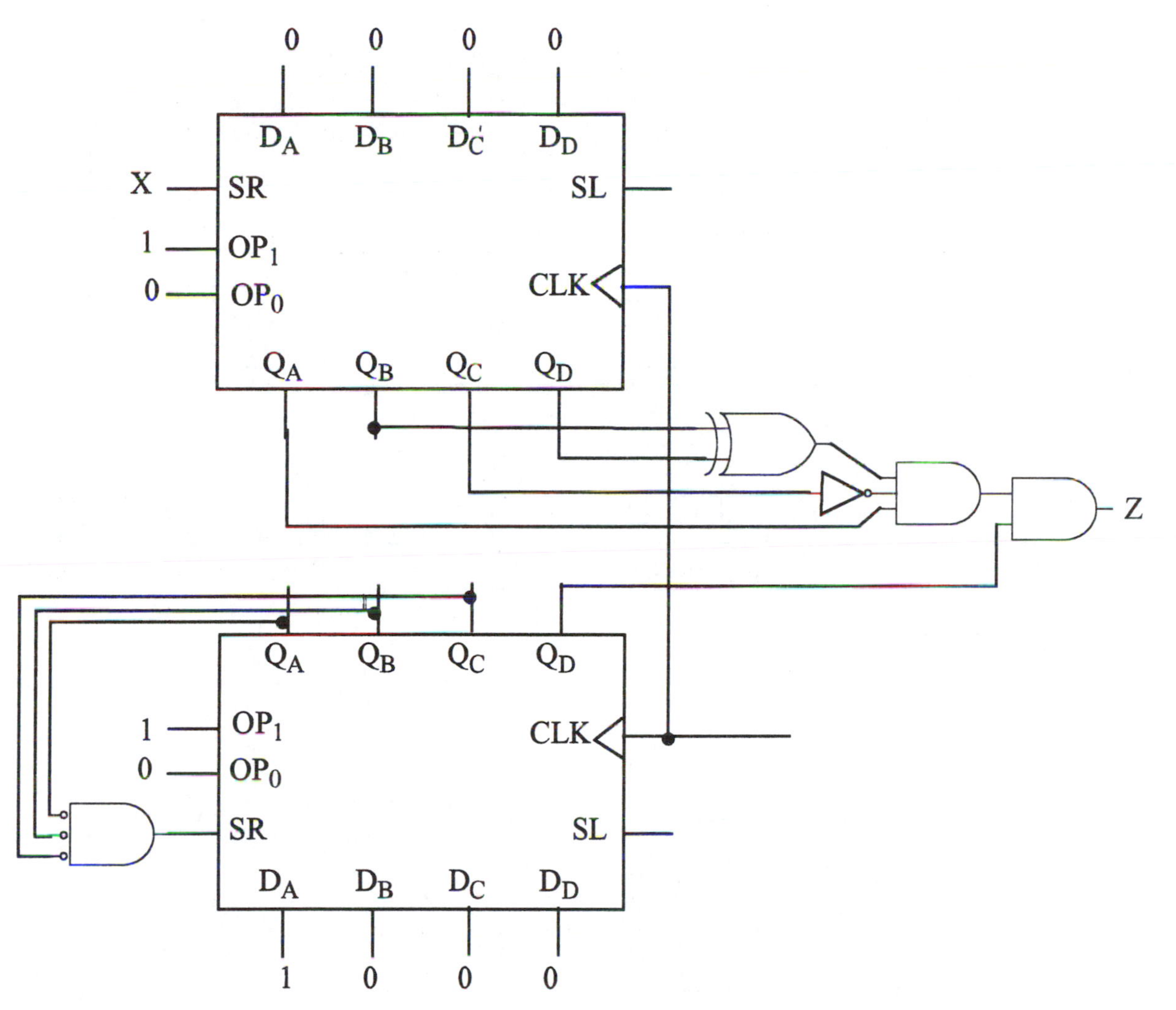

The top universal shift register detects the patterns 1001 or 1100. The bottom universal shift register clocks out the pattern detect on every fourth clock cycle.

10-16. Solution.

Load shift registers with the initial pattern $Q_A Q_B Q_C Q_D = 1000$.

a.

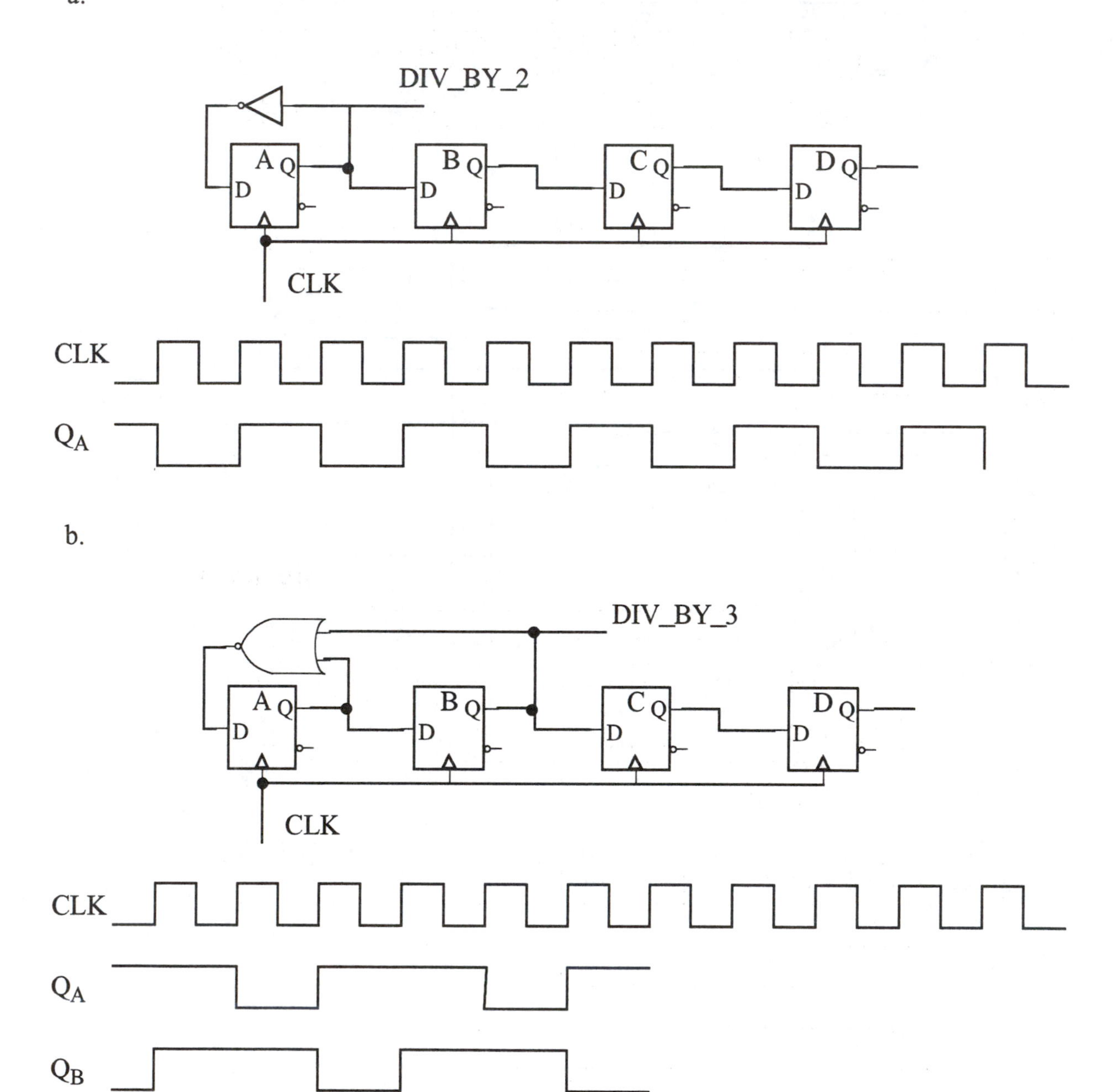

c.

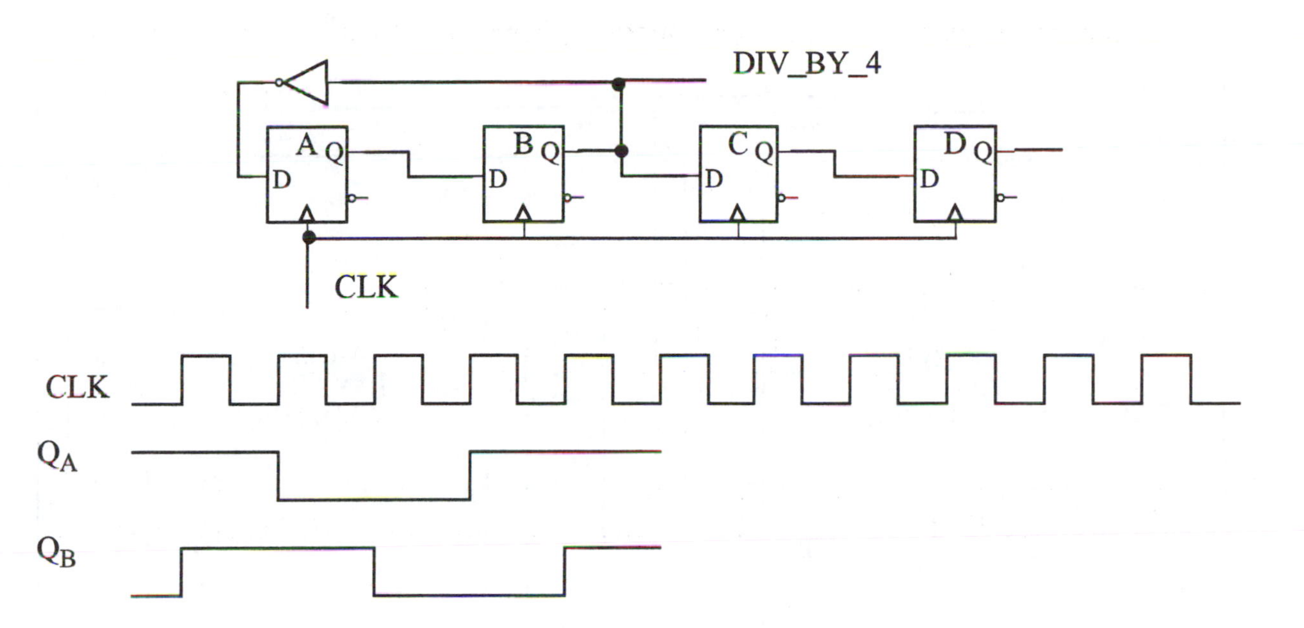
DIV_BY_4
A
Q
D
B
Q
D
C
Q
D
D
Q
D
CLK
CLK
Q_A
Q_B

d.

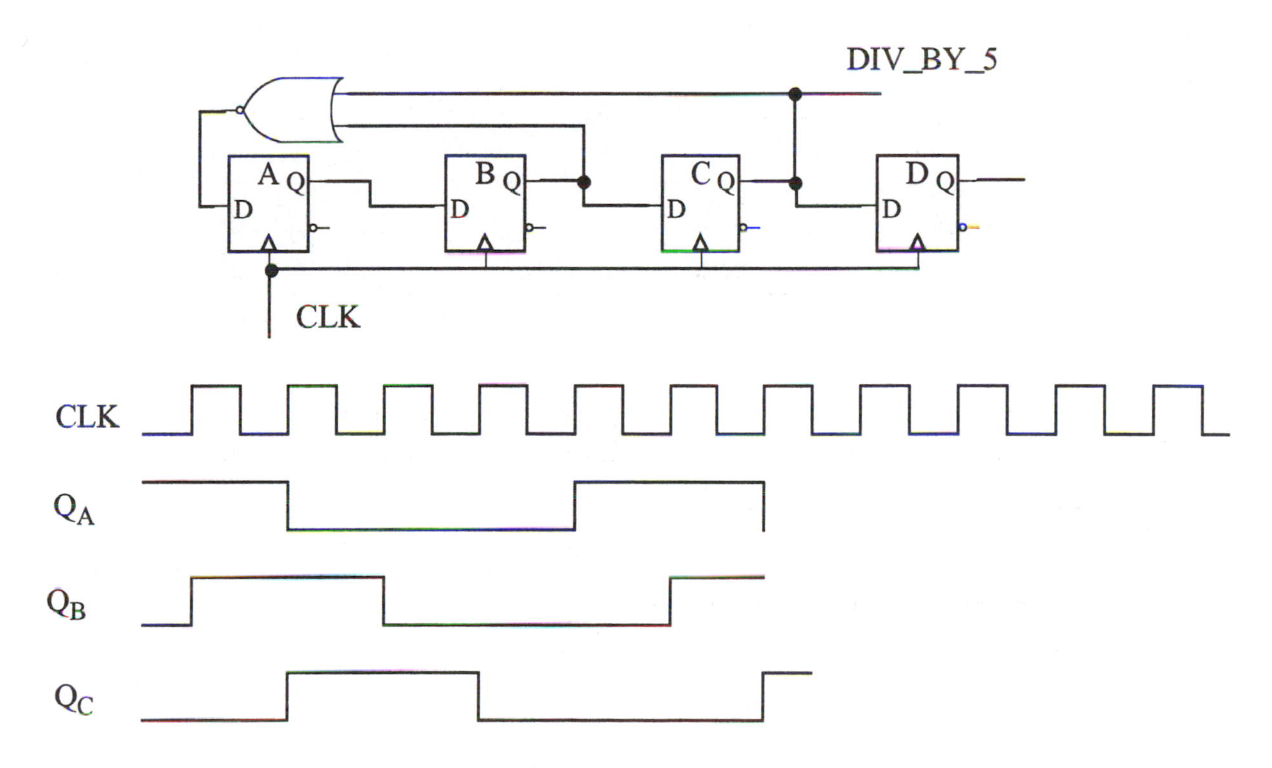
DIV_BY_5
A
Q
D
B
Q
D
C
Q
D
D
Q
D
CLK
CLK
Q_A
Q_B
Q_C

10-17. Solution.

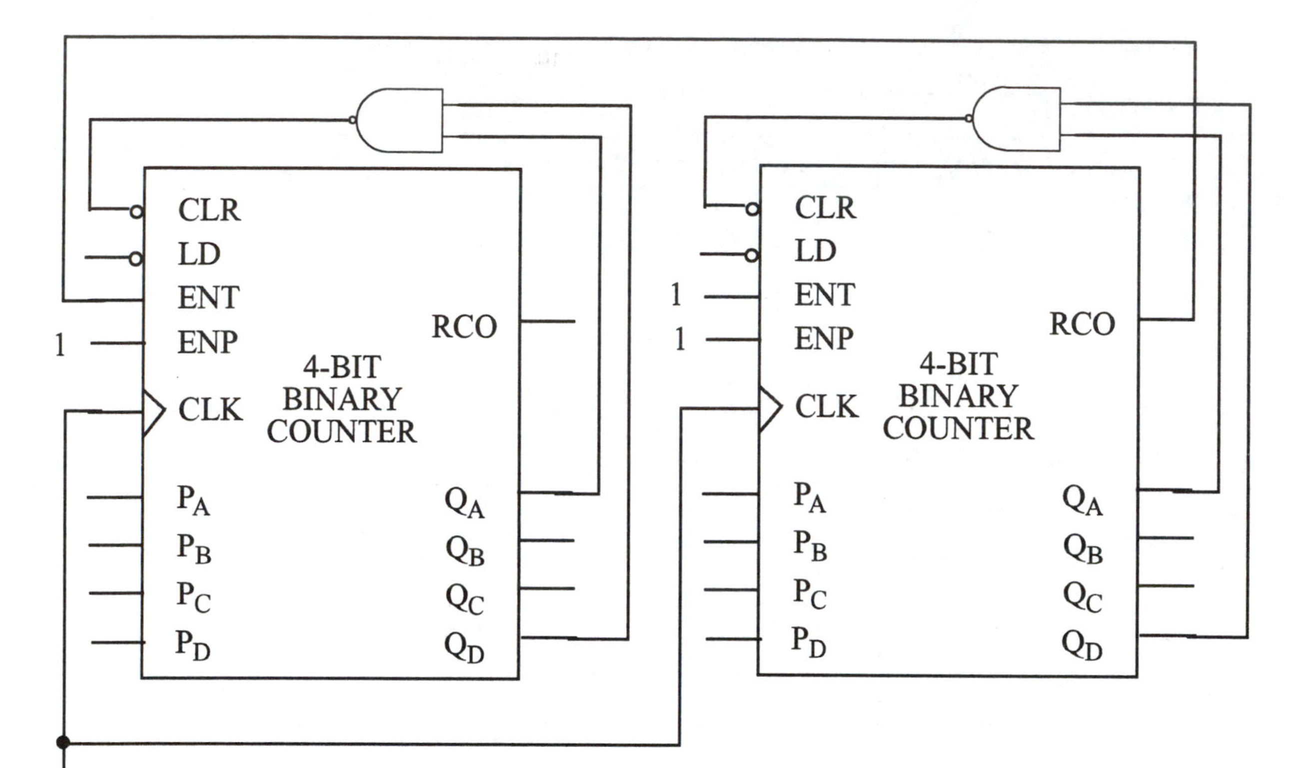

10-18. Solution.

Initially clear the counter and then load the counter on every clock cycle to generate desired even count sequence.

$Q_AQ_BQ_CQ_D$	LD_BAR	$P_AP_BP_CP_D$
0000	0	0010
0001	1	----
0010	0	0100
0011	1	----
0100	0	0110
0101	1	----
0110	0	1000
0111	1	----
1000	0	1010
1001	1	----
1010	0	1100
1011	1	----
1100	0	1110
1101	1	----
1110	0	0000
1111	1	----

LD_BAR

Q_AQ_B \ Q_CQ_D	00	01	11	10
00	0	1	1	0
01	0	1	1	0
11	0	1	1	0
10	0	1	1	0

P_A

Q_AQ_B \ Q_CQ_D	00	01	11	10
00	0	-	-	0
01	0	-	-	1
11	1	-	-	0
10	1	-	-	1

P_B

Q_AQ_B \ Q_CQ_D	00	01	11	10
00	0	-	-	1
01	1	-	-	0
11	1	-	-	0
10	0	-	-	1

P_C

Q_AQ_B \ Q_CQ_D	00	01	11	10
00	1	-	-	0
01	1	-	-	0
11	1	-	-	0
10	1	-	-	0

P_D

Q_AQ_B \ Q_CQ_D	00	01	11	10
00	0	-	-	0
01	0	-	-	0
11	0	-	-	0
10	0	-	-	0

$$P_A = (Q_A \bullet \overline{Q}_C) + (Q_A \bullet \overline{Q}_B) + (\overline{Q}_A \bullet Q_B \bullet Q_C)$$

$$P_B = (Q_B \bullet \overline{Q}_C) + (\overline{Q}_B \bullet Q_C) = Q_B \oplus Q_C$$

$$P_C = \overline{Q}_C$$

$$P_D = 0$$

$$\text{LD\_BAR} = Q_D$$

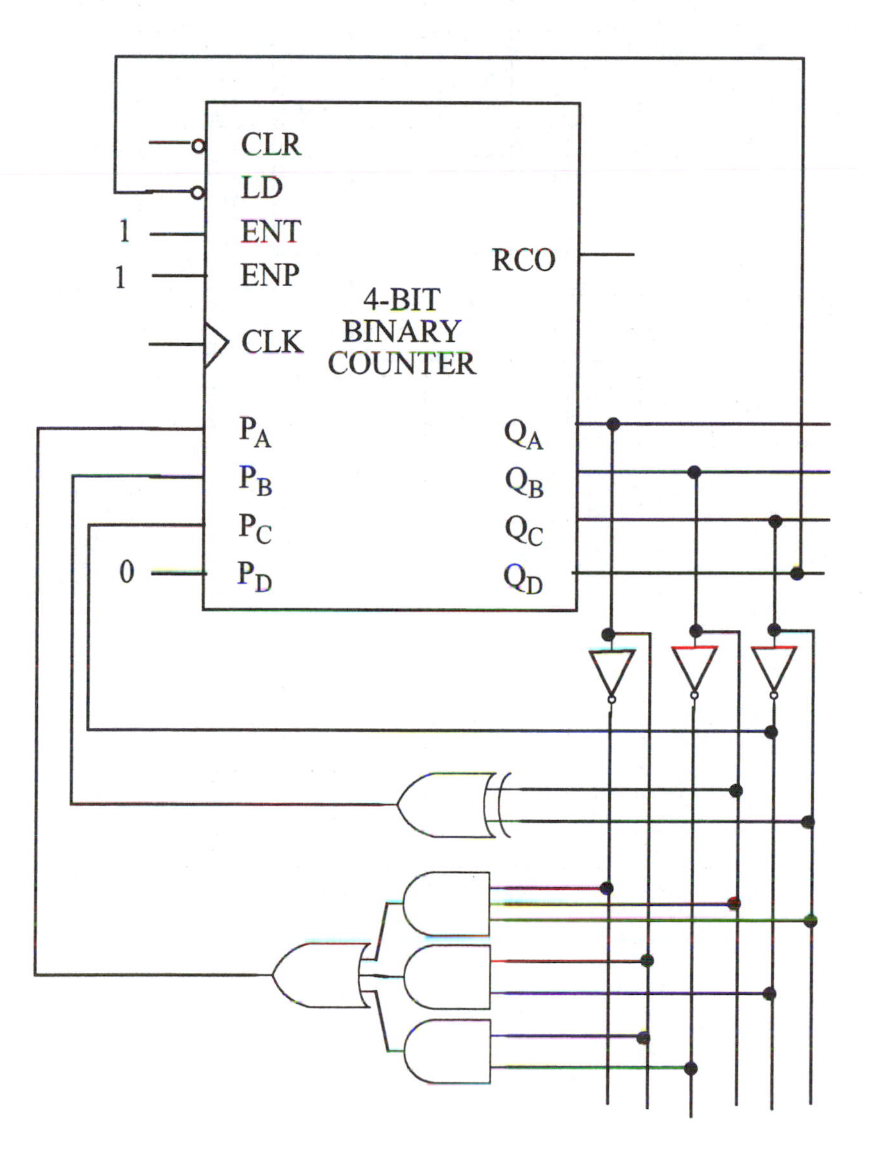

10-19. Solution.

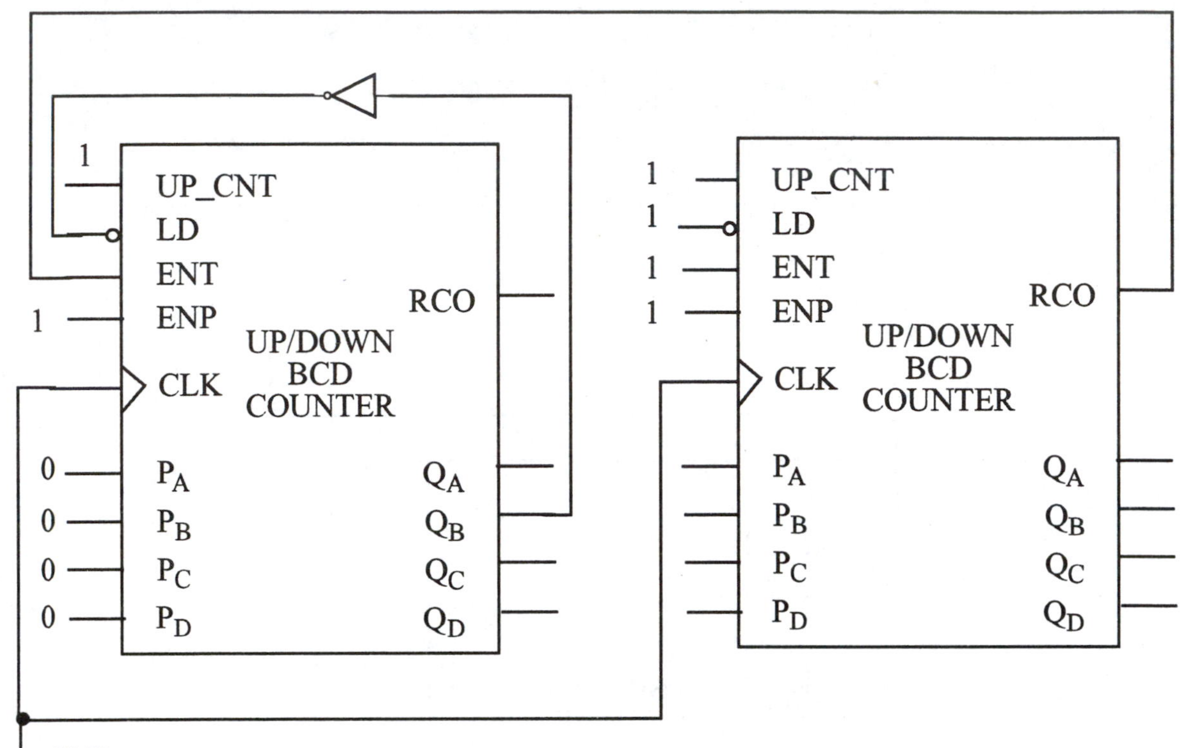

10-20. Solution.

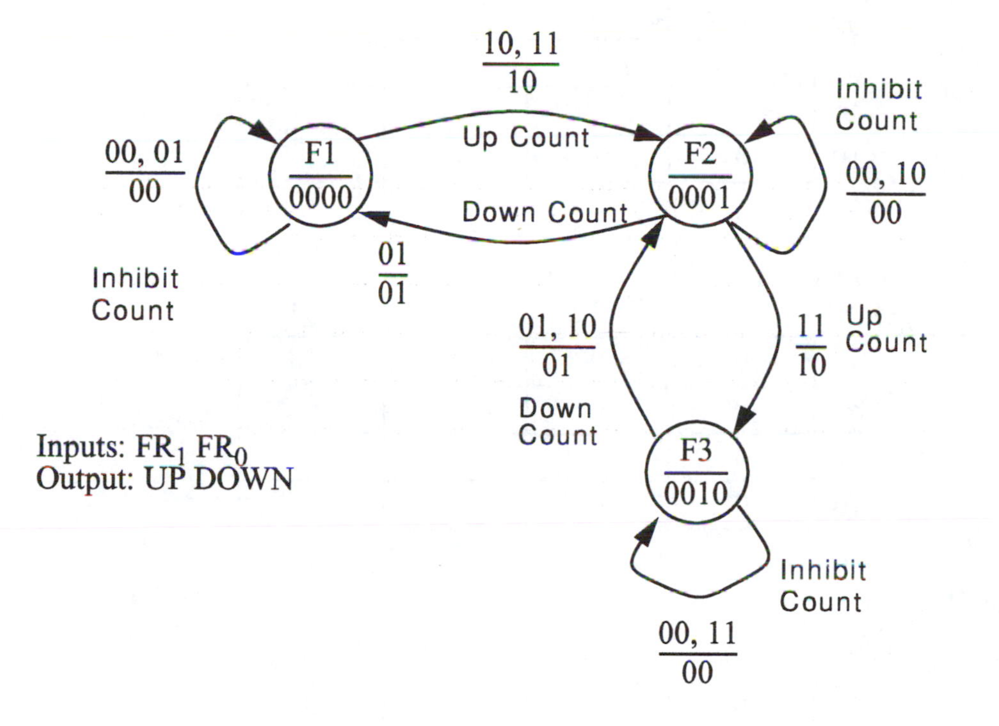

$Q_A Q_B Q_C Q_D$	UP_CNT (FR)				ENT/ENP (FR)				UP/DOWN (FR)			
	00	**01**	**10**	**11**	**00**	**01**	**10**	**11**	**00**	**01**	**10**	**11**
0000	-	-	1	1	0	0	1	1	00	00	10	10
0001	-	0	-	1	0	1	0	1	00	01	00	10
0010	-	0	0	-	0	1	1	0	00	01	01	00

UP_CNT

Q_AQ_B \ FR_1FR_0	00	01	11	10
00	-	-	1	1
01	-	0	1	-
11	-	-	-	-
10	-	0	-	0

$ENT = ENP$

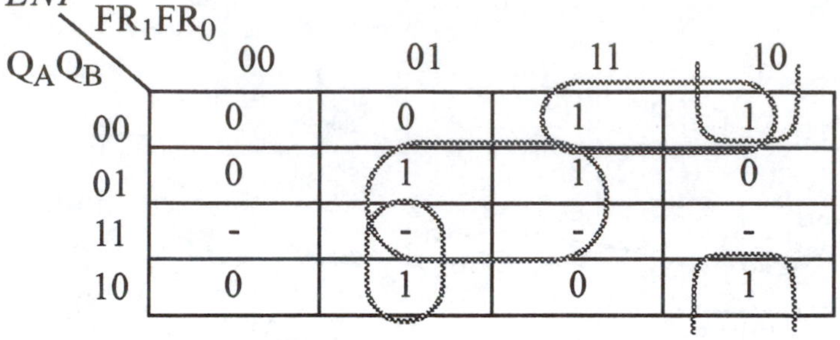

Q_AQ_B \ FR_1FR_0	00	01	11	10
00	0	0	1	1
01	0	1	1	0
11	-	-	-	-
10	0	1	0	1

UP

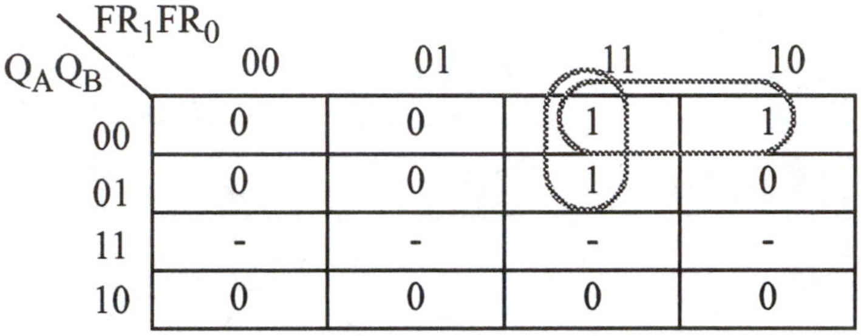

Q_AQ_B \ FR_1FR_0	00	01	11	10
00	0	0	1	1
01	0	0	1	0
11	-	-	-	-
10	0	0	0	0

DOWN

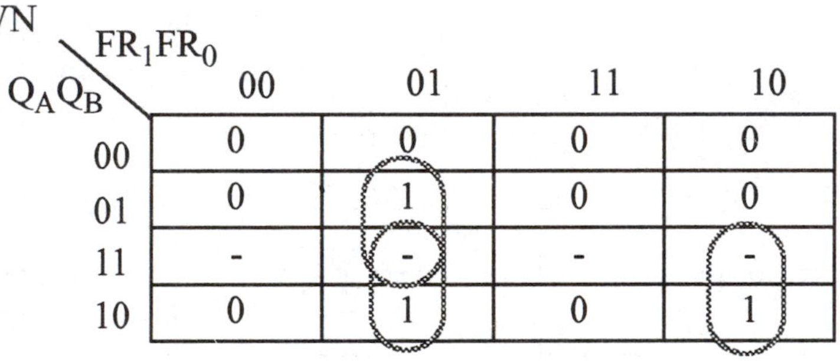

Q_AQ_B \ FR_1FR_0	00	01	11	10
00	0	0	0	0
01	0	1	0	0
11	-	-	-	-
10	0	1	0	1

$$\text{UP\_CNT} = \overline{Q}_C \bullet FR_1$$

$$ENT = ENP = \left(Q_C \bullet \overline{FR}_1 \bullet FR_0\right) + (Q_D \bullet FR_0) + (\overline{Q}_C \bullet \overline{Q}_D \bullet FR_1) + \left(\overline{Q}_D \bullet FR_1 \bullet \overline{FR}_0\right)$$

$$UP = (\overline{Q}_C \bullet FR_1 \bullet FR_0) + (\overline{Q}_C \bullet \overline{Q}_D \bullet FR_1)$$

$$DOWN = \left(Q_D \bullet \overline{FR}_1 \bullet FR_0\right) + \left(Q_C \bullet \overline{FR}_1 \bullet FR_0\right) + \left(Q_C \bullet FR_1 \bullet \overline{FR}_0\right)$$

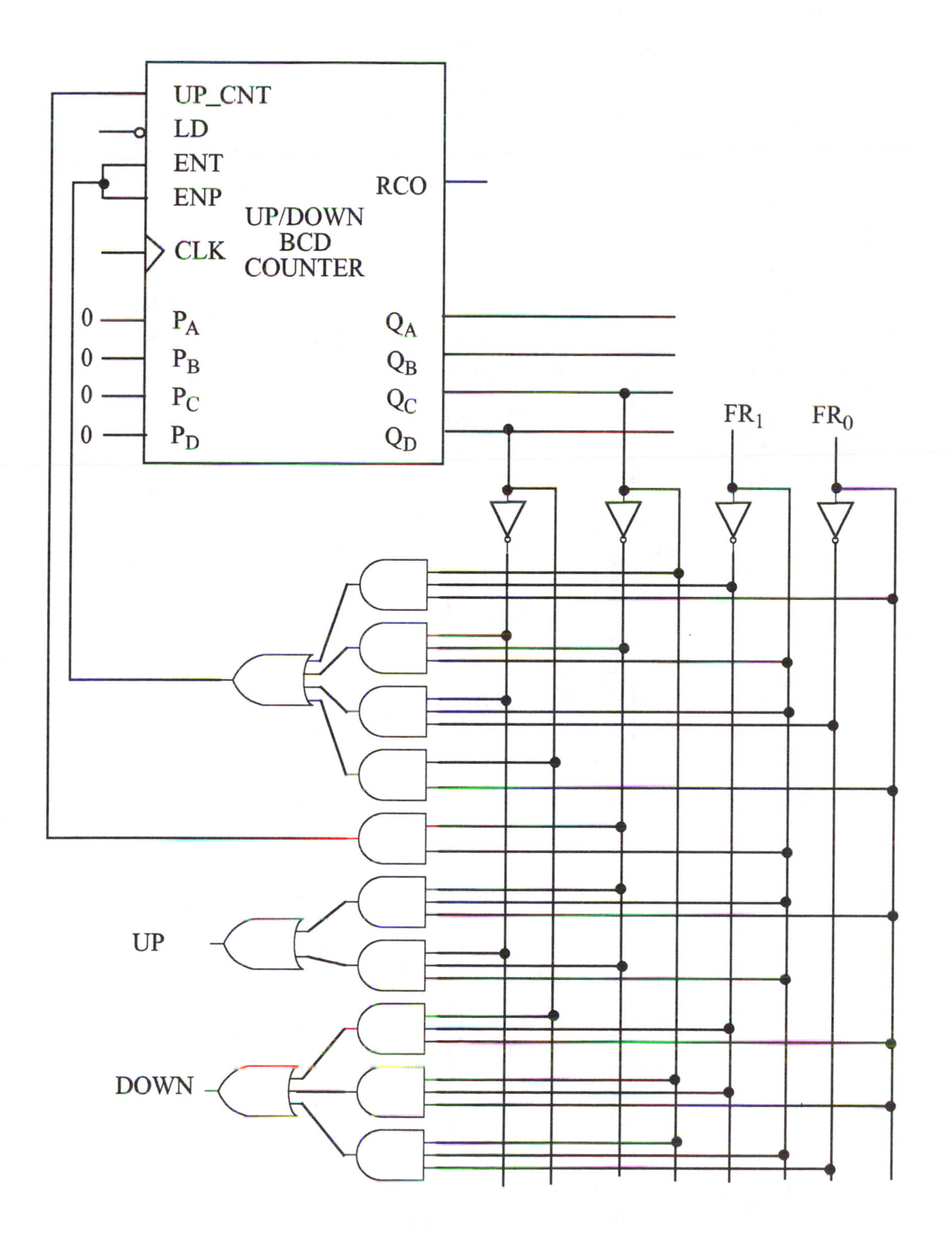
UP_CNT
LD
ENT
ENP
CLK
RCO
UP/DOWN
BCD
COUNTER
0
0
0
0
P_A
P_B
P_C
P_D
Q_A
Q_B
Q_C
Q_D
FR_1
FR_0
UP
DOWN

10-21. Solution.

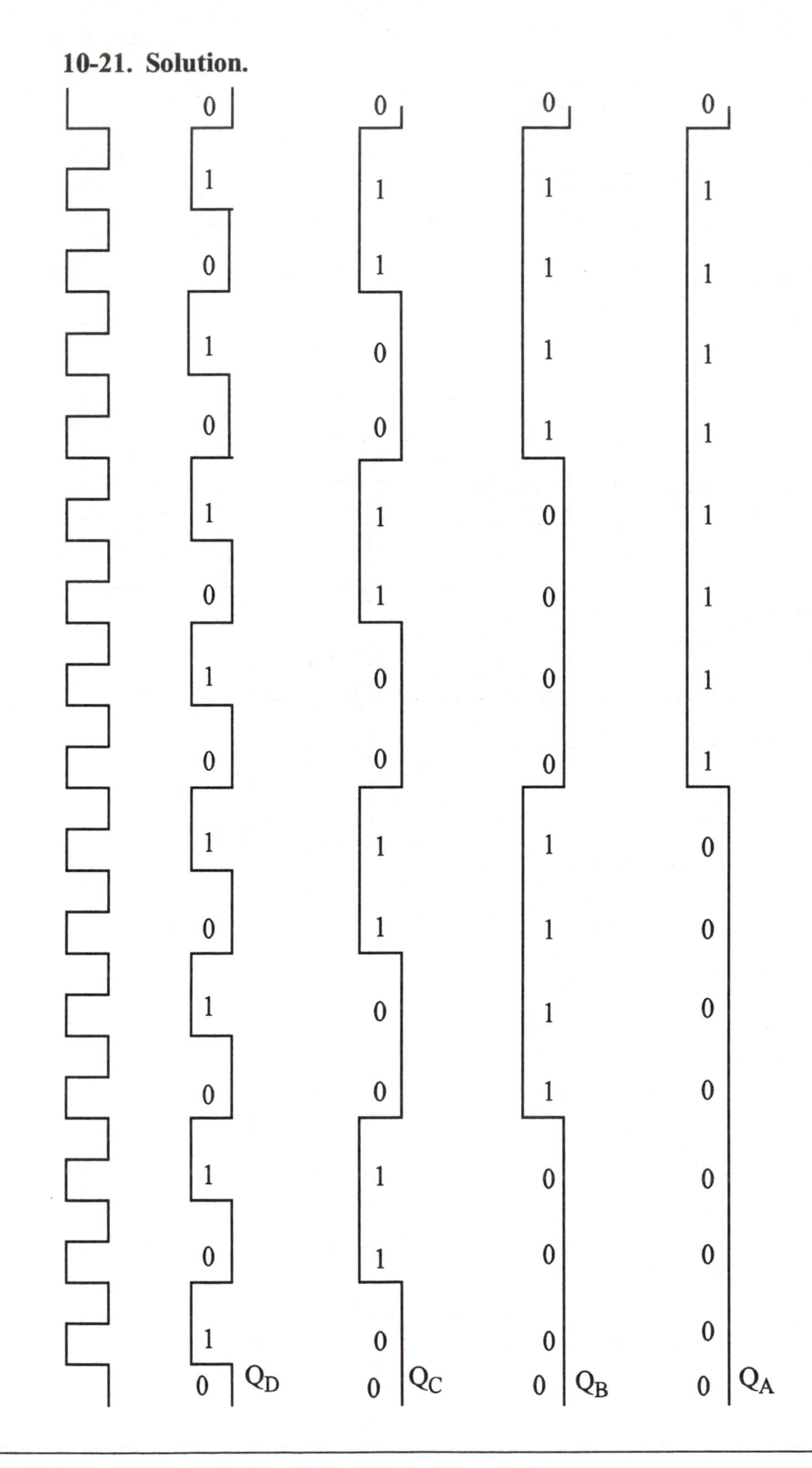

10-22. Solution.

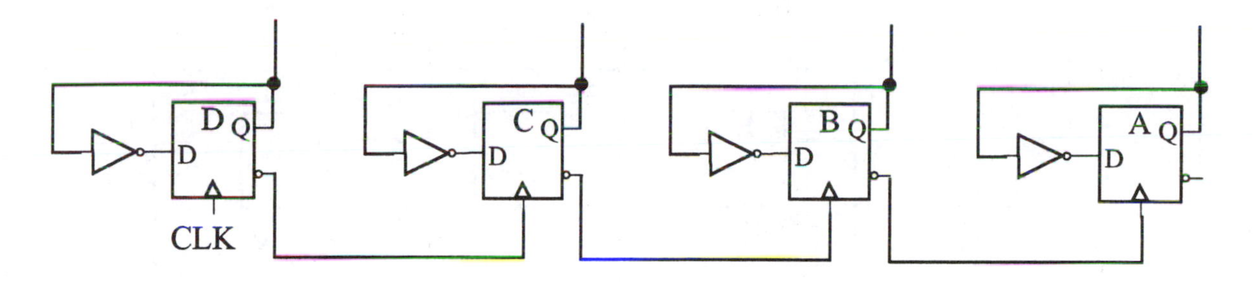

10-23. Solution.

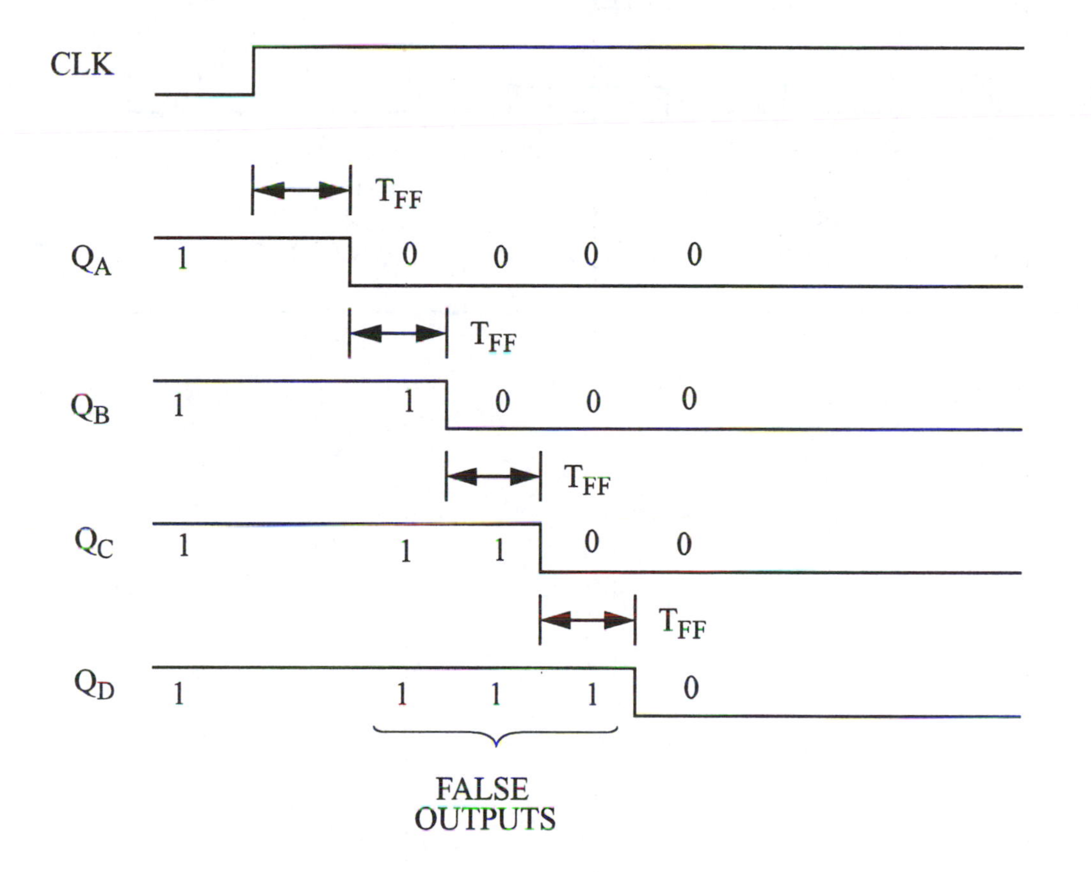

10-24. Solution.

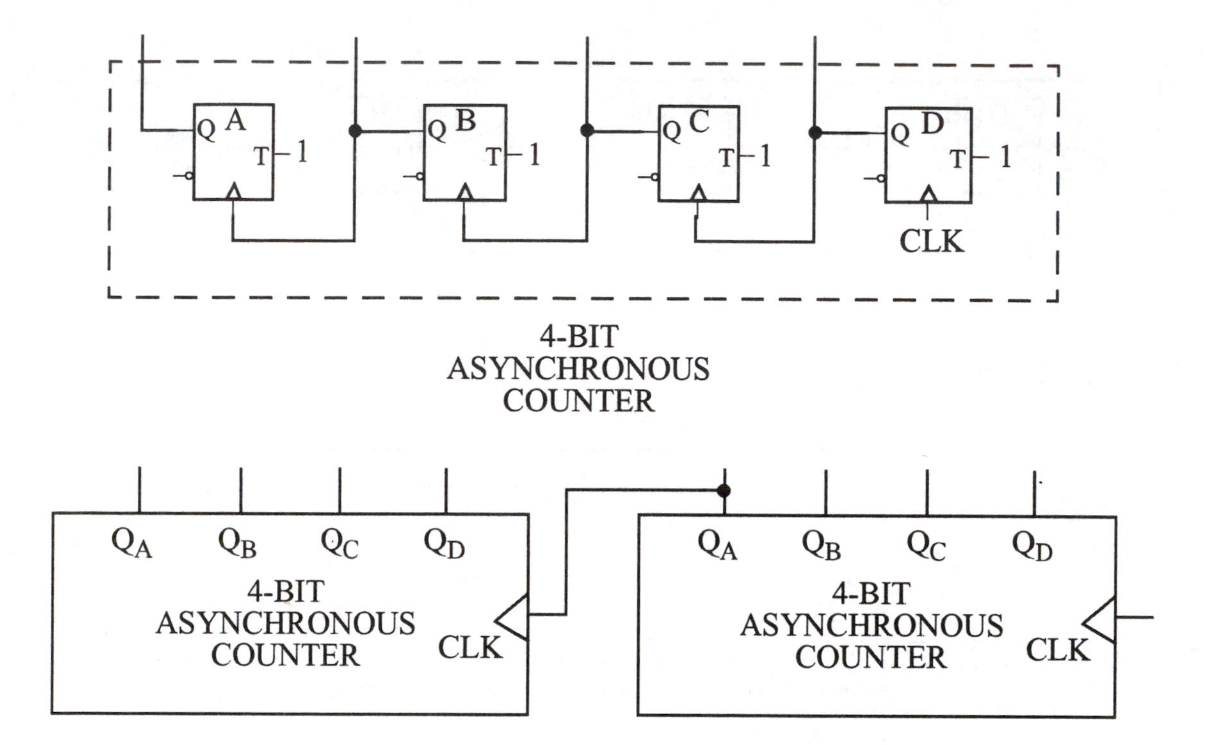

Chapter 11. VHDL: A First Look

11-1. Solution.

a. A design entity is the basic construct in VHDL for modeling a hardware component. A design entity is composed of an entity declaration (interface) and an architecture definition (body).

b. An interface or entity declaration declares the existence of a design entity and its name. An entity declaration also defines information accessible by the environment, such as signals that enter and/or leave the design entity. Input/output signals are called ports.

c. An architecture defines the behavior of the design entity, i.e., how the outputs respond to the inputs. An architecture contains a declaration section for declaring objects, such as signals, that are internal to the design entity and not accessible to the larger environment. An architecture also contains statements describing input signal to output signal transformations. For simple switching algebra operations, the architecture contains concurrent signal assignment statements describing logic operations and expressions.

d. A signal declaration may occur within an architecture and defines signals to be used within the architecture by associated statements. Such signals have names and types, but no mode.

e. A concurrent signal assignment statement describes a signal transformation. Anytime any signal appearing on the right-hand side of the assignment changes value, the concurrent signal assignment statement executes. The right-hand side logic expression containing logic operators and signals is evaluated and the resulting value is assigned to the target signal on the left-hand side of the assignment statement.

11-2. Solution.

a. A data type defines a set of values.

b. Data types rigorously define the range of values an object hold, thereby providing structured information modéling. Structured information modeling helps match the model to the domain, which facilitates generating, understanding, and using models. Structured information modeling also helps catch inadvertent modeling errors caused by incorrectly using objects.

c. The data type BIT represents the two character literals '0' and '1'.

11-3. Solution.

a. Illegal - keyword

b. Legal

c. Illegal - starts with a number

d. Illegal - two successive underscore characters

e. Legal

f. Illegal - # is not a legal character

g. Illegal - ends with an underscore character

11-4. Solution.

a. Illegal - keyword

b. Legal

c. Illegal - / is not a legal character

d. Legal

e. Illegal - starts with a number

f. Illegal - no trailing enclosing backslash

g. Illegal - # is not a legal character

11-5. Solution.

a.
```
entity INVERT is
  port (A : in BIT; A_BAR : out BIT);
end INVERT;
architecture INVERT of INVERT is
begin
  A_BAR <= not A;
end INVERT;
```

b.
```
entity AND2 is
  port (A, B : in BIT; Z : out BIT);
end AND2;
architecture AND2 of AND2 is
begin
  Z <= A and B;
end AND2;
```

c.
```
entity OR2 is
  port (A, B : in BIT; Z : out BIT);
end OR2;
architecture OR2 of OR2 is
begin
  Z <= A or B;
end OR2;
```

d.
```
entity XNOR2 is
   port (A, B : in BIT; Z : out BIT);
end XNOR2;
architecture XNOR2 of XNOR2 is
begin
   -- VHDL-93
   Z <= A xnor B;
end XNOR2;
```

11-6. Solution.

a.
```
entity BUF is
  port (A : in BIT; Z : out BIT);
end BUF;
architecture BUF of BUF is
begin
  Z <= A;
end BUF;
```

b.
```
entity NAND3 is
  port (A, B, C : in BIT; Z : out BIT);
end NAND3;
architecture NAND3 of NAND3 is
begin
  Z <= not (A and B and C);
end NAND3;
```

c.
```
entity AOI is
  port (A, B, C, D : in BIT; Z : out BIT);
end AOI;
architecture AOI of AOI is
begin
  Z <= (A and B) nor (C and D);
end AOI;
```

d.
```
entity NOR4 is
  port (A, B, C, D : in BIT; Z : out BIT);
end NOR4;
architecture NOR4 of NOR4 is
begin
  Z <= not (A or B or C or D);
end NOR4;
```

11-7. Solution.

a. $Z = (\overline{A} \bullet \overline{B} \bullet C) + (\overline{A} \bullet B \bullet \overline{C}) + (A \bullet \overline{B} \bullet \overline{C}) + (A \bullet B \bullet C)$

b.

A	B	C	Z
0	0	0	0
0	0	1	1
0	1	0	1
0	1	1	0
1	0	0	1
1	0	1	0
1	1	0	0
1	1	1	1

c. The output is asserted active-1 when an odd number of inputs are asserted active-1.

```
entity LOGIC_OPERATION is
  port (A, B, C : in BIT; Z : out BIT);
end LOGIC_OPERATION;
architecture NEW_MODEL of LOGIC_OPERATION is
begin
  Z <= A xor B xor C;
end NEW_MODEL;
```

11-8. Solution.

a. $Z = (A + B) \bullet (A + \bar{B})$

b.

A	B	Z
0	0	0
0	1	0
1	0	1
1	1	1

c. The output Z equals the input A

```
entity LOGIC_OPERATION is
  port (A, B : in BIT; Z : out BIT);
end LOGIC_OPERATION;
architecture NEW_MODEL of LOGIC_OPERATION is
begin
  Z <= A;
end NEW_MODEL;
```

11-9. Solution.

a.

A	B	Z
0	0	0
0	1	1
1	0	0
1	1	0

```
entity FA is
 port (A, B : in BIT; Z : out BIT);
end FA;
architecture FA of FA is
begin
 Z <= not A and B;
end FA;
```

b.

A	B	Z
0	0	0
0	1	0
1	0	1
1	1	0

```
entity FB is
  port (A, B : in BIT; Z : out BIT);
end FB;
architecture FB of FB is
begin
  Z <= not A nor B;
end FB;
```

c.

A	B	Z
0	0	1
0	1	1
1	0	0
1	1	1

```
entity FC is
 port (A, B : in BIT; Z : out BIT);
end FC;
architecture FC of FC is
begin
 Z <= A nand not B;
end FC;
```

d.

A	B	Z
0	0	1
0	1	0
1	0	0
1	1	1

```
entity FD is
 port (A, B : in BIT; Z : out BIT);
end FD;
architecture FD of FD is
begin
  Z <= not A xor B;
end FD;
```

11-10. Solution.

In switching algebra, conjunction has higher precedence than disjunction; however, in VH-DL, conjunction and disjunction have equal precedences.

```
entity NICE_TRY is
  port (A, B, C : in BIT; Z : out BIT);
end NICE_TRY;
architecture NEW_MODEL of NICE_TRY is
begin
  Z <= A or (B and C);
end NEW_MODEL;
```

11-11. Solution.

a. Anytime A and/or B change value.

b. 1) not A = RESULT1
   2) not B = RESULT2
   3) RESULT1 or RESULT2 = RESULT3
   4) not RESULT3

c. $Z = \overline{(\overline{A} + \overline{B})}$

d.

A	B	Z
0	0	0
0	1	0
1	0	0
1	1	1

e. The output is asserted active-1 when both inputs are asserted active-1.

```
entity LOGIC_OPERATION is
  port (A, B : in BIT; Z : out BIT);
end LOGIC_OPERATION;
architecture NEW_MODEL of LOGIC_OPERATION is
begin
  Z <= A and B;
end NEW_MODEL;
```

11-12. Solution.

a. Statement 2

   Statement 3

b.

Signal	Old Value	New Value
INT1	1	1
INT2	1	0
INT3	1	1
Z	0	0

11-13. Solution.

a. Statement 1 and Statement 2
   Statement 3
   Statement 4

b.

Signal	Old Value	New Value
INT1	1	0
INT2	1	0
INT3	1	0
Z	0	1

Chapter 12. Structural Modeling in VHDL - Part 1

12-1. Solution.

a.

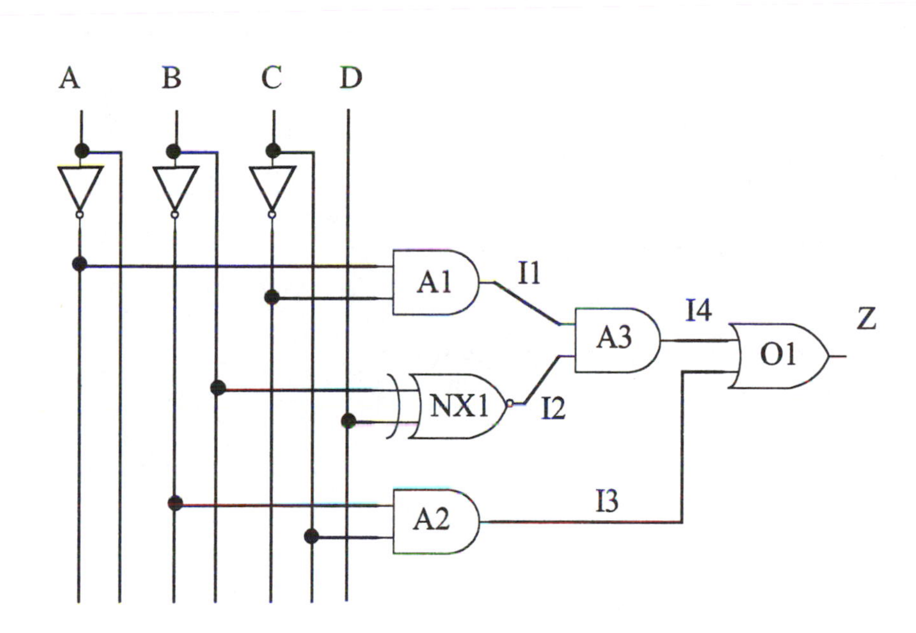

b. $Z = (\bar{A} \bullet \bar{C} \bullet \overline{(B \oplus D)}) + (\bar{B} \bullet C)$

c.

ABCD	Z
0000	1
0101	1
1010	1
1111	0

d.
```
entity AND2_OP is
 port
 (A, B : in BIT; Z : out BIT);
end AND2_OP;

architecture BEHAVIOR of AND2_OP is
begin
 Z <= A and B;
end BEHAVIOR;

entity OR2_OP is
 port
 (A, B : in BIT; Z : out BIT);
end OR2_OP;

architecture BEHAVIOR of OR2_OP is
begin
 Z <= A or B;
end BEHAVIOR;

entity XNOR2_OP is
 port
 (A, B : in BIT; Z : out BIT);
end XNOR2_OP;

architecture BEHAVIOR of XNOR2_OP is
begin
 Z <= not (A xor B);
end BEHAVIOR;

entity NOT_OP is
 port
 (A : in BIT; A_BAR: out BIT);
end NOT_OP;

architecture BEHAVIOR of NOT_OP is
begin
 A_BAR <= not A;
end BEHAVIOR;
```

12-2. Solution.

a.

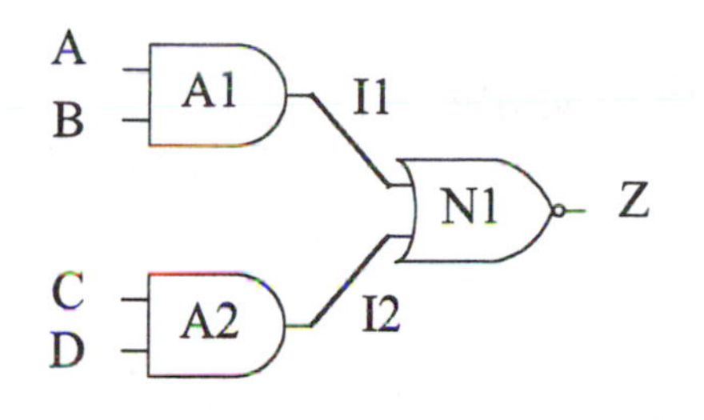

b. $Z = \overline{(A \bullet B) + (C \bullet D)}$

c.

ABCD	Z
0000	1
0001	1
0010	1
0011	0
0100	1
0101	1
0110	1
0111	0
1000	1
1001	1
1010	1
1011	0
1100	0
1101	0
1110	0
1111	0

d.
```
entity AND2_OP is
 port
 (A, B : in BIT; Z : out BIT);
end AND2_OP;

architecture BEHAVIOR of AND2_OP is
begin
 Z <= A and B;
end BEHAVIOR;

entity NOR2_OP is
 port
 (A, B : in BIT; Z : out BIT);
end NOR2_OP;

architecture BEHAVIOR of NOR2_OP is
begin
 Z <= A nor B;
end BEHAVIOR;
```

12-3. Solution.

a.

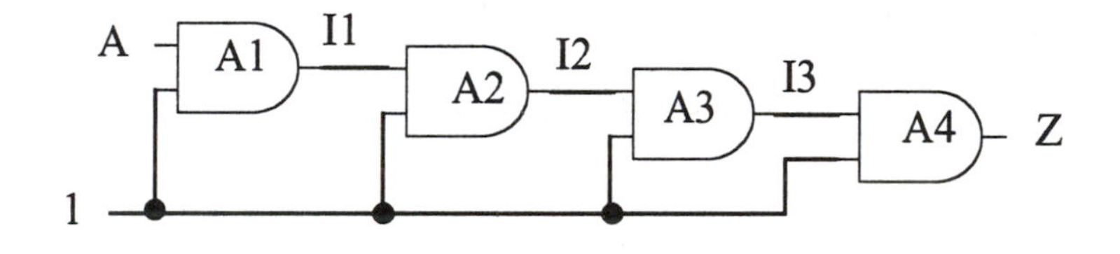

b. $Z = (1 \bullet (1 \bullet (1 \bullet (1 \bullet A)))) = A$

c.

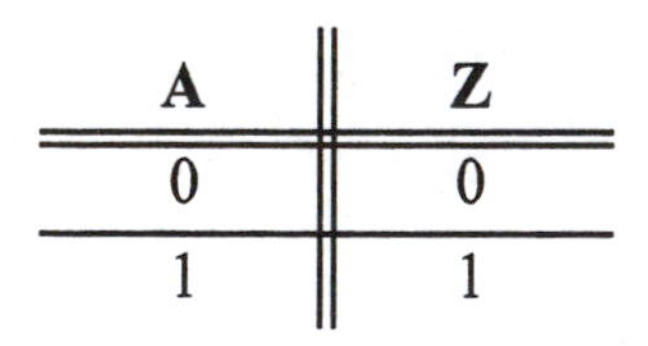

A	Z
0	0
1	1

d.
```
entity AND2_OP is
 port
 (A, B : in BIT; Z : out BIT);
end AND2_OP;

architecture BEHAVIOR of AND2_OP is
begin
 Z <= A and B;
end BEHAVIOR;
```

12-4. Solution.

Packages are containers for commonly used declarations. Other design units, such as packages, entities, and architectures, access the declarations contained within a package by a use clause. Packages avoid needlessly rewriting code. Packages help coordinate modeling styles and conventions across a design project. Packages help hide details of VHDL that an engineer may not need to know.

12-5. Solution.

a. AND2_OP : package PKG1 and package PKG2.

b. OR2_OP : package PKG1, entity SCOPE, and architecture STRUCTURE.

c. NAND2_OP : architecture STRUCTURE

12-6. Solution.

a.
```
entity PROBLEM_6 is
   port (A, B, C, D : in BIT;
         Z : out BIT);
end PROBLEM_6;

use WORK.LOGIC_PKG.all;
architecture POSITIONAL of PROBLEM_6 is
  signal A_BAR, B_BAR, C_BAR, I1, I2, I3 : BIT;
begin
 N1: NAND2_OP port map (A, A, A_BAR);
 N2: NAND2_OP port map (B, B, B_BAR);
 N3: NAND2_OP port map (C, C, C_BAR);

 A1: AND3_OP port map (A_BAR, B_BAR, D, I1);
 A2: AND3_OP port map (A, B, C, I2);
 A3: AND3_OP port map (B_BAR, C_BAR, C_BAR, I3);

 O1: OR4_OP port map (I1, I2, I3, I3, Z);
end POSITIONAL;
```

b.
```
entity PROBLEM_6 is
 port
 (A, B, C, D : in BIT;
 Z : out BIT);
end PROBLEM_6;

use WORK.LOGIC_PKG.all;
architecture NAMED of PROBLEM_6 is
  signal A_BAR, B_BAR, C_BAR, I1, I2, I3 : BIT;
begin
 N1: NAND2_OP port map (A=>A, B=>A, Z=>A_BAR);
 N2: NAND2_OP port map (A=>B, B=>B, Z=>B_BAR);
 N3: NAND2_OP port map (A=>C, B=>C, Z=>C_BAR);

 A1: AND3_OP port map (A=>A_BAR, B=>B_BAR, C=>D, Z=>I1);
 A2: AND3_OP port map (A=>A, B=>B, C=>C, Z=>I2);
 A3: AND3_OP port map (A=>B_BAR, B=>C_BAR, C=>C_BAR, Z=>I3);

 O1: OR4_OP port map (A=>I1, B=>I2, C=>I3, D=>I3, Z=>Z);
end NAMED;
```

c.

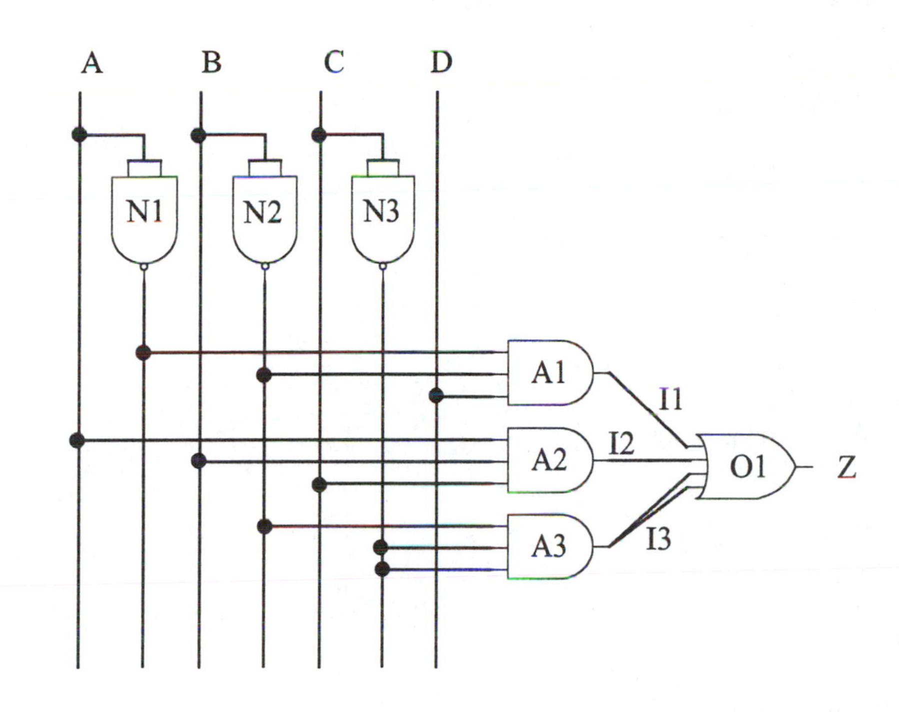

d. Z(ABCD = 0101) = 0

Z(ABCD = 0110) = 0

Z(ABCD = 1001) = 1

Z(ABCD = 1110) = 1

e.
```
entity AND3_OP is
 port
 (A, B, C : in BIT; Z : out BIT);
end AND3_OP;

architecture BEHAVIOR of AND3_OP is
begin
 Z <= A and B and C;
end BEHAVIOR;

entity NAND2_OP is
 port
 (A, B : in BIT; Z : out BIT);
end NAND2_OP;
```

```
architecture BEHAVIOR of NAND2_OP is
begin
 Z <= A nand B;
end BEHAVIOR;

entity OR4_OP is
 port
 (A, B, C, D : in BIT; Z : out BIT);
end OR4_OP;

architecture BEHAVIOR of OR4_OP is
begin
 Z <= A or B or C or D;
end BEHAVIOR;
```

12-7. Solution.

a.

```
entity PROBLEM_7 is
 port
 (A, B, C : in BIT;
  Z : out BIT);
end PROBLEM_7;

use WORK.LOGIC_PKG.all;
architecture POSITIONAL of PROBLEM_7 is

 signal A_BAR, B_BAR, C_BAR, I1, I2, I3 : BIT;
begin
 N1: NAND2_OP port map (A, A, A_BAR);
 N2: NAND2_OP port map (B, B, B_BAR);
 N3: NAND2_OP port map (C, C, C_BAR);

 O1: OR4_OP port map (A, B, C_BAR, C_BAR, I1);
 O2: OR4_OP port map (A_BAR, C, C, C, I2);
 O3: OR4_OP port map (A_BAR, B_BAR, B_BAR, B_BAR, I3);

 A1: AND3_OP port map (I1, I2, I3, Z);

end POSITIONAL;
```

b.
```
entity PROBLEM_7 is
 port
 (A, B, C : in BIT;
 Z : out BIT);
end PROBLEM_7;

use WORK.LOGIC_PKG.all;
architecture NAMED of PROBLEM_7 is

 signal A_BAR, B_BAR, C_BAR, I1, I2, I3 : BIT;
begin
 N1: NAND2_OP port map (A=>A, B=>A, Z=>A_BAR);
 N2: NAND2_OP port map (A=>B, B=>B, Z=>B_BAR);
 N3: NAND2_OP port map (A=>C, B=>C, Z=>C_BAR);

 O1: OR4_OP port map (A=>A, B=>B, C=>C_BAR, D=>C_BAR, Z=>I1);
 O2: OR4_OP port map (A=>A_BAR, B=>C, C=>C, D=>C, Z=>I2);
 O3: OR4_OP port map (A=>A_BAR, B=>B_BAR, C=>B_BAR, D=>B_BAR, Z=>I3);

 A1: AND3_OP port map (A=>I1, B=>I2, C=>I3, Z=>Z);

end NAMED;
```

c.

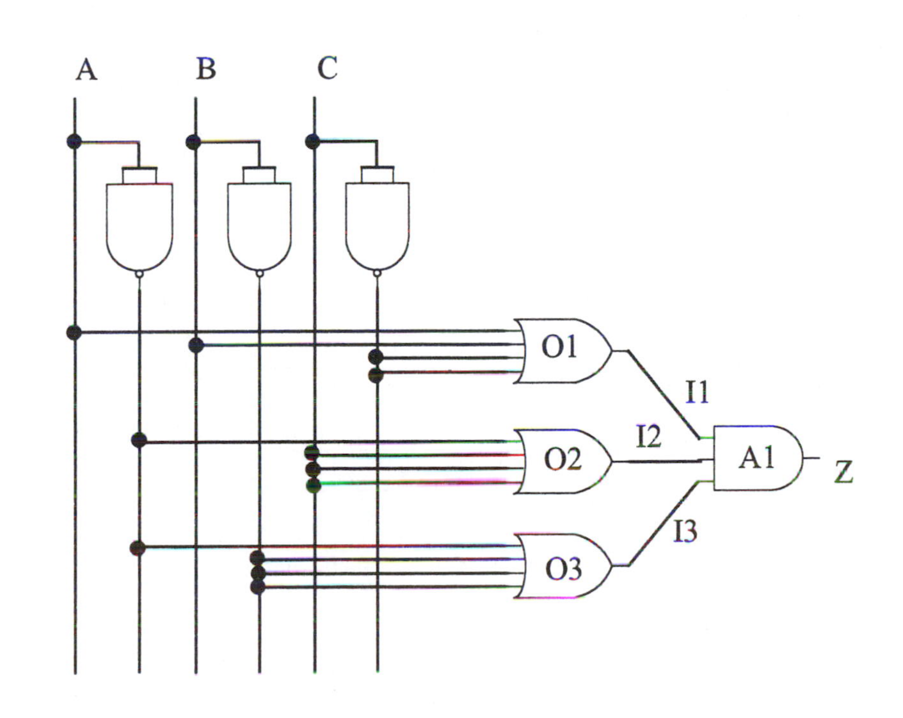

d. Z(ABC = 000) = 1

   Z(ABC = 100) = 0

   Z(ABC = 110) = 0

   Z(ABC = 111) = 0

e.
```
entity AND3_OP is
 port
 (A, B, C : in BIT; Z : out BIT);
end AND3_OP;

architecture BEHAVIOR of AND3_OP is
begin
 Z <= A and B and C;
end BEHAVIOR;

entity NAND2_OP is
 port
 (A, B : in BIT; Z : out BIT);
end NAND2_OP;

architecture BEHAVIOR of NAND2_OP is
begin
 Z <= A nand B;
end BEHAVIOR;

entity OR4_OP is
 port
 (A, B, C, D : in BIT; Z : out BIT);
end OR4_OP;

architecture BEHAVIOR of OR4_OP is
begin
 Z <= A or B or C or D;
end BEHAVIOR;
```

12-8. Solution.

Using DeMorgan's Theorem, $(\bar{A} + \bar{B}) = \overline{A \bullet B}$ and $(\bar{A} + C) = \overline{A \bullet \bar{C}}$

a.
```
entity PROBLEM_8 is
 port
 (A, B, C : in BIT;
 Z : out BIT);
end PROBLEM_8;

use WORK.LOGIC_PKG.all;
architecture POSITIONAL of PROBLEM_8 is
signal C_BAR, I1, I2, I3 : BIT;
begin
 N1: NAND2_OP port map (C, C, C_BAR);

 O1: OR4_OP port map (A, B, C_BAR, C_BAR, I1);
 N2: NAND2_OP port map (A, B, I2);
 N3: NAND2_OP port map (A, C_BAR, I3);

 A3: AND3_OP port map (I1, I2, I3, Z);
end POSITIONAL;
```

b.
```
entity PROBLEM_8 is
 port
 (A, B, C : in BIT;
 Z : out BIT);
end PROBLEM_8;

use WORK.LOGIC_PKG.all;
architecture NAMED of PROBLEM_8 is

 signal C_BAR, I1, I2, I3 : BIT;
begin
 N1: NAND2_OP port map (A=>C, B=>C, Z=>C_BAR);

 O1: OR4_OP port map (A=>A, B=>B, C=>C_BAR, D=>C_BAR, Z=>I1);
 N2: NAND2_OP port map (A=>A, B=>B, Z=>I2);
 N3: NAND2_OP port map (A=>A, B=>C_BAR, Z=>I3);

 A3: AND3_OP port map (A=>I1, B=>I2, C=>I3, Z=>Z);

end NAMED;
```

c.

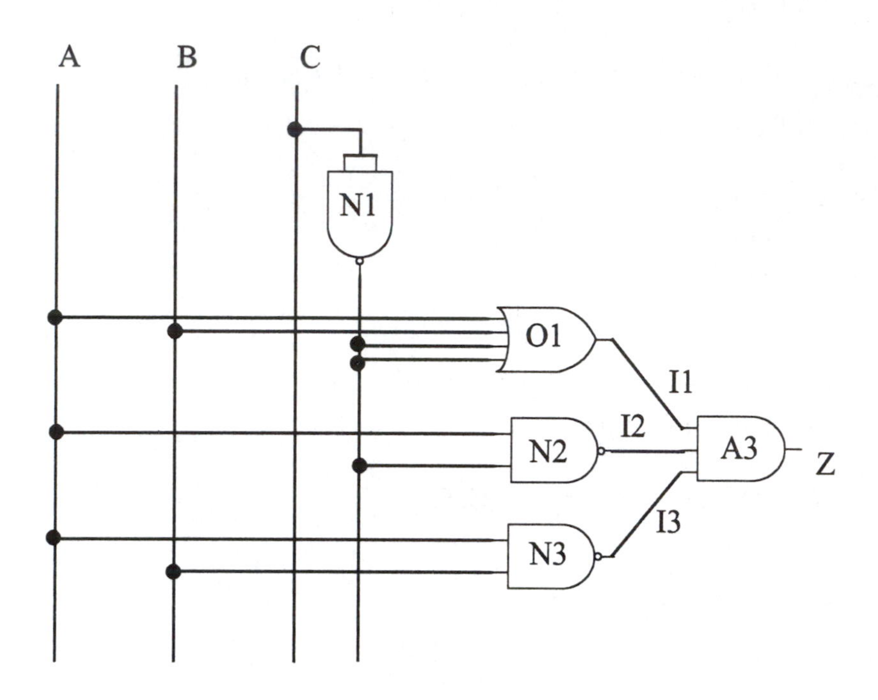

d. Z(ABC = 000) = 1

Z(ABC = 100) = 0

Z(ABC = 110) = 0

Z(ABC = 111) = 0

e.
```
entity AND3_OP is
 port
 (A, B, C : in BIT; Z : out BIT);
end AND3_OP;

architecture BEHAVIOR of AND3_OP is
begin
 Z <= A and B and C;
end BEHAVIOR;

entity NAND2_OP is
 port
 (A, B : in BIT; Z : out BIT);
end NAND2_OP;
```

```
architecture BEHAVIOR of NAND2_OP is
begin
 Z <= A nand B;
end BEHAVIOR;

entity OR4_OP is
 port
 (A, B, C, D : in BIT; Z : out BIT);
end OR4_OP;

architecture BEHAVIOR of OR4_OP is
begin
 Z <= A or B or C or D;
end BEHAVIOR;
```

12-9. Solution.

No. The last component instantiation statement should be

```
O1: OR2_OP port map (I1, I2, Z);
```

Chapter 13. Data Flow Modeling in VHDL

13-1. Solution.

$$F_1(A, B, C, D) = \sum m(4, 5, 6, 7, 11, 12, 13)$$

a.

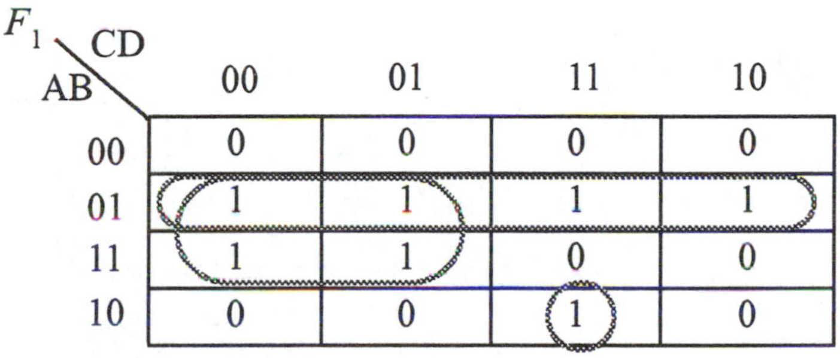

F_1 AB \ CD	00	01	11	10
00	0	0	0	0
01	1	1	1	1
11	1	1	0	0
10	0	0	1	0

$$F_1 = (\overline{A} \bullet B) + (B \bullet \overline{C}) + (A \bullet \overline{B} \bullet C \bullet D)$$

b.

```
entity AND2_OP is
  port
    (A, B : in BIT;    Z : out BIT);
end AND2_OP;

architecture AND2_OP of AND2_OP is
begin
 Z <= A and B;
end AND2_OP;
```

```
entity AND4_OP is
 port
 (A, B, C, D : in BIT;   Z : out BIT);
end AND4_OP;

architecture AND4_OP of AND4_OP is
begin
 Z <= A and B and C and D;
end AND4_OP;


entity OR3_OP is
 port
 (A, B, C : in BIT;
 Z : out BIT);
end OR3_OP;

architecture OR3_OP of OR3_OP is
begin
 Z <= A or B or C;
end OR3_OP;


entity NOT_OP is
 port
 (A : in BIT;
 A_BAR : out BIT);
end NOT_OP;

architecture NOT_OP of NOT_OP is
begin
 A_BAR <= not A;
end NOT_OP;
```

```
entity F1 is
 port
 (A, B, C, D : in BIT;  Z : out BIT);
end F1;

architecture STRUCTURE of F1 is
 component AND2_OP
 port (A, B : in BIT; Z : out BIT);
 end component;
 component AND4_OP
 port (A, B, C, D : in BIT; Z : out BIT);
 end component;
 component OR3_OP
 port (A, B, C : in BIT; Z : out BIT);
 end component;
 component NOT_OP
 port (A : in BIT; A_BAR : out BIT);
 end component;

 signal A_BAR, B_BAR, C_BAR, INT1, INT2, INT3 : BIT;
begin
 N1: NOT_OP port map (A, A_BAR);
 N2: NOT_OP port map (B, B_BAR);
 N3: NOT_OP port map (C, C_BAR);

 A1: AND2_OP port map (A_BAR, B, INT1);
 A2: AND2_OP port map (B, C_BAR, INT2);
 A3: AND4_OP port map (A, B_BAR, C, D, INT3);

 O1: OR3_OP port map (INT1, INT2, INT3, Z);

end STRUCTURE;
```

c.

```
entity F1 is
 port
 (A, B, C, D : in BIT;   Z : out BIT);
end F1;

architecture BEHAVIOR of F1 is
begin
 Z <= (not A and B) or (B and not C) or (A and not B and C and D);
end BEHAVIOR;
```

$F_2(A, B, C) = \sum m(1, 2, 4, 7)$

a.

$$F_2 = (\bar{A} \bullet \bar{B} \bullet C) + (A \bullet \bar{B} \bullet \bar{C}) + (A \bullet B \bullet C) + (\bar{A} \bullet B \bullet \bar{C})$$

$$F_2 = \bar{B} \bullet ((\bar{A} \bullet C) + (A \bullet \bar{C})) + B \bullet ((A \bullet C) + (\bar{A} \bullet \bar{C}))$$

$$F_2 = \bar{B} \bullet (A \oplus C) + B \bullet \overline{(A \oplus C)}$$

$$F_2 = A \oplus B \oplus C$$

b.

```
entity XOR3_OP is
 port
 (A, B, C : in BIT;
 Z : out BIT);
end XOR3_OP;

architecture XOR3_OP of XOR3_OP is
begin
 Z <= A xor B xor C;
end XOR3_OP;
```

```
entity F2 is
 port
 (A, B, C : in BIT;
 Z : out BIT);
end F2;

architecture STRUCTURE of F2 is
 component XOR3_OP
 port (A, B, C : in BIT; Z : out BIT);
 end component;
begin
 X1: XOR3_OP port map (A, B, C, Z);

end STRUCTURE;
```

c.

```
entity F2 is
 port
 (A, B, C : in BIT;
 Z : out BIT);
end F2;

architecture BEHAVIOR of F2 is
begin
 Z <= A xor B xor C;
end BEHAVIOR;
```

d.

e. Structural models describe the implementation of a design entity; the specification can be derived from the component behaviors and their interconnections. Behavioral models describe a specification; the implementation can be inferred from the concurrency of the signal assignment statements.

13-2. Solution.

a. 1. A changes

2. C<= A or B; and E <= D or A; execute

3. D <= C or B; executes

4. E <= D or A executes

b. 1. A and B change

2. C <= A or B; D <= C or B; and E <= D or A; execute

3. D <= C or B; and E <= D or A; execute

4. E <= D or A; executes

c. No.

13-3. Solution.

a.

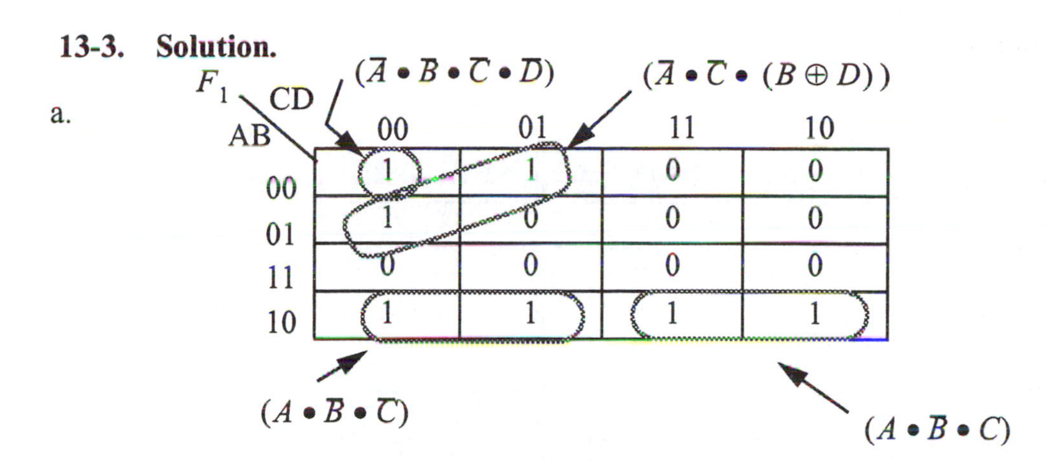

ABCD	Z
0000	1
0001	1
0010	0
0011	0
0100	1
0101	0
0110	0
0111	0
1000	1
1001	1
1010	1
1011	1
1100	0
1101	0
1110	0
1111	0

b.

13-4. Solution.

a.

SWAP	OLD_SIG	NEW_SIG
0	00	00
0	01	01
0	10	10
0	11	11
1	00	00
1	01	10
1	10	01
1	11	11

13-5. Solution.

a.

M_1 M_0	D_1 D_0	S_0	S_1	S_2	S_3	S_4	S_5	S_6
00	00	0	0	0	0	0	0	0
00	01	0	0	0	0	0	0	0
00	10	0	0	0	0	0	0	0
00	11	0	0	0	0	0	0	0
01	00	1	1	1	1	1	1	1
01	01	1	1	1	1	1	1	1
01	10	1	1	1	1	1	1	1
01	11	1	1	1	1	1	1	1
10	00	0	0	1	0	0	0	0
10	01	1	0	1	0	1	1	1
10	10	0	0	0	1	0	0	1
10	11	0	0	0	0	0	1	1
11	00	1	1	0	1	1	1	1
11	01	0	1	0	1	0	0	0
11	10	1	1	1	0	1	1	0
11	11	1	1	1	1	1	0	0

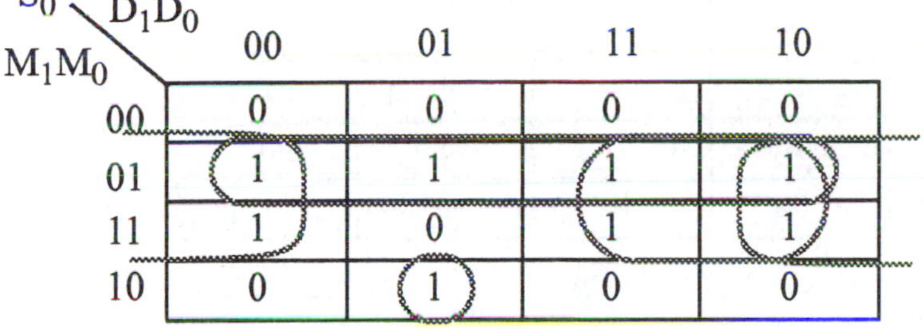

S_0 (M_1M_0 \ D_1D_0)	00	01	11	10
00	0	0	0	0
01	1	1	1	1
11	1	0	1	1
10	0	1	0	0

$$S_0 = \left(M_0 \bullet \overline{D_0}\right) + (M_0 \bullet D_1) + \left(\overline{M_1} \bullet M_0\right) + \left(M_1 \bullet \overline{M_0} \bullet \overline{D_1} \bullet D_0\right)$$

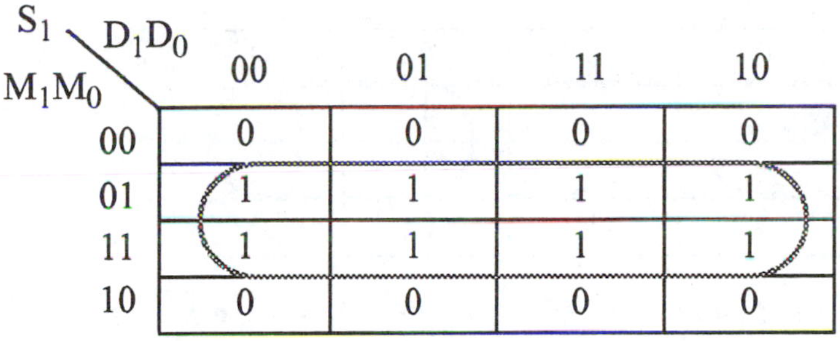

S_1 (M_1M_0 \ D_1D_0)	00	01	11	10
00	0	0	0	0
01	1	1	1	1
11	1	1	1	1
10	0	0	0	0

$$S_1 = M_0$$

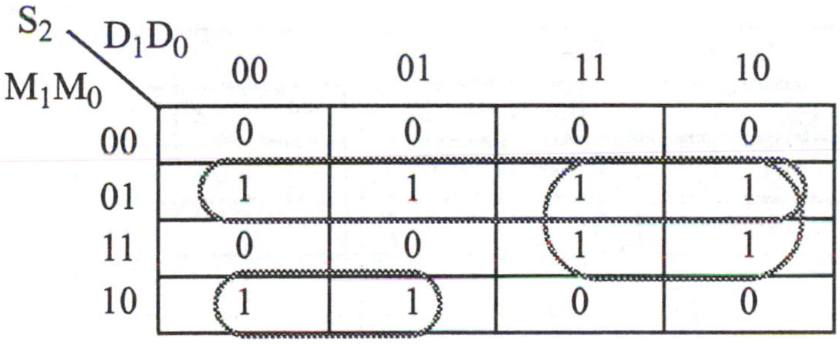

S_2 (M_1M_0 \ D_1D_0)	00	01	11	10
00	0	0	0	0
01	1	1	1	1
11	0	0	1	1
10	1	1	0	0

$$S_2 = \left(\overline{M_1} \bullet M_0\right) + (M_0 \bullet D_1) + \left(M_1 \bullet \overline{M_0} \bullet \overline{D_1}\right)$$

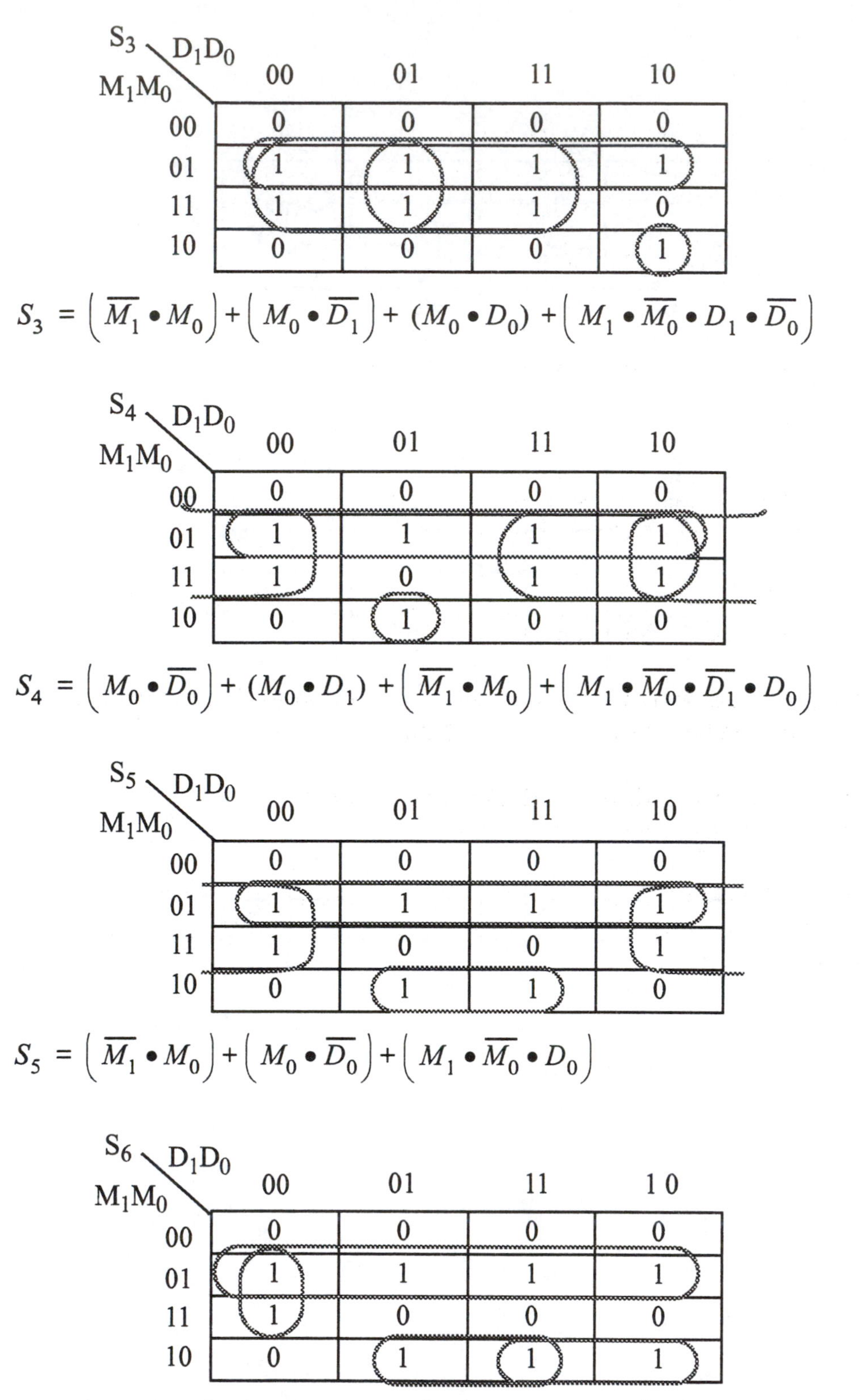

S_3 (M_1M_0 \ D_1D_0)	00	01	11	10
00	0	0	0	0
01	1	1	1	1
11	1	1	1	0
10	0	0	0	1

$$S_3 = \left(\overline{M_1} \bullet M_0\right) + \left(M_0 \bullet \overline{D_1}\right) + (M_0 \bullet D_0) + \left(M_1 \bullet \overline{M_0} \bullet D_1 \bullet \overline{D_0}\right)$$

S_4 (M_1M_0 \ D_1D_0)	00	01	11	10
00	0	0	0	0
01	1	1	1	1
11	1	0	1	1
10	0	1	0	0

$$S_4 = \left(M_0 \bullet \overline{D_0}\right) + (M_0 \bullet D_1) + \left(\overline{M_1} \bullet M_0\right) + \left(M_1 \bullet \overline{M_0} \bullet \overline{D_1} \bullet D_0\right)$$

S_5 (M_1M_0 \ D_1D_0)	00	01	11	10
00	0	0	0	0
01	1	1	1	1
11	1	0	0	1
10	0	1	1	0

$$S_5 = \left(\overline{M_1} \bullet M_0\right) + \left(M_0 \bullet \overline{D_0}\right) + \left(M_1 \bullet \overline{M_0} \bullet D_0\right)$$

S_6 (M_1M_0 \ D_1D_0)	00	01	11	10
00	0	0	0	0
01	1	1	1	1
11	1	0	0	0
10	0	1	1	1

$$S_6 = \left(\overline{M_1} \bullet M_0\right) + \left(M_0 \bullet \overline{D_1} \bullet \overline{D_0}\right) + \left(M_1 \bullet \overline{M_0} \bullet D_0\right) + \left(M_1 \bullet \overline{M_0} \bullet D_1\right)$$

b.

```
entity LED_DRIVER is
port
(M, D : in BIT_VECTOR(1 downto 0);
      S : out BIT_VECTOR(6 downto 0));
end LED_DRIVER;

architecture SEVEN of LED_DRIVER is
begin
 S(0) <= (M(0) and not D(0)) or (M(0) and D(1)) or (not M(1) and M(0)) or
           (M(1) and not M(0) and not D(1) and D(0));

 S(1) <= M(0);

 S(2) <= (not M(1) and M(0)) or (M(0) and D(1)) or
           (M(1) and not M(0) and not D(1));

 S(3) <= (not M(1) and M(0)) or (M(0) and not D(1)) or (M(0) and D(0)) or
           (M(1) and not M(0) and D(1) and not D(0));

 S(4) <= (M(0) and not D(0)) or (M(0) and D(1)) or (not M(1) and M(0)) or
           (M(1) and not M(0) and not D(1) and D(0));

 S(5) <= (not M(1) and M(0)) or (M(0) and not D(0)) or
           (M(1) and not M(0) and D(0));

 S(6) <= (not M(1) and M(0)) or (M(0) and not D(1) and not D(0)) or
           (M(1) and not M(0) and D(0)) or (M(1) and not M(0) and D(1));
end SEVEN;
```

c. architecture ONE of LED_DRIVER is
constant ZERO_BAR : BIT_VECTOR := B"0000100";
constant ONE_BAR : BIT_VECTOR := B"1110101";
constant TWO_BAR : BIT_VECTOR := B"1001000";
constant THREE_BAR : BIT_VECTOR := B"1100000";
begin
S <= B"0000000" when (M=B"00") else
B"1111111" when (M=B"01") else

ZERO_BAR when (M=B"10" and D=B"00") else
not ZERO_BAR when (M=B"11" and D=B"00") else

ONE_BAR when (M=B"10" and D=B"01") else
not ONE_BAR when (M=B"11" and D=B"01") else

TWO_BAR when (M=B"10" and D=B"10") else
not TWO_BAR when (M=B"11" and D=B"10") else

THREE_BAR when (M=B"10" and D=B"11") else
not THREE_BAR;
end ONE;

13-6. Solution.

a. 1. not A = '1' not C = 1 => '1' and B or '1' xor D

2. '1' and B = '1' => '1' or '1' xor D

3. '1' or '1' = '1' => '1' xor D

4. '1' xor D = '0'

b. 1. not A = '1' not C = '1' => ('1' and B) or ('1' xor D)

2. '1' and B = '1' '1' xor D = '0' => '1' or '0'

3. '1' or '0' = '1'

13-7. Solution.

a. 1. not A = '1' => '1' < B and C = D

2. '1' < B = FALSE   C = D = FALSE => FALSE and FALSE

3. FALSE and FALSE = FALSE

b. 1. A < B = TRUE   C=D = FALSE => not (TRUE and FALSE)

2. TRUE and FALSE = FALSE = > not (FALSE)

3. not FALSE = TRUE

13-8. Solution.

a.

OP_SEL	A	B	Z
00	0	0	0
00	0	1	0
00	1	0	0
00	1	1	1
01	0	0	1
01	0	1	1
01	1	0	1
01	1	1	0
10	0	0	0
10	0	1	1
10	1	0	1
10	1	1	1
11	0	0	1
11	0	1	0
11	1	0	0
11	1	1	0

b. Simulate.

13-9. Solution.

a.
```
entity BUF_INV is
 port
 (MODE_CNTRL : in BIT;
 SIG_IN : in BIT;
 SIG_OUT : out BIT);
end BUF_INV;

architecture PART_A of BUF_INV is
begin
 SIG_OUT <= not (SIG_IN xor MODE_CNTRL);
end PART_A;
```

b.
```
architecture PART_B of BUF_INV is
begin
 SIG_OUT <= not SIG_IN when (MODE_CNTRL='0') else
                SIG_IN;
end PART_B;
```

c.
```
architecture PART_C of BUF_INV is
begin
 with MODE_CNTRL select
 SIG_OUT <= not SIG_IN when '0',
                SIG_IN     when '1';
end PART_C;
```

d. The model given in a. is more suggestive of an implementation. The models given in parts b. and c. are more readable. The model given in part c. emphasizes the control aspect of the signal MODE_CNTRL.

13-10. Solution.

a.

VECA

7	6	5	4	3	2	1	0
1	1	1	0	0	1	0	1

VECB

10	11	12	13	14	15
0	1	1	1	1	0

b.

VECA(6 downto 2)

6	5	4	3	2
1	1	0	0	1

VECB(10 to 13)

10	11	12	13
0	1	1	1

MSB

7
1

OCTET

10	11	12
0	1	1

c. VECA'LEFT = 7

VECA'RIGHT = 0

VECA'HIGH = 7

VECA'LOW = 0

VECA'RANGE = 7 downto 0

VECA'REVERSE_RANGE = 0 to 7

VECA'LENGTH = 8

VECB'LEFT = 10

VECB'RIGHT = 15

VECB'HIGH = 15

VECB'LOW = 10

VECB'RANGE =10 to 15

VECB'REVERSE_RANGE = 15 downto 10

VECB'LENGTH = 6

13-11. Solution.

a. VECA and B"1111_0000" = B"1110_0000"

VECA xor X"FF" = B"00011010"

VECA >= VECB = TRUE

b. The first logic operation provides a way to "mask" bits, i.e., keeping some and resetting others to 0. The second logic operation provides a way to selectively invert bits.

Chapter 14. Structural Modeling in VHDL - Part 2

14-1. Solution.

Component Port	Design Entity Port		
	in	**out**	**inout**
in	Y	N	Y
out	N	Y	Y
inout	N	N	Y

14-2. Solution.

a.
```
entity SMODEL is
port
(P1 : in BIT;
 P2 : out BIT;
 P3 : inout BIT);
end SMODEL;

architecture STRUCTURE of SMODEL is
 component UNIT
 port (C1, C2 : in BIT; C3 : out BIT);
 end component;

begin
 U1: UNIT port map (C1=> P1 , C2=> P3 , C3=> P2);
end STRUCTURE;
```

b. YES

14-3. Solution.

a.

BDAC	Z
0000	1
0001	1
0010	1
0011	0
0100	0
0101	0
0110	0
0111	1
1000	0
1001	0
1010	1
1011	1
1100	0
1101	0
1110	1
1111	1

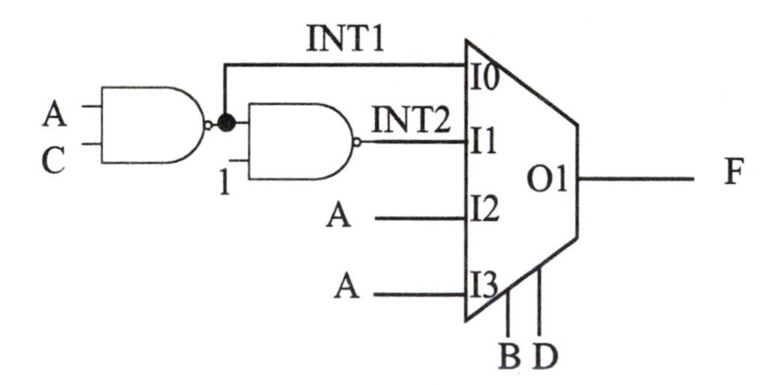

b. entity F is

```
 port
 (A, B, C, D : in BIT;
  F          : out BIT);
end F;

use WORK.LOGIC_COMPONENTS.all;
architecture STRUCTURE of F is
 signal INT1, INT2 : BIT;
 signal TIE_ONE : BIT := '1';
begin
 N1: NAND2_OP port map (A, C, INT1);
 N2: NAND2_OP port map (INT1, TIE_ONE, INT2);
 M1: MUX port map (INT1, INT2, A, A, B, D, F);
end STRUCTURE;
```

c. entity NAND2_GATE is

```
 port
 (A, B : in BIT;
  Z    : out BIT);
end F;

architecture NAND2_GATE of NAND2_GATE is
begin
 Z <= A nand B;
end NAND2_GATE;

entity MUX is
 port (I0, I1, I2, I3, S1, S0 : in BIT;
       O1                     : out BIT);
end MUX;

architecture MUX of MUX is
begin
 with (S1&S0) select
  O1 <= I0 when B"00",
        I1 when B"01",
        I2 when B"10",
        I3 when B"11;
end MUX;
```

d. entity F is

```
port
(A, B, C, D : in BIT;
 F          : out BIT);
end F;

use WORK.LOGIC_COMPONENTS.all;
architecture STRUCTURE of F is
 signal INT1, INT2 : BIT;
begin
 N1: NAND2_OP port map (A, C, INT1);
 N2: NAND2_OP port map (INT1, open, INT2);
 M1: MUX      port map (INT1, INT2, A, A, B, D, F);
end STRUCTURE;
```

e. entity NAND2_GATE is

```
port
(A, B : in BIT := '1';
  Z : out BIT);
end F;

architecture NAND2_GATE of NAND2_GATE is
begin
 Z <= A nand B;
end NAND2_GATE;
```

14-4. Solution.

a.

Minterm	ABCD	F
0	0000	1
1	0001	1
2	0010	1
3	0011	1
4	0100	0
5	0101	1
6	0110	0
7	0111	0
8	1000	1
9	1001	0
10	1010	0
11	1011	0
12	1100	0
13	1101	1
14	1110	0
15	1111	0

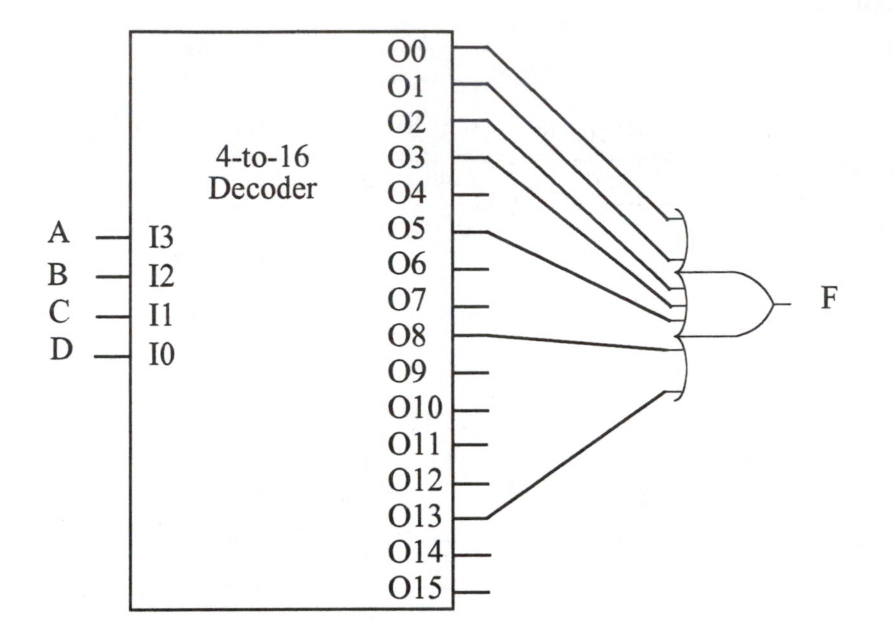

b. entity FIBONACCI is

```
port
(A, B, C, D : in BIT;
F : out BIT);
end FIBONACCI;

use WORK.LOGIC_COMPONENTS.all;
architecture STRUCTURE of FIBONACCI is
 signal M0, M1, M2, M3, M5, M8, M13 : BIT;
 signal OR1, OR2, OR3 : BIT;
begin
 D1: DECODE4_16 port map (I(3)=>A, I(2)=>B, I(1)=>C, I(0)=>D,
                          O0=>M0, O1=>M1, O2=>M2, O3=>M3,
                          O4=>open, O5=>M5, O6=>open, O7=>open,
                          O8=>M8, O9=>open, O10=>open, O11=>open,
                          O12=>open, O13=>M13, O14=>open, O15=>open);

 O1: OR2_GATE port map (M0, M1, M2, OR1);
 O2: OR2_GATE port map (M3, M5, M8, OR2);
 O3: OR2_GATE port map (M13, OR1, OR2, F);
end STRUCTURE;
```

c. entity OR3_GATE is

```
entity OR3_GATE is
 port
 (A, B, C : in BIT;
 Z : out BIT);
end OR3_GATE;

architecture OR3_GATE of OR3_GATE is
begin
 Z <= A or B or C;
end OR3_GATE;

entity DECODE4_16 is
 port (I : in BIT_VECTOR(3 downto 0));
      O0,O1,O2,O3,O4,O5,O6,O7,O8,O9,O10,O11,O12,O13,O14,O15 : out BIT);
end DECODE4_16;

architecture DECODE4_16 of DECODE4_16 is
begin
O0  <= '1' when (I = X"0") else '0';
O1  <= '1' when (I = X"1") else '0';
O2  <= '1' when (I = X"2") else '0';
O3  <= '1' when (I = X"3") else '0';
O4  <= '1' when (I = X"4") else '0';
O5  <= '1' when (I = X"5") else '0';
O6  <= '1' when (I = X"6") else '0';
O7  <= '1' when (I = X"7") else '0';
O8  <= '1' when (I = X"8") else '0';
O9  <= '1' when (I = X"9") else '0';
O10 <= '1' when (I = X"A") else '0';
O11 <= '1' when (I = X"B") else '0';
O12 <= '1' when (I = X"C") else '0';
O13 <= '1' when (I = X"D") else '0';
O14 <= '1' when (I = X"E") else '0';
O15 <= '1' when (I = X"F") else '0';
end DECODE4_16;
```

d. Simulate.

14-5. Solution.

a.
```
entity RIP_ADD8 is
 port
 (A, B : in BIT_VECTOR(7 downto 0);
   CI   : in BIT;
   SUM : out BIT_VECTOR(7 downto 0);
   CO  : out BIT);
end RIP_ADD8;

use WORK.LOGIC_COMPONENTS.all;
architecture LONG_STRUCTURE of RIP_ADD8 is
 signal CARRIES : BIT_VECTOR(6 downto 0);
begin
 ADD0: FULL_ADDER port map (A(0), B(0), CI, SUM(0), CARRIES(0));
 ADD1: FULL_ADDER port map (A(1), B(1), CARRIES(0), SUM(1), CARRIES(1));
 ADD2: FULL_ADDER port map (A(2), B(2), CARRIES(1), SUM(2), CARRIES(2));
 ADD3: FULL_ADDER port map (A(3), B(3), CARRIES(2), SUM(3), CARRIES(3));
 ADD4: FULL_ADDER port map (A(4), B(4), CARRIES(3), SUM(4), CARRIES(4));
 ADD5: FULL_ADDER port map (A(5), B(5), CARRIES(4), SUM(5), CARRIES(5));
 ADD6: FULL_ADDER port map (A(6), B(6), CARRIES(5), SUM(6), CARRIES(6));
 ADD7: FULL_ADDER port map (A(7), B(7), CARRIES(6), SUM(7), CO);
end LONG_STRUCTURE;
```

b.
```
architecture SHORT_STRUCTURE of RIP_ADD8 is
 signal CARRIES : BIT_VECTOR(6 downto 0);
begin
 LP: for I in 0 to 7 generate
 CASE1: if (I=0) generate
 ADD0: FULL_ADDER port map (A(I), B(I), CI, SUM(I), CARRIES(I));
 end generate CASE1;

 CASEX: if (I>0 and I<7) generate
 ADD1: FULL_ADDER port map (A(I), B(I), CARRIES(I-1), SUM(I), CARRIES(I));
 end generate CASEX;

 CASE7: if (I=7) generate
 ADD0: FULL_ADDER port map (A(I), B(I), CARRIES(I-1), SUM(I), CO);
 end generate CASE7;
 end generate LP;

end SHORT_STRUCTURE;
```

c. entity FULL_ADDER is

```
entity FULL_ADDER is
 port
 (A, B, CIN : in BIT;
  S, COUT : out BIT);
end FULL_ADDER;

architecture FULL_ADDER of FULL_ADDER is
begin
 S <= A xor B xor CIN;
 COUT <= (A and B) or ((A xor B) and CIN);
end FULL_ADDER;
```

14-6. Solution.

a.

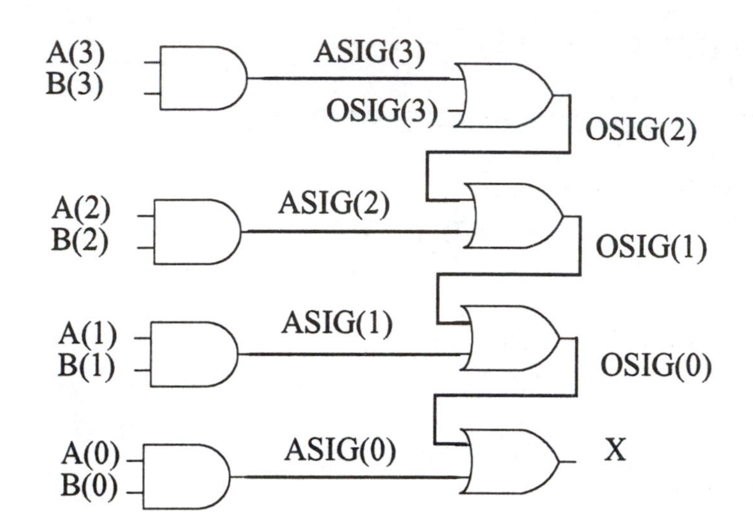

b.

$$X = (A(0) \bullet B(0)) + (A(1) \bullet B(1)) + (A(2) \bullet B(2)) + (A(3) \bullet B(3))$$

c.

A(0)B(0)	A(1)B(1)	A(2)B(2)	A(3)B(3)	X
11	--	--	--	1
--	11	--	--	1
--	--	11	--	1
--	--	--	11	1

14-7. Solution.

a.

```
entity LOGIC is
 port
 (A, B : in BIT_VECTOR(3 downto 0);
 X : out BIT);
end LOGIC;

architecture BEHAVIOR of LOGIC is

 signal ASIG, OSIG : BIT_VECTOR(3 downto 0) := X"0";
begin
 R: for COL in 1 to 2 generate
   C: for ROW in 3 downto 0 generate
     R1: if (COL=1) generate
       ASIG(ROW) <= A(ROW) and B(ROW);
     end generate R1;

     R1C: if (COL=2 and ROW/=0) generate
       OSIG(ROW-1) <= ASIG(ROW) or OSIG(ROW);
     end generate R1C;

     R1C0: if (COL=2 and ROW=0) generate
       X <= ASIG(ROW) or OSIG(ROW);
     end generate R1C0;
   end generate C;
 end generate R;

end BEHAVIOR;
```

b.

```
ASIG(3) <= A(3) and B(3);
ASIG(2) <= A(2) and B(2);
ASIG(1) <= A(1) and B(1);
ASIG(0) <= A(0) and B(0);

OSIG(2) <= ASIG(3) or OSIG(3);
OSIG(1) <= ASIG(2) or OSIG(2);
OSIG(0) <= ASIG(1) or OSIG(1);

X <= ASIG(0) or OSIG(0);
```

14-8. Solution.

a.
```
entity RAND is
 port (SIG_IN : in BIT_VECTOR; SIG_OUT : out BIT);
end RAND;
architecture RAND of RAND is
 component RAND
 port
    (SIG_IN : in BIT_VECTOR; SIG_OUT : out BIT);
 end component;

 signal SIG_INT : BIT;
begin
 RECURSE: if (SIG_IN'LENGTH > 2) generate
 C: RAND port map(SIG_IN=>SIG_IN(SIG_IN'HIGH-1 downto SIG_IN'LOW),
                  SIG_OUT=>SIG_INT);
 end generate RECURSE;

 SIG_OUT <= SIG_IN(SIG_IN'HIGH) and SIG_IN(SIG_IN'LOW)
                                     when SIG_IN'LENGTH=2 else
            SIG_IN(SIG_IN'HIGH) and SIG_INT;

end RAND;
```

b.
```
entity TOP_LEVEL is
 port (A : in BIT_VECTOR(3 downto 0);
       B : out BIT);
end TOP_LEVEL;
architecture TOP_LEVEL of TOP_LEVEL is
 component RAND
 port (SIG_IN : in BIT_VECTOR; SIG_OUT : out BIT);
 end component;
begin
 CA: RAND port map (SIG_IN=>A, SIG_OUT=>B);
end TOP_LEVEL;
```

14-9. Solution.

a.
```
entity BIT_REVERSE is
 port
 (VEC_IN : in BIT_VECTOR;
  VEC_REV : out BIT_VECTOR);
end BIT_REVERSE;

architecture DATA_FLOW of BIT_REVERSE is
begin
 -- Assume both VEC_IN and VEC_REV have same index constraint
 LP: for I in VEC_REV'LOW to VEC_REV'HIGH generate
   VEC_REV(I) <= VEC_IN(VEC_IN'HIGH - (I - VEC_IN'LOW));
 end generate LP;
end DATAFLOW;
```

b.
```
entity TOP_LEVEL is
 port
 (A : in BIT_VECTOR(3 downto 0);
  Z : out BIT_VECTOR(3 downto 0));
end TOP_LEVEL;

architecture STRUCTURE of TOP_LEVEL is
 component BIT_REVERSE
 port
  (VEC_IN : in BIT_VECTOR;
   VEC_REV : out BIT_VECTOR);
 end component;

begin
 C1: BIT_REVERSE port map (A, Z);
end STRUCTURE;
```

c. Simulate

14-10. Solution.

a.
```
entity LOGIC is
 port
 (A, B : in BIT_VECTOR;
  X : out BIT);
end LOGIC;

architecture STRUCTURE of LOGIC is
 component AND2_GATE
 port
 (A, B : in BIT; Z : out BIT);
 end component;

 component OR2_GATE
 port
 (A, B : in BIT; Z : out BIT);
 end component;

 signal ASIG, OSIG : BIT_VECTOR(A'RANGE) := X"0";
begin
 R: for COL in 1 to 2 generate
  C: for ROW in A'RANGE generate
   R1: if (COL=1) generate
     AX: AND2_GATE port map (A(ROW), B(ROW), ASIG(ROW));
   end generate R1;

   R1C: if (COL=2 and ROW/=0) generate
     OX: OR2_GATE port map (ASIG(ROW), OSIG(ROW), OSIG(ROW-1));
   end generate R1C;

   R1C0: if (COL=2 and ROW=0) generate
     O0: OR2_GATE port map (ASIG(ROW), OSIG(ROW), X);
   end generate R1C0;
  end generate C;
 end generate R;

end STRUCTURE;
```

b. entity LOGIC is

```
entity LOGIC is
 generic
  (N : POSITIVE);
 port
 (A, B : in BIT_VECTOR(N-1 downto 0);
  X : out BIT);
end LOGIC;

architecture STRUCTURE of LOGIC is
 component AND2_GATE
 port
 (A, B : in BIT; Z : out BIT);
 end component;

 component OR2_GATE
 port
 (A, B : in BIT; Z : out BIT);
 end component;

 signal ASIG, OSIG : BIT_VECTOR(N-1 downto 0) := X"0";
begin
 R: for COL in 1 to 2 generate
  C: for ROW in N-1 downto 0 generate
    R1: if (COL=1) generate
      AX: AND2_GATE port map (A(ROW), B(ROW), ASIG(ROW));
    end generate R1;

    R1C: if (COL=2 and ROW/=0) generate
      OX: OR2_GATE port map (ASIG(ROW), OSIG(ROW), OSIG(ROW-1));
    end generate R1C;

    R1C0: if (COL=2 and ROW=0) generate
      O0: OR2_GATE port map (ASIG(ROW), OSIG(ROW), X);
    end generate R1C0;
  end generate C;
 end generate R;

end STRUCTURE;
```

c. The generic parameter approach makes it clear that the input signals A and B must have the same length and indexing scheme, whereas the unconstrained BIT_VECTOR approach does not enforce such requirements.

14-11. Solution.

a. -11

b. Error, the converted value -11 is not a legal value of the target type POSITIVE.

c. -10

d. Either 10 or 11 depending on the particular implementation software.

e. 2

14-12. Solution.

a. 5#1130#   6#433#

b. 2#10111110#   8#276#   16#BE#

c. 100_000   1E6   1e+6   1E+6

14-13. Solution.

a. 0.96

b. 28.75

c. 136.5

14-14. Solution.

a. Illegal - Cannot take an integer to a negative power.

b. Legal - 900

c. Legal - 256.0

d. Legal - 0

e. Legal 1/16**2 = 1/256

f. Illegal - mod operator accepts only integer operands

g. Illegal - cannot take a floating point number to a noninteger power

14-15. Solution.

a.

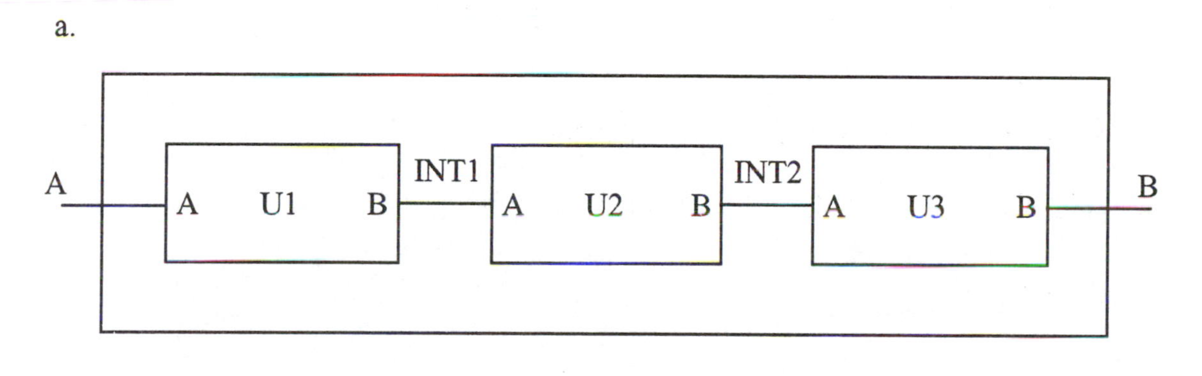

b.

":element(structure):cascade(2):output_case:u3@unit(unit):a"

":element(structure):cascade(1):middle_case:u2@unit(unit):a"

":element(structure):cascade(0):input_case:u1@unit(unit):a"

Chapter 15. VHDL Technology Information - Part 1

15-1. Solution.

a. 96×10^{-6} sec

b. 600 sec

c. 152 sec

d. 5×10^{-3} sec

e. 9 sec

f. 390 sec

g. 6×10^{-15} sec

15-2. Solution.

a. 1.2 min = 72×10^{15} fs

b. 0 fs

c. 999,999 fs

15-3. Solution.

a.
```
type CURRENT is range 0 to INTEGER'HIGH
 units
  uA;                  -- microamps
  mA = 1000 uA;        -- milliamps
  A  = 1000 mA;        -- amps
 end units;
```

b.
```
type LENGTH range 0 to INTEGER'HIGH
 units
  u;                   -- micron
  mm = 1000 u;         -- millimeter
  cm  = 1000 mm;       -- centimeter
  mil = 394 u;         -- milliinch
  in  = 1000 mil;      -- inch
 end units;
```

c.
```
type CELSIUS range 0 to INTEGER'HIGH
 units
  C;                   -- Celsius
 end units;
```

15-4. Solution.

a.
```
entity LOGIC is
  port
   (A, B, C, D : in BIT;
    Z          : out BIT);
end LOGIC;

architecture DATA_FLOW of LOGIC is
 signal INT1, INT2 : BIT;
begin
 INT1 <= A and B after 5 ns;
 INT2 <= INT1 or C after 5 ns;
 Z    <= INT2 and D after 5 ns;
end DATA_FLOW;
```

b. Simulate.

15-5. Solution.

a.

C4

N4N3 \ N2N1	00	01	11	10
00	1	1	0	0
01	0	0	0	0
11	-	-	-	-
10	0	0	-	-

C3

N4N3 \ N2N1	00	01	11	10
00	0	0	1	1
01	1	1	0	0
11	-	-	-	-
10	0	0	-	-

C2

N4N3 \ N2N1	00	01	11	10
00	0	0	1	1
01	0	0	1	1
11	-	-	-	-
10	0	0	-	-

C1

N4N3 \ N2N1	00	01	11	10
00	1	0	0	1
01	1	0	0	1
11	-	-	-	-
10	1	0	-	-

$$C4 = \overline{N4} \bullet \overline{N3} \bullet \overline{N2}$$
$$C3 = N2 \oplus N3$$
$$C2 = N2$$
$$C1 = \overline{N1}$$

b.
```
entity LOGIC is
  port
   (N4, N3, N2, N1 : in BIT;
    C4, C3, C2, C1 : out BIT);
end LOGIC;

architecture DATA_FLOW of LOGIC is
 signal N4_BAR, N3_BAR, N2_BAR, N1_BAR : BIT;
begin
 N4_BAR <= not N4 after 5 ns;
 N3_BAR <= not N3 after 5 ns;
 N2_BAR <= not N2 after 5 ns;
 N1_BAR <= not N1 after 5 ns;

 C4 <= N4_BAR and N3_BAR and N2_BAR after 7 ns;
 C3 <= N2 xor N3 after 7 ns;
 C2 <= N2;
 C1 <= N1_BAR;
end DATA_FLOW;
```

c. Simulate.

15-6. Solution.

a.
```
entity LOGIC is
  port
   (A, B, C, D : in BIT;
    CNTRL      : in BIT;
    Z          : out BIT);
end LOGIC;

architecture DATA_FLOW of LOGIC is
 signal INT1, INT2 : BIT;
begin
 INT1 <= A and B after 7 ns;
 INT2 <= C nand D after 5 ns;

 with CNTRL select
 Z <= INT1 after 12 ns when '0',
      INT2 after 12 ns when '1';
end DATA_FLOW;
```

b. Simulate.

15-7. Solution.

a.
```
entity LOGIC is
  port
   (A : in BIT;
    C  : out BIT);
end LOGIC;

architecture DATA_FLOW of LOGIC is
 signal B : BIT;
begin
  B <= A after 5 ns;
  C <= B after 10 ns;
end DATA_FLOW;
```

b.

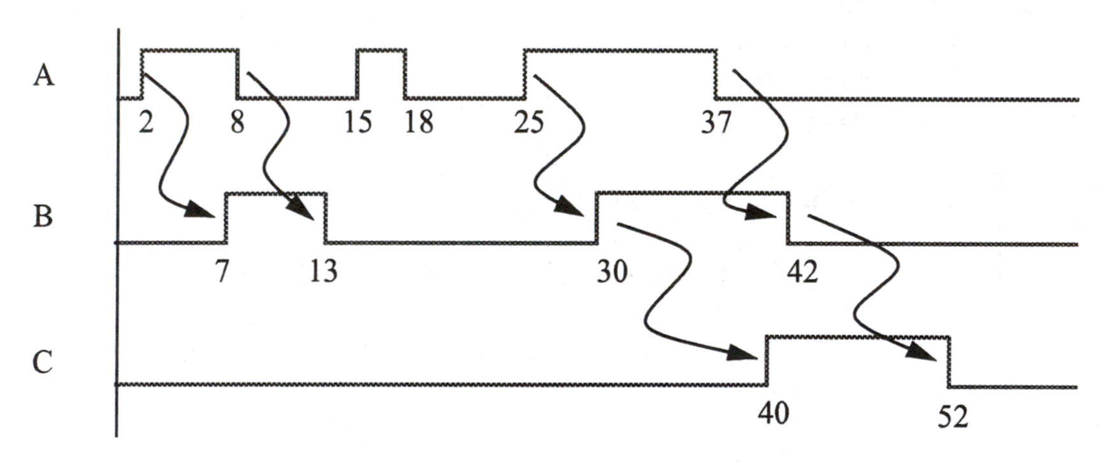

15-8. Solution.

a.
```
entity LOGIC is
  port
   (A : in BIT;
    C  : out BIT);
end LOGIC;

architecture DATA_FLOW of LOGIC is
 signal B : BIT;
begin
  B <= transport A after 5 ns;
  C <= transport B after 10 ns;
end DATA_FLOW;
```

b.

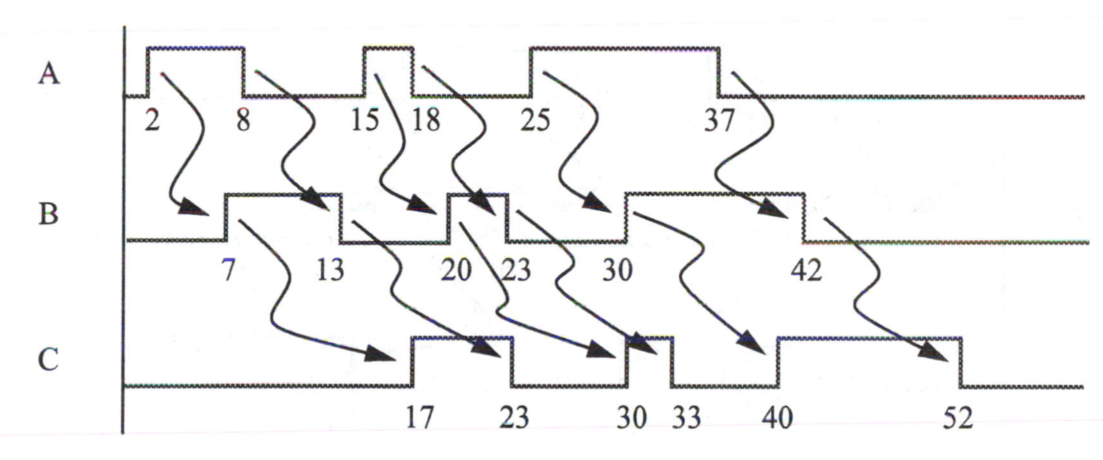

15-9. Solution.

a.
```
entity LOGIC is
  port
   (A : in BIT;
    C  : out BIT);
end LOGIC;

architecture DATA_FLOW of LOGIC is
 signal B : BIT;
begin
  B <= A after 5 ns;
  C <= reject 5 ns inertial B after 10 ns;
end DATA_FLOW;
```

b.

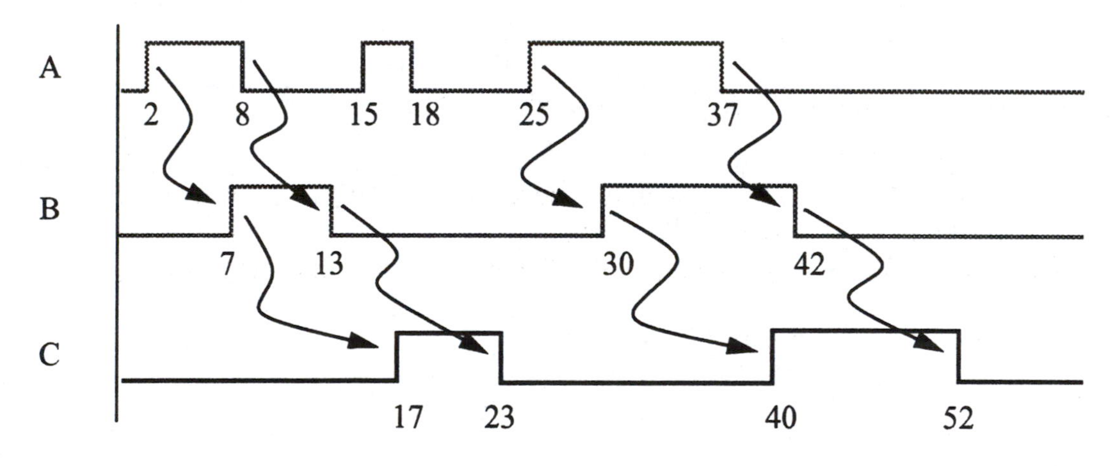

15-10. Solution.

Cycle 1: Time advances to 2 nanoseconds

Signal A changes from 0 to 1
First concurrent signal assignment statement executes and schedules B to change from 0 to 1 in 5 nanoseconds

Cycle 2: Time advances to 7 nanoseconds

Signal B changes from 0 to 1
Second concurrent signal assignment statement executes and schedules C to change from 0 to 1 in 10 nanoseconds

Cycle 3: Time advances to 8 nanoseconds

Signal A changes from 1 to 0
First concurrent signal assignment statement executes and schedules B to change from 1 to 0 in 5 nanoseconds

Cycle 4: Time advances to 13 nanoseconds

Signal B changes from 1 to 0
Second concurrent signal assignment statement executes and schedules C to change from 1 to 0 in 10 nanoseconds. The 1 did not persist for more than the inertial delay of 10 nanoseconds, so the events scheduled for 17 and 23 nanoseconds are erased.

Cycle 5: Time advances to 15 nanoseconds

Signal A changes from 0 to 1
First concurrent signal assignment statement executes and schedules B to change from 0 to 1 in 5 nanoseconds.

Cycle 6: Time advances to 18 nanoseconds

Signal A changes from 1 to 0
First concurrent signal assignment statement executes and schedules B to change from 1 to 0 in 5 nanoseconds. The 1 did not persist for more than the inertial delay of 5 nanoseconds, so the events scheduled for 20 and 23 nanoseconds are erased.

Cycle 7: Time advances to 25 nanoseconds

Signal A changes from 0 to 1
First concurrent signal assignment statement executes and schedules B to change from 0 to 1 in 5 nanoseconds

Cycle 8: Time advances to 30 nanoseconds

Signal B changes from 0 to 1
Second concurrent signal assignment statement executes and schedules C to change from 0 to 1 in 10 nanoseconds

Cycle 9: Time advances to 37 nanoseconds

Signal A changes from 1 to 0
First concurrent signal assignment statement executes and schedules B to change from 1 to 0 in 5 nanoseconds

Cycle 10: Time advances to 40 nanoseconds

Signal C changes from 0 to 1

Cycle 11: Time advances to 42 nanoseconds

Signal B changes from 1 to 0
Second concurrent signal assignment statement executes and schedules C to change from 1 to 0 in 10 nanoseconds

Cycle 12: Time advances to 52 nanoseconds

Signal C changes from 1 to 0

15-11. Solution.

Cycle 1: Time advances to 2 nanoseconds

Signal A changes from 0 to 1
First concurrent signal assignment statement executes and schedules B to change from 0 to 1 in 5 nanoseconds

Cycle 2: Time advances to 7 nanoseconds

Signal B changes from 0 to 1
Second concurrent signal assignment statement executes and schedules C to change from 0 to 1 in 10 nanoseconds

Cycle 3: Time advances to 8 nanoseconds

Signal A changes from 1 to 0
First concurrent signal assignment statement executes and schedules B to change from 1 to 0 in 5 nanoseconds

Cycle 4: Time advances to 13 nanoseconds

Signal B changes from 1 to 0
Second concurrent signal assignment statement executes and schedules C to change from 1 to 0 in 10 nanoseconds.

Cycle 5: Time advances to 15 nanoseconds

Signal A changes from 0 to 1
First concurrent signal assignment statement executes and schedules B to change from 0 to 1 in 5 nanoseconds.

Cycle 6: Time advances to 17 nanoseconds

Signal C changes from 0 to 1

Cycle 7: Time advances to 18 nanoseconds

Signal A changes from 1 to 0
First concurrent signal assignment statement executes and schedules B to change from 1 to 0 in 5 nanoseconds.

Cycle 8: Time advances to 20 nanoseconds

Signal B changes from 1 to 0
Second concurrent signal assignment statement executes and schedules C to change from 1 to 0 in 10 nanoseconds.

Cycle 9: Time advances to 23 nanoseconds

Signals B and C change from 1 to 0
Second concurrent signal assignment statement executes and schedules C to change from 1 to 0 in 10 nanoseconds.

Cycle 10: Time advances to 25 nanoseconds

Signal A changes from 0 to 1
First concurrent signal assignment statement executes and schedules B to change from 0 to 1 in 5 nanoseconds

Cycle 11: Time advances to 30 nanoseconds

Signals B and C change from 0 to 1
Second concurrent signal assignment statement executes and schedules C to change from 0 to 1 in 10 nanoseconds

Cycle 12: Time advances to 33 nanoseconds

Signal C changes from 1 to 0

Cycle 13: Time advances to 37 nanoseconds

Signal A changes from 1 to 0
First concurrent signal assignment statement executes and schedules B to change from 1 to 0 in 5 nanoseconds

Cycle 14: Time advances to 40 nanoseconds

Signal C changes from 0 to 1

Cycle 15: Time advances to 42 nanoseconds

Signal B changes from 1 to 0
Second concurrent signal assignment statement executes and schedules C to change from 1 to 0 in 10 nanoseconds

Cycle 16: Time advances to 52 nanoseconds

Signal C changes from 1 to 0

15-12. Solution.

```
entity NOR2 is
 generic
  (TPLH, TPHL : TIME);
port
  (A, B : in  BIT;
   Z    : out BIT);
end NOR2;

architecture NOR2 of NOR2 is
 signal Z_INT : BIT;
begin
  Z_INT <= A nor B;

  Z <= Z_INT after TPLH when (Z_INT='1') else
        Z_INT after TPHL;
end NOR2;
```

15-13. Solution.

```
entity OSCILLATOR is
 generic
  (PERIOD : TIME);
port
  (CLK : inout BIT);
end OSCILLATOR;

architecture OSCILLATOR of OSCILLATOR is
begin
   CLK <= not CLK after (PERIOD/2);
end OSCILLATOR;
```

15-14. Solution.

a.
```
package TEMP_PKG is
  type TEMPERATURE is range 0 to 150
   units
     C;
   end units;
end TEMP_PKG;

use WORK.TEMP_PKG.all;
entity NOR2T is
 generic
  (TEMP : TEMPERATURE);
port
  (A, B : in  BIT;
   Z    : out BIT);
end NOR2T;

architecture NOR2T of NOR2T is
begin
  Z <= A nor B after (((10*TEMP)/20 C) * 1 ns);
end NOR2T;
```

b. Simulate.

15-15. Solution.

a.
```
package BCD_PKG is
  function BCD_BAD (BCD_IN : BIT_VECTOR(3 downto 0)) return BIT;
end TEMP_PKG;

package body BCD_PKG is
  function BCD_BAD (BCD_IN : BIT_VECTOR(3 downto 0)) return BIT is
  begin
   if (BCD_IN >= X"A") then
     return ('1');
   else
     return ('0');
   end if;
  end BCD_BAD;
end BCD_PKG;
```

b.
```
entity BCD_CK is
port
  (BCD     : in  BIT_VECTOR(3 downto 0);
   BCD_OUT : out BIT);
end BCD_CK;

use WORK.BCD_PKG.all;
architecture BCD_CK of BCD_CK is
begin
  BCD_OUT <= BCD_BAD(BCD);
end BCD_CK;
```

c. Simulate.

15-16. Solution.

```
package UTILS_PKG is
  function ONES (ARG : BIT_VECTOR) return NATURAL;
end UTILS_PKG;

package body UTILS_PKG is
  function ONES (ARG : BIT_VECTOR) return NATURAL is
   variable TOTAL_CNT : NATURAL := 0;
  begin
   for I in ARG'RANGE loop
    if (ARG(I)='1') then
     TOTAL_CNT := TOTAL_CNT + 1;
    end if;
   end loop;
   return (TOTAL_CNT);
  end ONES;
end UTILS_PKG;
```

15-17. Solution.

a.

```
package UTILS_PKG is
  function AND_OP (ARG : BIT_VECTOR) return BIT;
end UTILS_PKG;

package body UTILS_PKG is
  function AND_OP (ARG : BIT_VECTOR) return BIT is
    -- assume index scheme is n-1 downto 0
  begin
    if (ARG'LENGTH=2) then
     return (ARG(1) and ARG(0));
    else
     return (ARG(HIGH) and ARG_OP(HIGH-1 downto 0));
    end if;
 end AND_OP;
end UTILS_PKG;
```

b.
```
entity ANDN_OP is
 port
  (A : in BIT_VECTOR;  Z : out BIT);
end ANDN_OP;

use WORK.UTILS_PKG.all;
architecture ANDN_OP of ANDN_OP is
begin
  Z <= AND_OP(A);
end ANDN_OP;

entity TOP_LEVEL is
 port
  (A : in BIT_VECTOR(3 downto 0);  Z : out BIT);
end TOP_LEVEL;

architecture TOP_LEVEL of TOP_LEVEL is
 component ANDN_OP
  port
   (A : in BIT_VECTOR;  Z : out BIT);
 end component;
begin
  U1 : ANDN_OP port map (A, Z);
end TOP_LEVEL;
```

c. Simulate.

15-18. Solution.

a.

```
package UTILS_PKG is
  function WIRED_AND (SOURCES : BIT_VECTOR) return BIT;
end UTILS_PKG;

package body UTILS_PKG is
  function WIRED_AND (SOURCES : BIT_VECTOR) return BIT is
    variable RESULT : BIT := '1';
  begin
    for I in SOURCES'RANGE loop
     RESULT := RESULT and SOURCES(I);
    end loop;
    return (RESULT);
  end WIRED_AND;
end UTILS_PKG;

use WORK.UTILS_PKG.all;
entity NOR2_IIL is
 port
  (A, B : in  BIT;
   Z    : out WIRED_AND BIT);
end NOR2_IIL;

architecture NOR2_IIL of NOR2_IIL is
 component IIL_GATE
  port
   (A_IN : in BIT;
    Z1_OUT, Z2_OUT, Z3_OUT : out BIT);
 end component;
begin
  I1 : IIL_GATE port map (A, Z, open, open);
  I2 : IIL_GATE port map (B, Z, open, open);
end NOR2_IIL;
```

b. use WORK.UTILS_PKG.all;

```
entity NAND2_IIL is
 port
  (A, B : in  BIT;
   Z    : out BIT);
end NAND2_IIL;

architecture NAND2_IIL of NAND2_IIL is
 component IIL_GATE
  port
   (A_IN : in BIT;
    Z1_OUT, Z2_OUT, Z3_OUT : out BIT);
 end component;

 signal A_BAR, B_BAR : BIT;
 signal INT : WIRED_AND BIT;
begin
  I1 : IIL_GATE port map (A, A_BAR, open, open);
  I2 : IIL_GATE port map (B, B_BAR, open, open);

  I3 : IIL_GATE port map (A_BAR, INT, open, open);
  I4 : IIL_GATE port map (B_BAR, INT, open, open);
  I5 : IIL_GATE port map (INT, Z, open, open);
end NAND2_IIL;
```

15-19. Solution.

Signals: Signals have a time history of values, involving past, present, and future values. Past and present signals values can be queried, but not updated. Future signal values can be updated, but not queried. Signals are assigned a value/time pair, where the time is the additional time from the present into the future that the value will take affect. Signal assignments always affect future values, never the present value. Even a time element of zero, implies a delta delay.

Variables: Variables have only one value, their present values. The present value of a variable may be queried and changed. Variable assignments execute in zero time during the present simulation cycle or, equivalently simulation time step.

Constants: Constants have only one value, their present values. The present value of a constant is set at constant declaration and thereafter may not be updated.

Chapter 16. VHDL Technology Information - Part 2

16-1. Solution.

a. 'U' - Uninitialize. Used as initial, "undriven" values for signals.

'0' - Logical 0 state with a forcing strength from active source.

'1' - Logical 1 state with a forcing strength from active source.

'X' - Logical unknown state with a forcing strength. Result of '0' and '1' conflict.

'L' - Logical 0 state with a resistive strength from passive source.

'H' - Logical 1 state with a resistive strength from passive source.

'W' - Logical unknown state with a resistive strength. Result of 'L' and 'W' conflict.

'Z' - Disconnect from any source.

'-' - Don't care.

b. TTL - Values having forcing strengths, uninitialized, and don't care: 'U', 'X', '0', '1', and '-'.

c. Open Collector TTL - Values having forcing strengths, resistive pull-up, unitialized, and don't care: 'U', 'X', '0', '1', '-', and 'H'.

d. Static CMOS - Values having forcing strengths, uninitialized, and don't care: 'U', 'X', '0', '1', and '-'. Also, high impedance 'Z', if pass transistor logic is used.

16-2. Solution.

a. No. The 1164 not operator assumes an active output logic family that refreshes logic values. The 1164 not operator does not support passive (resistive strength) logic output values.

b. Making a conservative assumption that input uncertainty yields weak output uncertainty.

```
library IEEE;
use IEEE.STD_LOGIC_1164.all;
entity NMOS_INVERTER is
 port
  (A : in STD_ULOGIC;  A_BAR : out STD_ULOGIC);
end NMOS_INVERTER;
architecture BEHAVIOR of NMOS_INVERTER is
begin
 with A select
  A_BAR <= '0' when '1' | 'H',          -- invert
           'H' when '0' | 'L',          -- invert
           'U' when 'U',                -- pass through
           'H' when 'Z',                -- float to high
           'W' when 'X' | 'W' | '-';    -- weak unknown
end BEHAVIOR;
```

16-3. Solution.

a. type STD_LOGIC_TABLE is array (STD_ULOGIC, STD_ULOGIC) of STD_ULOGIC;

```
constant AND_TABLE : STD_LOGIC_TABLE :=
('U' => ('U','U','0','U','U','U','0','U','U'),
 'X' => ('U','X','0','X','X','X','0','X','X'),
 '0' => ('0','0','0','0','0','0','0','0','0'),
 '1' => ('U','X','0','1','X','X','0','1','X'),
 'Z' => ('U','X','0','X','X','X','0','X','X'),
 'W' => ('U','X','0','X','X','X','0','X','X'),
 'L' => ('0','0','0','0','0','0','0','0','0'),
 'H' => ('U','X','0','1','X','X','0','1','X'),
 '-' => ('U','X','0','X','X','X','0','X','X'));

function "and" (L,R : STD_ULOGIC) return STD_ULOGIC is
begin
  return (AND_TABLE(L,R));
end "and";
```

b.
```
library IEEE;
use IEEE.STD_LOGIC_1164.all;
entity AND2_OP is
 port (A, B : in STD_ULOGIC; Z : out STD_ULOGIC);
end AND2_OP;

architecture BEHAVIOR of AND2_OP is
begin
 Z <= A and B;
end BEHAVIOR;
```

c.
```
type STD_LOGIC_TABLE is array (STD_ULOGIC, STD_ULOGIC) of STD_ULOGIC;

constant AND_TABLE : STD_LOGIC_TABLE :=
 ('U'  => ('U','U','0','U','U','U','0','U','U'),
  'X'  => ('U','X','0','X','X','X','0','X','X'),
  '0'  => ('0','0','0','0','0','0','0','0','0'),
  '1'  => ('U','X','0','1','X','X','0','1','X'),
  'Z'  => ('U','X','0','X','X','X','0','X','X'),
  'W' => ('U','X','0','X','X','X','0','X','X'),
  'L'  => ('0','0','0','0','0','0','0','0','0'),
  'H'  => ('U','X','0','1','X','X','0','1','X'),
  '-'  => ('U','X','0','X','X','X','0','X','X'));

function "and" (L,R : STD_ULOGIC) return STD_ULOGIC is
begin
 if (L='0' or L='L') then
   return('0');
 else
   return (AND_TABLE(L,R));
 end if;
end "and";
```

d. library IEEE;

```
use IEEE.STD_LOGIC_1164.all;
entity AND2_OP is
  port (A, B : in STD_ULOGIC; Z : out STD_ULOGIC);
end AND2_OP;

architecture BEHAVIOR of AND2_OP is
type STD_LOGIC_TABLE is array (STD_ULOGIC, STD_ULOGIC) of STD_ULOGIC;

constant AND_TABLE : STD_LOGIC_TABLE :=
 ('U'  => ('U','U','0','U','U','U','0','U','U'),
  'X'  => ('U','X','0','X','X','X','0','X','X'),
  '0'  => ('0','0','0','0','0','0','0','0','0'),
  '1'  => ('U','X','0','1','X','X','0','1','X'),
  'Z'  => ('U','X','0','X','X','X','0','X','X'),
  'W' => ('U','X','0','X','X','X','0','X','X'),
  'L'  => ('0','0','0','0','0','0','0','0','0'),
  'H'  => ('U','X','0','1','X','X','0','1','X'),
  '-'  => ('U','X','0','X','X','X','0','X','X'));

function "and" (L,R : STD_ULOGIC) return STD_ULOGIC is
begin
 if (L='0' or L='L') then
   return('0');
 else
   return (AND_TABLE(L,R));
 end if;
end "and";

begin
 Z <= A and B;
end BEHAVIOR;
```

16-4. Solution.

a. type STD_LOGIC_TABLE is array (STD_ULOGIC, STD_ULOGIC) of STD_ULOGIC;

```
constant OR_TABLE : STD_LOGIC_TABLE :=
 ('U'  => ('U','U','U','1','U','U','U','1','U'),
  'X'  => ('U','X','X','1','X','X','X','1','X'),
  '0'  => ('U','X','0','1','X','X','0','1','X'),
  '1'  => ('1','1','1','1','1','1','1','1','1'),
  'Z'  => ('U','X','X','1','X','X','X','1','X'),
  'W'  => ('U','X','X','1','X','X','X','1','X'),
  'L'  => ('U','X','0','1','X','X','0','1','X'),
  'H'  => ('1','1','1','1','1','1','1','1','1'),
  '-'  => ('U','X','X','1','X','X','X','1','X'));

function "or" (L,R : STD_ULOGIC) return STD_ULOGIC is
begin
 return (OR_TABLE(L,R));
end "or";
```

b.
```
library IEEE;
use IEEE.STD_LOGIC_1164.all;
entity OR2_OP is
 port (A, B : in STD_ULOGIC; Z : out STD_ULOGIC);
end OR2_OP;

architecture BEHAVIOR of OR2_OP is
begin
 Z <= A or B;
end BEHAVIOR;
```

c. type STD_LOGIC_TABLE is array (STD_ULOGIC, STD_ULOGIC) of STD_ULOGIC;

```
constant OR_TABLE : STD_LOGIC_TABLE :=
 ('U'  => ('U','U','U','1','U','U','U','1','U'),
  'X'  => ('U','X','X','1','X','X','X','1','X'),
  '0'  => ('U','X','0','1','X','X','0','1','X'),
  '1'  => ('1','1','1','1','1','1','1','1','1'),
  'Z'  => ('U','X','X','1','X','X','X','1','X'),
  'W'  => ('U','X','X','1','X','X','X','1','X'),
  'L'  => ('U','X','0','1','X','X','0','1','X'),
  'H'  => ('1','1','1','1','1','1','1','1','1'),
  '-'  => ('U','X','X','1','X','X','X','1','X'));
```

```
function "or" (L,R : STD_ULOGIC) return STD_ULOGIC is
begin
 if (L='1' or L='H') then
   return('1');
 else
   return (OR_TABLE(L,R));
 end if;
end "or";
```

d.
```
library IEEE;
use IEEE.STD_LOGIC_1164.all;
entity OR2_OP is
  port (A, B : in STD_ULOGIC; Z : out STD_ULOGIC);
end OR2_OP;

architecture BEHAVIOR of OR2_OP is
type STD_LOGIC_TABLE is array (STD_ULOGIC, STD_ULOGIC) of STD_ULOGIC;

constant OR_TABLE : STD_LOGIC_TABLE :=
 ('U'  => ('U','U','U','1','U','U','U','1','U'),
  'X'  => ('U','X','X','1','X','X','X','1','X'),
  '0'  => ('U','X','0','1','X','X','0','1','X'),
  '1'  => ('1','1','1','1','1','1','1','1','1'),
  'Z'  => ('U','X','X','1','X','X','X','1','X'),
  'W'  => ('U','X','X','1','X','X','X','1','X'),
  'L'  => ('U','X','0','1','X','X','0','1','X'),
  'H'  => ('1','1','1','1','1','1','1','1','1'),
  '-'  => ('U','X','X','1','X','X','X','1','X'));

function "or" (L,R : STD_ULOGIC) return STD_ULOGIC is
begin
 if (L='1' or L='H') then
   return('1');
 else
   return (OR_TABLE(L,R));
 end if;
end "or";

begin
 Z <= A or B;
end BEHAVIOR;
```

16-5. Solution.

a. Bit vector literals are defined to be composed of enumeration literals ‘0’ and ‘1’ of type BIT.

b.
```
library IEEE;
use IEEE.STD_LOGIC_1164.all;
entity AND_MASK is
 port
  (A_IN : in   STD_ULOGIC_VECTOR(63 downto 0);
   Z     : out STD_ULOGIC_VECTOR(63 downto 0));
end AND_MASK;

architecture BEHAVIOR of AND_MASK is
begin
  Z <= A_IN and
        “0000000100100011010001010110011110001001101010111100110111101111”;
end BEHAVIOR;
```

16-6. Solution.

a. number of truth table entries = 9^3

b.

A_IN	B_IN	C_IN	$\overline{A\_IN} \cdot B\_IN \cdot C\_IN$	$A\_IN \cdot \overline{B\_IN} \cdot C\_IN$	$A\_IN \cdot B\_IN \cdot \overline{C\_IN}$	$A\_IN \cdot B\_IN \cdot C\_IN$	Z_OUT
‘U’	‘U’	’U’	‘U’	‘U’	‘U’	‘U’	’U’
’U’	’U’	’0’	‘0’	‘0’	‘U’	‘0’	’U’
’U’	’U’	’1’	‘U’	‘U’	‘U’	‘U’	’U’
’U’	’0’	’U’	‘0’	‘U’	‘0’	‘0’	’U’
’U’	’0’	’0’	‘0’	‘0’	‘0’	‘0’	’0’
’U’	’0’	’1’	‘0’	‘U’	‘0’	‘0’	’U’
’U’	’1’	’U’	‘U’	‘0’	‘U’	‘U’	’U’
’U’	’1’	’0’	‘0’	‘0’	‘U’	‘0’	’U’
’U’	’1’	’1’	‘U’	‘0’	‘0’	‘U’	’U’
’0’	’U’	’U’	‘U’	‘0’	‘0’	‘0’	’U’
’0’	’U’	’0’	‘0’	‘0’	‘0’	‘0’	’0’
’0’	’U’	’1’	‘U’	‘0’	‘0’	‘0’	’U’
’0’	’0’	’U’	‘0’	‘0’	‘0’	‘0’	’0’
’0’	’0’	’0’	‘0’	‘0’	‘0’	‘0’	’0’
’0’	’0’	’1’	‘0’	‘0’	‘0’	‘0’	’0’
’0’	’1’	’U’	‘U’	‘0’	‘0’	‘0’	’U’
’0’	’1’	’0’	‘0’	‘0’	‘0’	‘0’	’0’
’0’	’1’	’1’	‘1’	‘0’	‘0’	‘0’	’1’
’1’	’U’	’U’	‘0’	‘U’	‘U’	‘U’	’U’
’1’	’U’	’0’	‘0’	‘0’	‘U’	‘0’	’U’
’1’	’U’	’1’	‘0’	‘U’	‘0’	‘U’	’U’
’1’	’0’	’U’	‘0’	‘U’	‘0’	‘0’	’U’
’1’	’0’	’0’	‘0’	‘0’	‘0’	‘0’	’0’
’1’	’0’	’1’	‘0’	‘1’	‘0’	‘0’	’1’
’1’	’1’	’U’	‘0’	‘0’	‘U’	‘U’	’U’
’1’	’1’	’0’	‘0’	‘0’	‘1’	‘0’	’1’
’1’	’1’	’1’	‘0’	‘0’	‘0’	‘1’	’1’

c. ‘U’

16-7. Solution.

a.

```
type STD_ULOGIC_VECTOR is array (NATURAL range <>) of STD_ULOGIC;
type STD_LOGIC_TABLE is array (STD_ULOGIC, STD_ULOGIC) of STD_ULOGIC;

constant RESOLUTION_TABLE : STD_LOGIC_TABLE := (
  --    ------------------------------------------------------
  --    | U   X   0   1   Z   W   L   H   -       | |
  --    ------------------------------------------------------
        ( 'U', 'U', 'U', 'U', 'U', 'U', 'U', 'U', 'U' ), -- | U |
        ( 'U', 'X', 'X', 'X', 'X', 'X', 'X', 'X', 'X' ), -- | X |
        ( 'U', 'X', '0', 'X', '0', '0', '0', '0', 'X' ), -- | 0 |
        ( 'U', 'X', 'X', '1', '1', '1', '1', '1', 'X' ), -- | 1 |
        ( 'U', 'X', '0', '1', 'Z', 'W', 'L', 'H', 'X' ), -- | Z |
        ( 'U', 'X', '0', '1', 'W', 'W', 'W', 'W', 'X' ), -- | W |
        ( 'U', 'X', '0', '1', 'L', 'W', 'L', 'W', 'X' ), -- | L |
        ( 'U', 'X', '0', '1', 'H', 'W', 'W', 'H', 'X' ), -- | H |
        ( 'U', 'X', 'X', 'X', 'X', 'X', 'X', 'X', 'X' )  -- | - |
    );

  function RESOLVED ( ARG : STD_ULOGIC_VECTOR ) return STD_ULOGIC is
    variable RESULT : STD_ULOGIC := 'Z';  -- weakest state default
  begin
    if (ARG'LENGTH = 1) then
      return ARG(ARG'LOW);
    else
      for I in ARG'RANGE loop
        RESULT := RESOLUTION_TABLE(RESULT, ARG(I));
      end loop;
    end if;
    return (RESULT);
  end RESOLVED;
```

16-8. Solution.

a.

A_IN	B_IN	C_IN	B_IN + C_IN	A_IN + B_IN	A_IN + C_IN	INT_SIG	ZOUT_BAR
'0'	'0'	'0'	'0'	'0'	'0'	'0'	'1'
'0'	'0'	'1'	'1'	'0'	'1'	'X'	'X'
'0'	'1'	'0'	'1'	'1'	'0'	'X'	'X'
'0'	'1'	'1'	'1'	'1'	'1'	'1'	'0'
'1'	'0'	'0'	'0'	'1'	'1'	'X'	'X'
'1'	'0'	'1'	'1'	'1'	'1'	'1'	'0'
'1'	'1'	'0'	'1'	'1'	'1'	'1'	'0'
'1'	'1'	'1'	'1'	'1'	'1'	'1'	'0'

b. Simulate.

c. No. The resolution function RESOLVED does not provide wired-**and** properties

16-9. Solution.

a.
```
library IEEE;
use IEEE.STD_LOGIC_1164.all;
entity TRI_DECODER_3to8 is
 port
  (A, B, C            : in  STD_ULOGIC;
   ENABLE_BAR : in  STD_ULOGIC;
   DECODE           : out STD_ULOGIC_VECTOR(7 downto 0));
end TRI_DECODER_3to8;

architecture BEHAVIOR of TRI_DECODER_3to8 is
begin
 DECODE <= "UUUUUUUU" when (ENABLE_BAR = 'U') else
           "XXXXXXXX" when (ENABLE_BAR = 'W' or ENABLE_BAR = 'X' or
                                  ENABLE_BAR = 'Z') else
           "ZZZZZZZZ" when (ENABLE_BAR = '1' or ENABLE_BAR = 'H') else
           (     A and     B and     C) &      -- Z(7)
           (     A and     B and not C) &      -- Z(6)
           (     A and not B and     C) &      -- Z(5)
           (     A and not B and not C) &      -- Z(4)
           (not A and     B and     C) &      -- Z(3)
           (not A and     B and not C) &      -- Z(2)
           (not A and not B and     C) &      -- Z(1)
           (not A and not B and not C);       -- Z(0)

end BEHAVIOR;
```

16-10. Solution.

a.

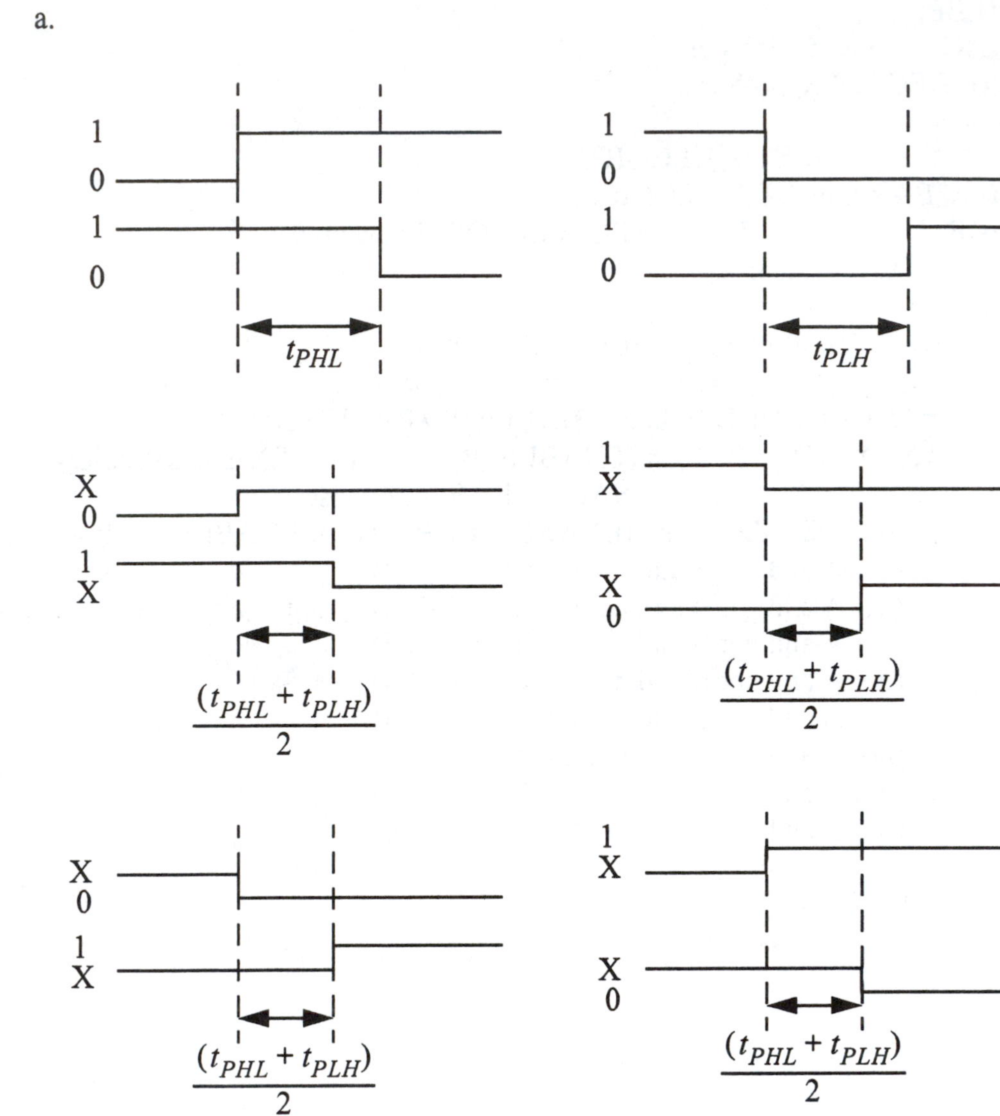

```
library IEEE;
  use IEEE.STD_LOGIC_1164.all;
  entity INVERTER is
    generic
      (TPHL, TPLH : TIME);
    port
      (A        : in  X01;
       A_BAR : out X01);
end INVERTER;

architecture BEHAVIOR of INVERTER is
begin
 A_BAR <= not A after TPHL when (RISING_EDGE(A)) else       -- 0->1
                  not A after TPLH when (FALLING_EDGE(A)) else -- 1->0
                  not A after (TPHL + TPLH)/2;                       -- 0/1->X, X->0/1

end BEHAVIOR;
```

If input transitions to/from unknown, output is complement with average of tphl and tplh delay.

b. Simulate.

16-11. Solution.

a.
```
library IEEE;
use IEEE.STD_LOGIC_1164.all;
entity INVERTER is
 generic
  (TPHL, TPLH : TIME;
   TR, TF     : TIME);
 port
  (A     : in  X01;
   A_BAR : out X01);
end INVERTER;

architecture BEHAVIOR of INVERTER is
begin
 A_BAR <= 'X' after TPHL, not A after (TPHL + TF/2) when (RISING_EDGE(A)) else
                'X' after TPLH, not A after (TPLH + TR/2) when (FALLING_EDGE(A)) else
                not A after ((TPHL + TPLH + TR + TF)/2);

end BEHAVIOR;
```

b. Simulate.

c.

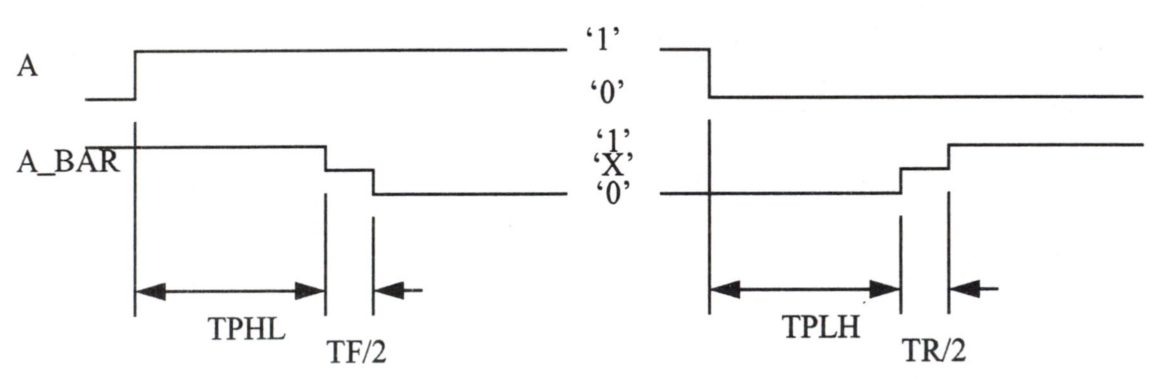

16-12. Solution.

a. 0

b. LOW

c. 2

d. LOW

16-13. Solution.

a. No. The position numbers associated with each enumeration literal do not yield the expected ordering relationship of NONE<LOW<NORM<HIGH

b. No. MOD is a reserved keyword.

16-14. Solution.

a. variable WORD : WORD_TYPE;

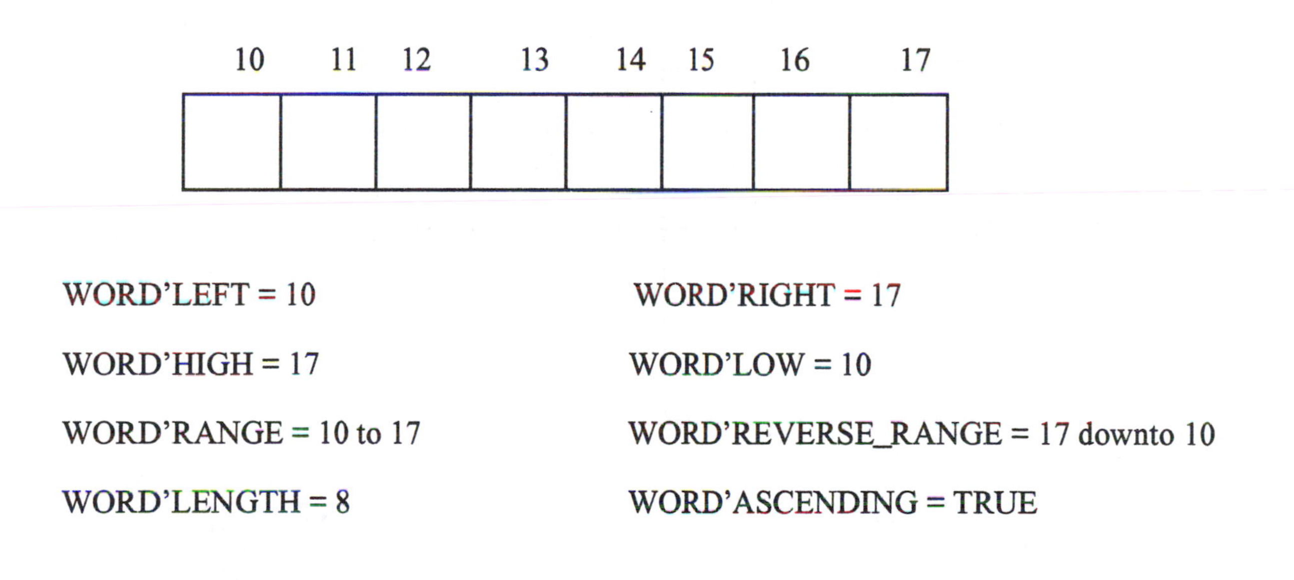

WORD'LEFT = 10 | WORD'RIGHT = 17

WORD'HIGH = 17 | WORD'LOW = 10

WORD'RANGE = 10 to 17 | WORD'REVERSE_RANGE = 17 downto 10

WORD'LENGTH = 8 | WORD'ASCENDING = TRUE

b. variable ELEMENT : ELEMENT_TYPE;

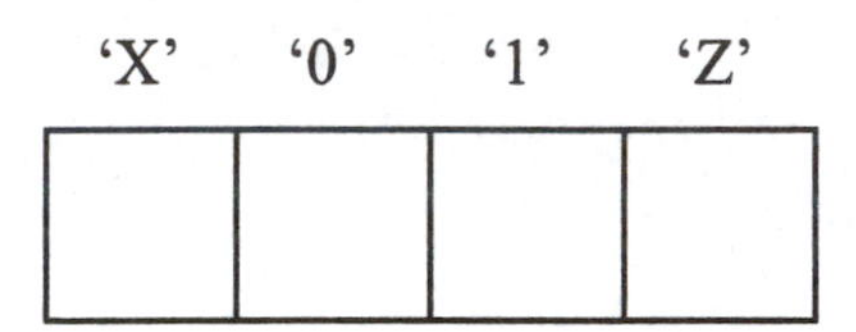

ELEMENT'LEFT = 'X' ELEMENT'RIGHT = 'Z'

ELEMENT'HIGH = 'Z' ELEMENT'LOW = 'X'

ELEMENT'RANGE = 'X' to 'Z' ELEMENT'REVERSE_RANGE = 'Z' downto 'X'

ELEMENT'LENGTH = 4 ELEMENT'ASCENDING = TRUE

c. variable TABLE : TABLE_TYPE;

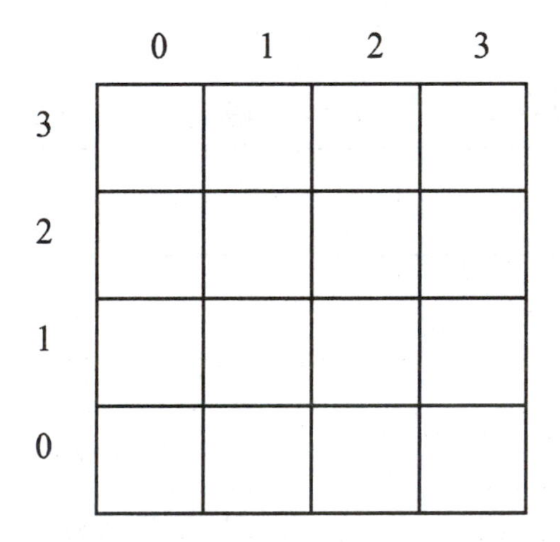

TABLE'LEFT(1) = 3 TABLE'RIGHT(1) = 0

TABLE'HIGH(1) = 3 TABLE'LOW(1) = 0

TABLE'RANGE(1) = 3 downto 0 TABLE'REVERSE_RANGE(1) = 0 to 3

TABLE'LENGTH(1) = 4 TABLE'ASCENDING(1) = FALSE

TABLE'LEFT(2) = 0

TABLE'RIGHT(2) = 3

TABLE'HIGH(2) = 3

TABLE'LOW(2) = 0

TABLE'RANGE(2) = 0 to 3

TABLE'REVERSE_RANGE(2) = 3 downto 0

TABLE'LENGTH(2) = 4

TABLE'ASCENDING(2) = TRUE

16-15. Solution.

a. variable LIST : LIST_TYPE;

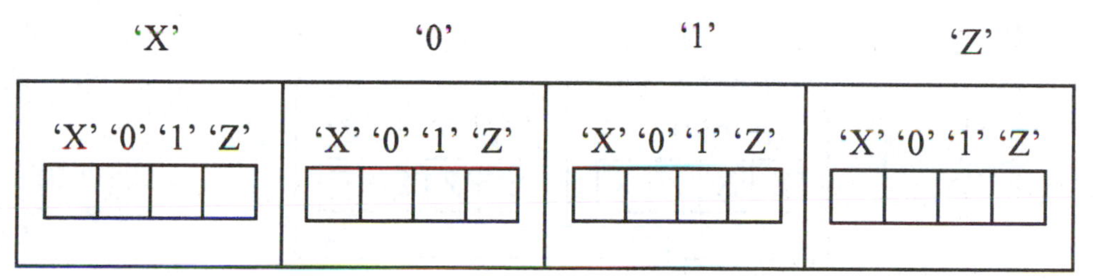

LIST'LEFT = 'X'

LIST'RIGHT = 'Z'

LIST'HIGH = 'Z'

LIST'LOW = 'X'

LIST'RANGE = 'X' to 'Z'

LIST'REVERSE_RANGE = 'Z' downto 'X'

LIST'LENGTH = 4

LIST'ASCENDING = TRUE

variable SUBLIST : SUBLIST_TYPE;

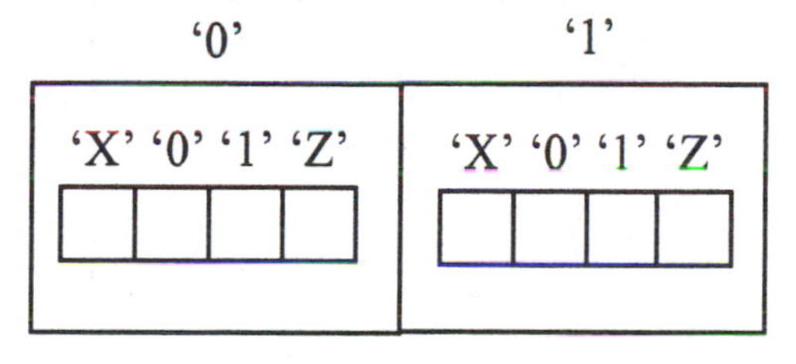

SUBLIST'LEFT = '0'

SUBLIST'RIGHT = '1'

SUBLIST'HIGH = '1'

SUBLIST'LOW = '0'

SUBLIST'RANGE = '0' to '1'

SUBLIST'REVERSE_RANGE = '1' downto '0'

SUBLIST'LENGTH = 2

SUBLIST'ASCENDING = TRUE

16-16. Solution.

a. ELEMENTA

'X'	'0'	'1'	'Z'
'X'	'1'	'1'	'0'

b. SUBLISTA

'0'	'1'
'X' '0' '1' 'Z' 'X' 'X' 'X' 'X'	'X' '0' '1' 'Z' 'Z' 'Z' 'Z' 'Z'

c. LISTA

'X'	'0'	'1'	'Z'

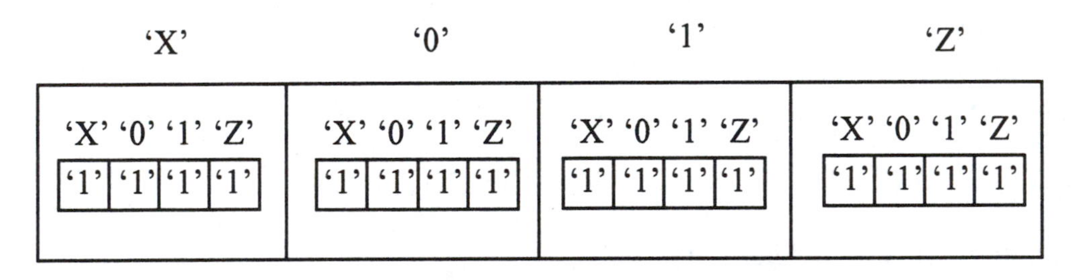

16-17. Solution.

```
package BIT_UTILS is
 type BIT_ARRAY is array (NATURAL <>, NATURAL <>) of BIT;
end BIT_UTILS;

use WORK.BIT_UTILS.all;
entity AND_MATRIX is
 port
  (A, B  : in  BIT_VECTOR(0 to 2);
   Z     : out BIT_ARRAY(0 to 2, 0 to 2));
end AND_MATRIX;

architecture BEHAVIOR of AND_MATRIX is
begin
ROW_RANGE: for ROW in B'RANGE generate
  COL_RANGE: for COL in A'RANGE generate
    Z(ROW, COL) <= B(ROW) and A(COL);
  end generate COL_RANGE;
 end generate ROW_RANGE;

end BEHAVIOR;
```

16-18. Solution.

a.
```
package MOS_TECH is
type DIMENSION is range 1 to INTEGER'HIGH
 units
  um;
  mm = 1000 um;
  cm  = 1000 mm;
  m   = 100   cm;
 end units;

 -- optional
 subtype GATE_DIMENSION is DIMENSION range 1 um to 3 um;

 type TIME_TABLE_TYPE is array (GATE_DIMENSION, GATE_DIMENSION) of
        TIME;
end MOS_TECH;

library IEEE;
use IEEE.STD_LOGIC_1164,all;
use WORK.MOS_TECH.all;
entity MOS_AND is
 generic
  (LENGTH, WIDTH : GATE_DIMENSION);
 port
  (A, B  : in  STD_ULOGIC;
   Z     : out STD_ULOGIC);
end MOS_AND;

architecture BEHAVIOR of MOS_AND is
  constant DELAY_TABLE : TIME_TABLE :=
   ((10 ns, 12 ns, 15 ns),
    ( 7 ns, 10 ns, 11 ns),
    ( 5 ns,  8 ns, 10 ns));
begin
 Z <= A and B after DELAY_TABLE(WIDTH, LENGTH);
end BEHAVIOR;
```

b. library IEEE;

```
use IEEE.STD_LOGIC_1164,all;
use WORK.MOS_TECH.all;
entity TOP_LEVEL is
 port
  (AIN, BIN   : in  STD_ULOGIC;
   Z1, Z2, Z3 : out STD_ULOGIC);
end TOP_LEVEL;

architecture STRUCTURE of TOP_LEVEL is
  component MOS_AND
   generic
    (LENGTH, WIDTH : GATE_DIMENSION);
   port
    (A, B  : in  STD_ULOGIC;
     Z     : out STD_ULOGIC);
  end component;
begin
  A1: MOS_AND generic map (1 um, 2 um)  port map(AIN, BIN, Z1);
  A2: MOS_AND generic map (2 um, 2 um)  port map(AIN, BIN, Z2);
  A1: MOS_AND generic map (3 um, 2 um)  port map(AIN, BIN, Z3);
end STRUCTURE;
```

16-19. Solution.

```
package MOS_TECH is
 type DIMENSION is range 1 to INTEGER'HIGH
  units
   um;
   mm = 1000 um;
   cm  = 1000 mm;
   m   = 100   cm;
  end units;

  -- optional
  subtype GATE_DIMENSION is DIMENSION range 1 um to 3 um;

  type TIME_TABLE_TYPE is array (GATE_DIMENSION, GATE_DIMENSION) of
        TIME;

  type PKG_TYPE is (DIL,     -- dual-in-line
                    PGA,    -- pin grid array
                    LCC);   -- leadless chip carrier
  type TEST_VEC_TYPE is array (1 to 3) of STD_ULOGIC;
  type TEST_ARRAY_TYPE is array (NATURAL range <>) of TEST_VEC_TYPE;

  attribute PKG : PKG_TYPE;
  attribute PIN_NAME : STRING;
  attribute TEST_VECTORS : TEST_ARRAY_TYPE;
end MOS_TECH;
```

```
library IEEE;
use IEEE.STD_LOGIC_1164,all;
use WORK.MOS_TECH.all;
entity MOS_AND is
 generic
  (LENGTH, WIDTH : GATE_DIMENSION);
 port
  (A, B  : in  STD_ULOGIC;
   Z     : out STD_ULOGIC);

 attribute PIN_NAME of A : signal is "PIN_1";
 attribute PIN_NAME of B : signal is "PIN_2";
 attribute PIN_NAME of Z : signal is "PIN_4";

 attribute PKG of MOS_AND : entity is DIL;

 attribute TEST_VECTORS of MOS_AND : entity is
    (('U', 'U', 'U'), ('U', 'X', 'U'), ('U', '0', '0'), .....); -- input, input, output, ...
end MOS_AND;

architecture BEHAVIOR of MOS_AND is
  constant DELAY_TABLE : TIME_TABLE :=
   ((10 ns, 12 ns, 15 ns),
    ( 7 ns, 10 ns, 11 ns),
    ( 5 ns,  8 ns, 10 ns));
begin
 Z <= A and B after DELAY_TABLE(WIDTH, LENGTH);
end BEHAVIOR;
```

16-20. Solution.

a. Illegal. Cannot redefine the METAL_LEVEL attribute of SIG_C.

b. Illegal. Cannot declare additional signals after an attribute specification for signals using the keywords all or others.

c. Illegal. The integer 0 is not a legal value of the attribute METAL_LEVEL of type POSITIVE.

16-21. Solution.

a. A'WIRE_CAP = 8 pF and X'WIRE_CAP = Undefined

b. A'WIRE_CAP = 8 pF and X'WIRE_CAP = 10 pF

16-22. Solution.

```
library IEEE;
use IEEE.STD_LOGIC_1164;all;
entity AND2_GATE is
 port
  (A, B  : in  STD_ULOGIC;
   Z     : out STD_ULOGIC);

 group PORT_GRP_TYPE is (signal <>);
 group PERMUTE_PORTS : PORT_GRP_TYPE(A, B);
end AND2_GATE;

architecture BEHAVIOR of AND2_GATE is
begin
 Z <= A and B;
end BEHAVIOR;
```

Chapter 17. Describing Synchronous Behavior in VHDL

17-1. Solution.

Assume the state of the latch is J=0, K=0, Q=0 and Q_BAR=1. When J=K=1, the first concurrent signal assignment statement executes and schedules Q to take on the complement of Q on the next delta cycle. On the next delta cycle, Q=Q_BAR=1 and both concurrent signal assignment statements execute. The first concurrent signal assignment statement schedules Q to take on the complement of Q on the next delta cycle. The second concurrent signal assignment statement schedules Q_BAR to take on the complement of Q on the next delta cycle. On the next delta cycle, Q=Q_BAR=0 and the complementary relationship between Q and Q_BAR is violated. As long as J=K=1, the JK_LATCH will oscillate between Q=Q_BAR=0 and Q=Q_BAR=1.

17-2. Solution.

a.
```
entity T_LATCH is
  port
   (T     : in BIT;
    Q     : inout BIT := '0';
    Q_BAR : inout BIT := '1');
end T_LATCH;

architecture TRUTH_TABLE of T_LATCH is
begin
 Q <= Q     when (T='0') else
      Q_BAR;

 Q_BAR <= not Q;
end TRUTH_TABLE;
```

b.

```
entity T_LATCH is
  port
    (T     : in BIT;
     Q     : inout BIT := '0';
     Q_BAR : inout BIT := '1');
end T_LATCH;

architecture CHAR_EQ of T_LATCH is
begin
 Q <= T xor not Q_BAR;

 Q_BAR <= not Q;
end CHAR_EQ;
```

17-3. Solution.

a.

```
entity T_GLATCH is
  port
    (T     : in BIT;
     CLK   : in BIT;
     Q     : inout BIT := '0';
     Q_BAR : inout BIT := '1');
end T_GLATCH;

architecture TRUTH_TABLE of T_GLATCH is
begin
 Q <= Q_BAR  when (CLK='1' and T='1') else
        Q;

 Q_BAR <= not Q;
end TRUTH_TABLE;
```

b. entity T_GLATCH is

```
port
  (T     : in BIT;
   CLK   : in BIT;
   Q     : inout BIT := '0';
   Q_BAR : inout BIT := '1');
end T_LATCH;

architecture TRUTH_TABLE of T_LATCH is
begin
 CLKED: block (CLK='1')
 begin
  Q <= guarded Q           when (T='0') else
                 Q_BAR;

  Q_BAR <= not Q;
 end block CLKED;
end TRUTH_TABLE;
```

c. The VHDL model in a. encodes the controlling functionality in an asynchronous fashion. The first signal assignment statement will still execute in response to changes on T, even when CLK=0, but will not change the value of the target signal when CLK=0. This description is closer to the actual gating hardware.

The VHDL model in b. encodes the controlling functionality in a synchronous fashion, making the control explicit. This description is more simulation efficient because the guarded signal assignment state will not execute in response to changes on T when CLK=0.

17-4. Solution.

The guard expression must yield a value of type BOOLEAN, i.e., either TRUE or FALSE. The object CONTROL_SIGNAL is of type BIT.

17-5. Solution.

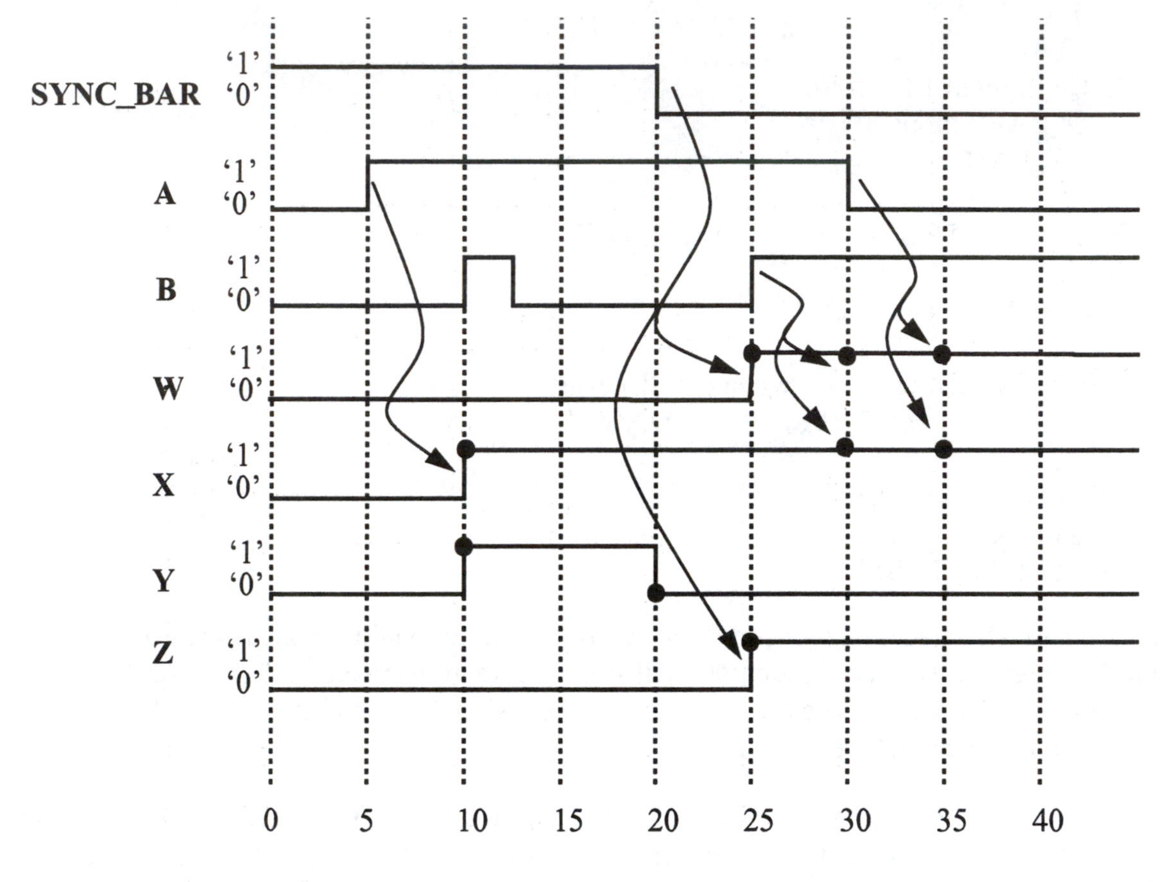

● - signal update

17-6. Solution.

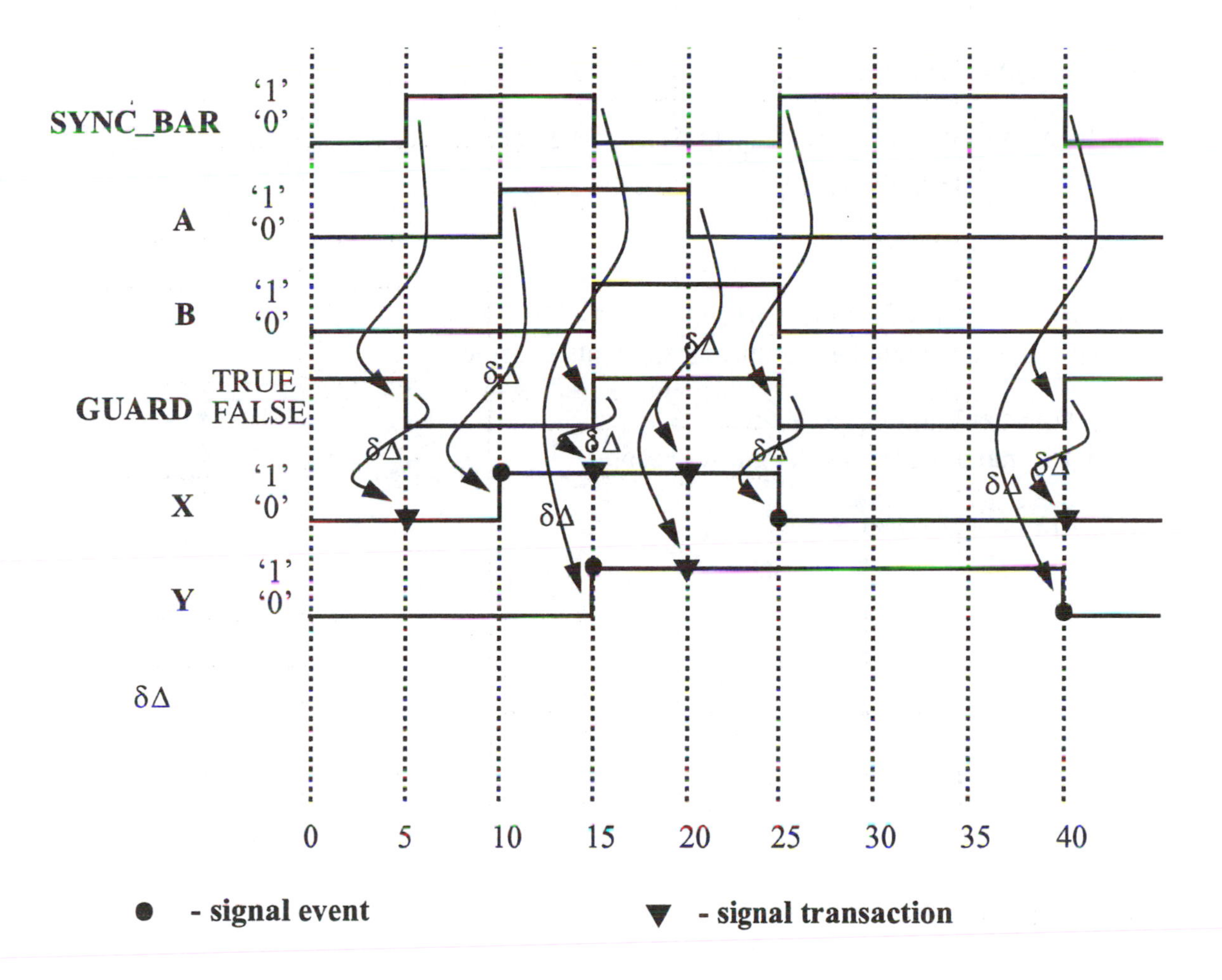

● - signal event ▼ - signal transaction

#1: At t=5, SYNC_BAR->1 and GUARD->FALSE
1st signal assignment statement executes

#2: At t=5+δ, and X=0

#3: At t=10, A->1
1st signal assignment statement executes

#4: At t=10+δ, and X->1

#5: At t=15, SYNC_BAR->0, GUARD->TRUE, B->1
1st and 2nd signal assignment statements execute

#6: At t=15+δ, and X=1, Y->1

#7: At t=20, A->0
1st and 2nd signal assignment statements execute

#8: At t=20+δ, and X=1, Y=1

#9: At t=25, SYNC_BAR->1, GUARD->TRUE, B->0
1st signal assignment statement execute

#10: At t=25+δ, and X->0

#11: At t=40, SYNC_BAR->0, GUARD->TRUE
1st and 2nd signal assignment statements execute

#11: At t=40+δ, and X=0, Y->0
1st signal assignment statement executes

#12: At t=40+2δ, X=0

17-7. Solution.

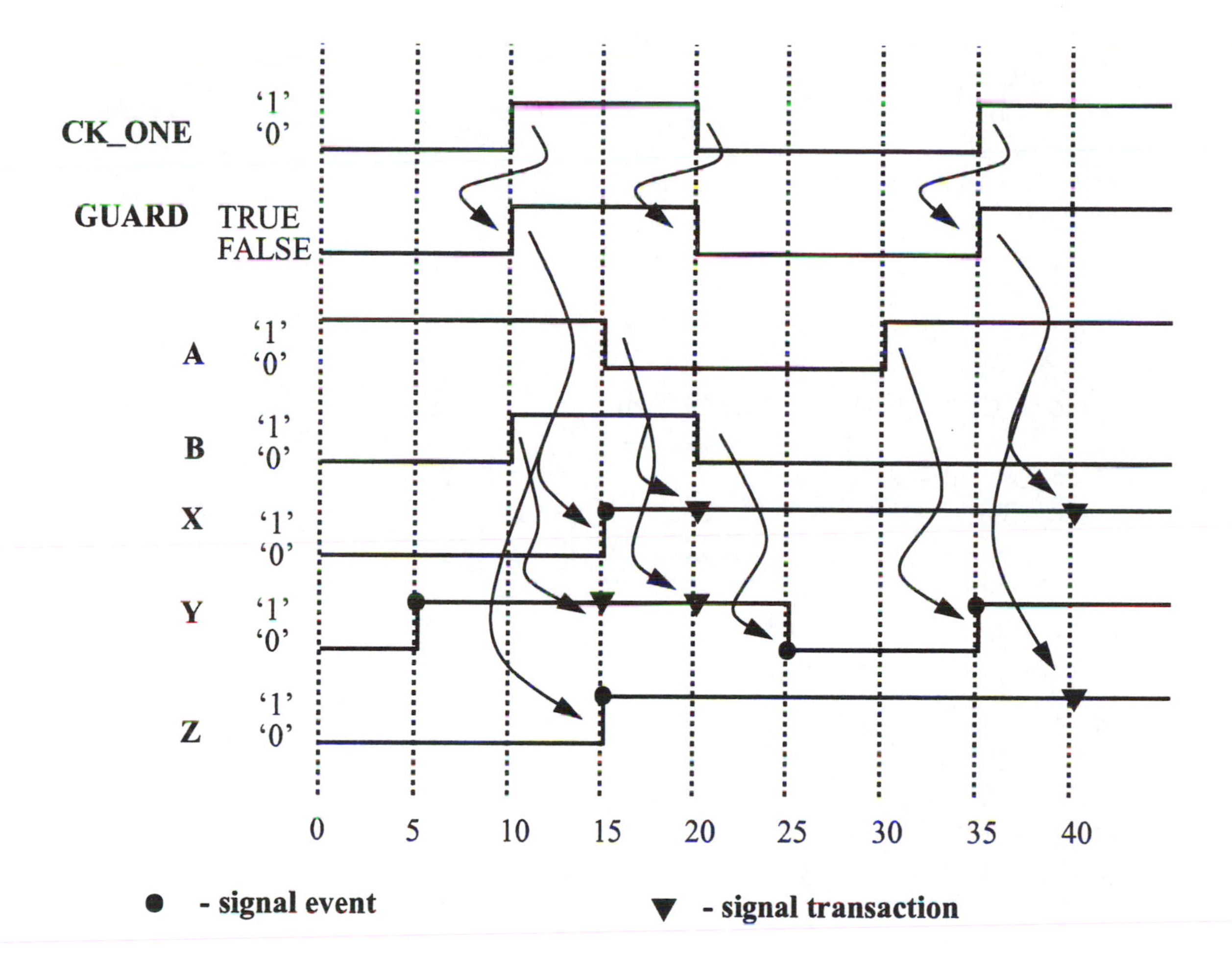

17-8. Solution.

a.
```
entity T_FF is
port
  (T     : in BIT;
   CLK : in BIT
   Q     : inout BIT := '0';
   Q_BAR : inout BIT := '1');
end T_FF;

architecture BEHAVIOR of T_FF is
begin
 B1: block (CLK='1' and not CLK'STABLE)
 begin
   Q <= guarded T xor Q;
 end block B1;

 Q_BAR <= not Q;
end BEHAVIOR;
```

b.
```
entity T_FF is
port
  (T     : in BIT;
   CLK : in BIT
   Q     : inout BIT := '0';
   Q_BAR : inout BIT := '1');
end T_FF;

architecture BEHAVIOR of T_FF is
begin
 B1: block (CLK='0' and not CLK'STABLE)
 begin
   Q <= guarded T xor Q;
 end block B1;

 Q_BAR <= not Q;
end BEHAVIOR;
```

17-9. Solution.

a. CLK'EVENT is a value, not a signal. Thus, CLK'EVENT changing value does not trigger the evaluation of the guard expression. When CLK changes from 0 to 1, CLK'EVENT is TRUE and the guard expression changes from FALSE to TRUE. However, when CLK'EVENT changes from TRUE to FALSE one delta cycle later, the guard expression is not reevaluated and thus stays TRUE until CLK changes from 1 to 0.

b. Level-sensitive D flip-flop.

17-10. Solution.

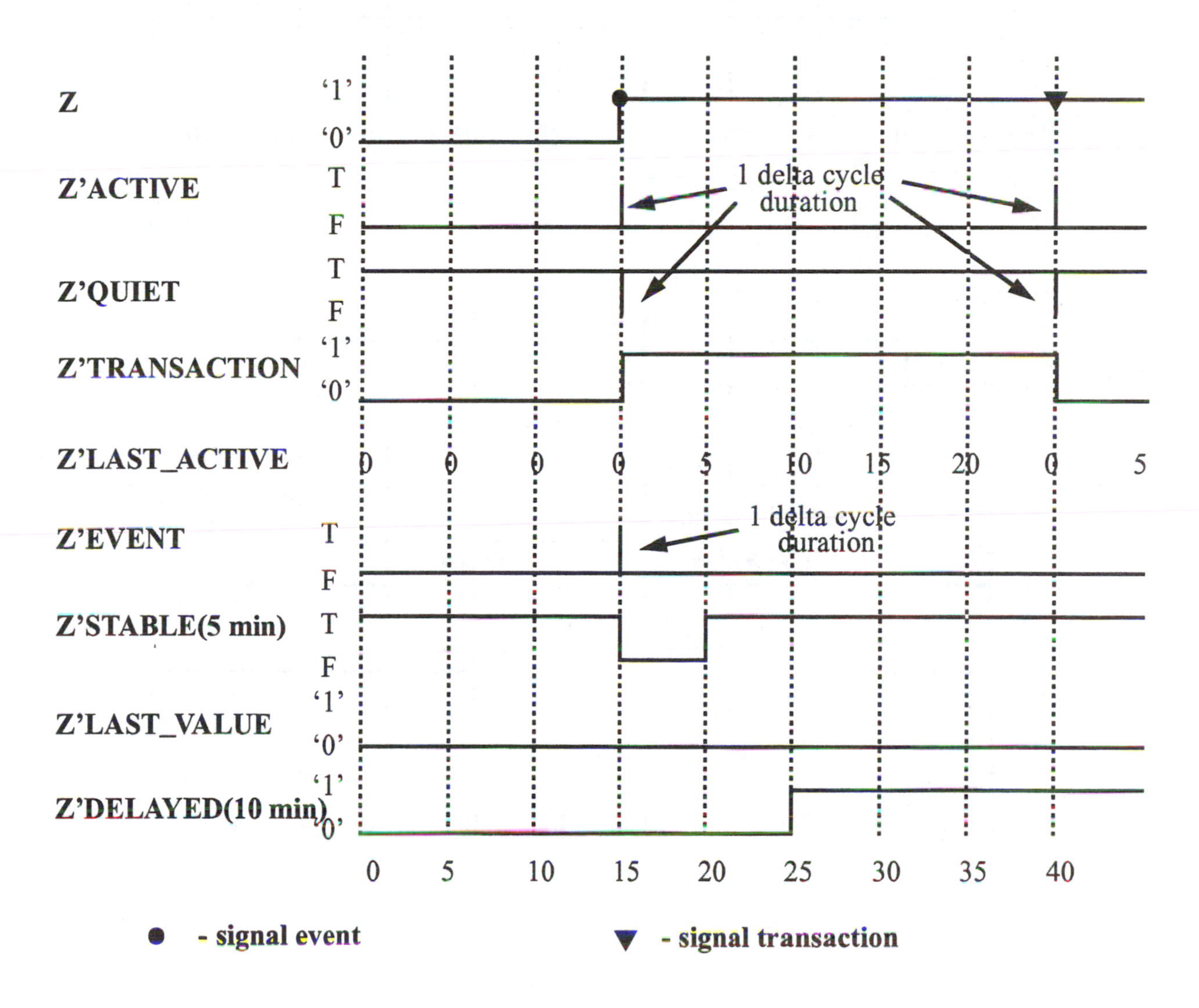

17-11. Solution.

a.

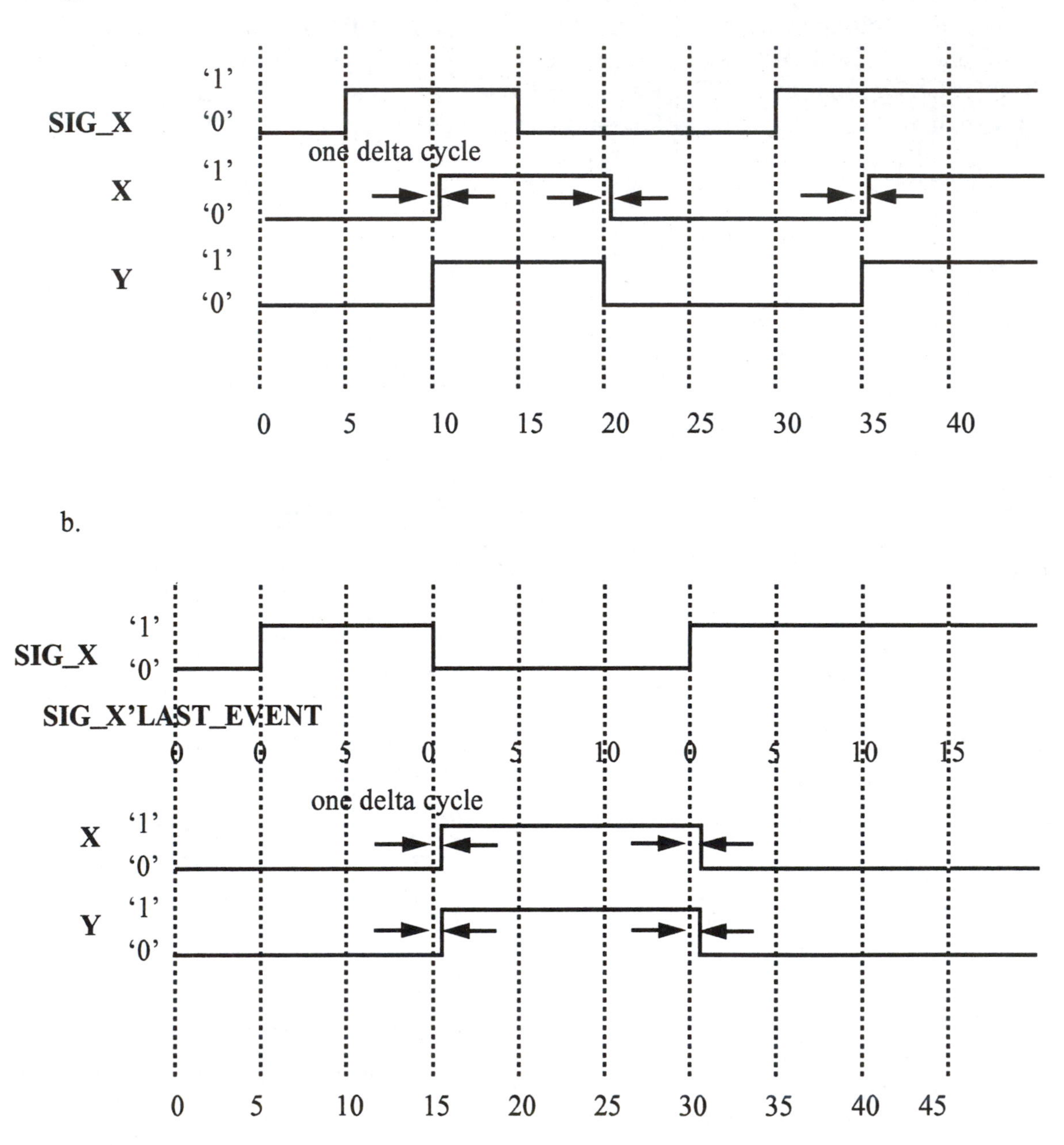

c.

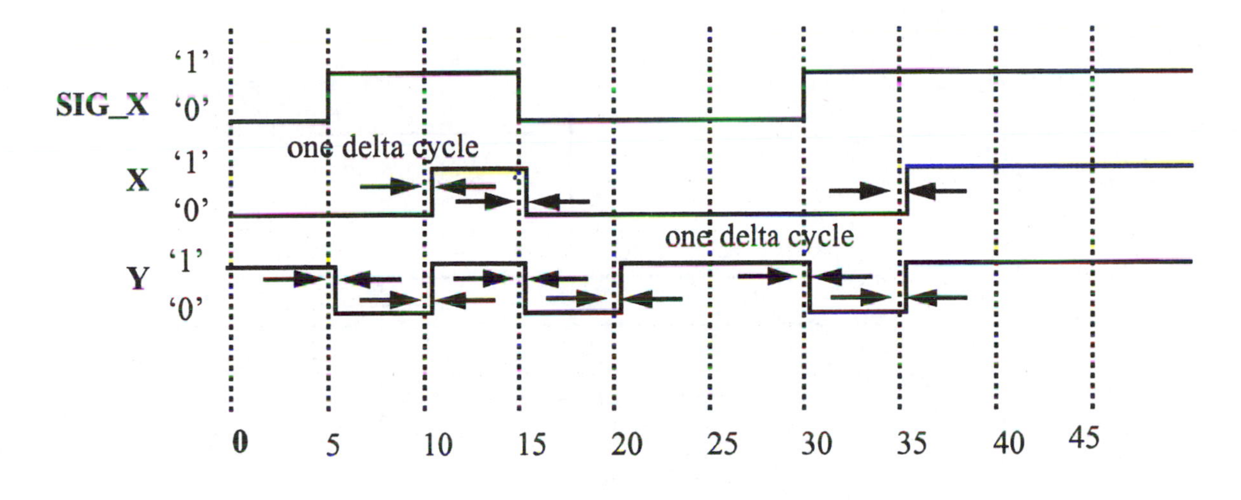

SIG_X
X
Y
'1'
'0'
'1'
'0'
'1'
'0'
one delta cycle
one delta cycle
0
5
10
15
20
25
30
35
40
45

17-12. Solution.

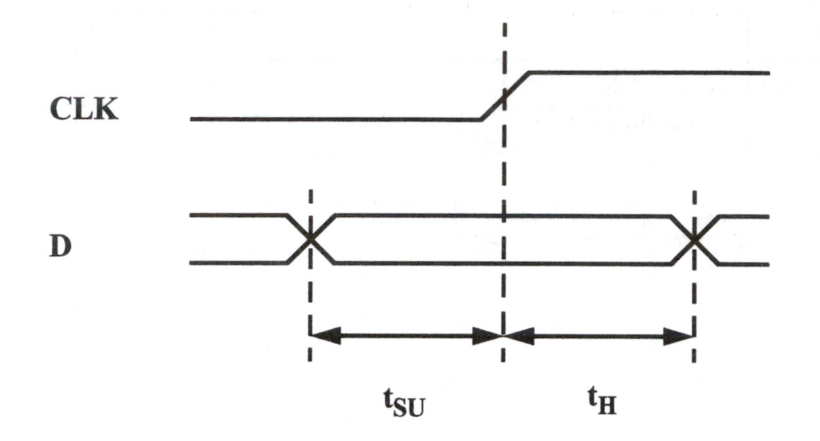


```
entity D_FF is
 generic
  (SETUP, HOLD : TIME);
 port
  (D, CLK   : in BIT;
   Q, Q_BAR : inout BIT);
end D_FF;

architecture DATA_FLOW of D_FF is
begin
 -- assertion statements defining setup and hold timing requirements
 assert not (CLK='1' and not CLK'STABLE and not D'STABLE(SETUP))
  report "Setup Timing Violation";

 assert not (not D'STABLE and CLK='1' and not CLK'STABLE(HOLD))
  report "Hold Timing Violation";

 CLKED: block (CLK = '1' and not CLK'STABLE)
 begin
  -- guarded concurrent signal assignment
  Q <= guarded D;
  Q_BAR <= not Q;
 end block CLKED;
end DATA_FLOW;
```

17-13. Solution.

```
package UTILS is
 function VALUE (signal ARG : BIT) return STRING;
end UTILS;
package body UTILS is
 function VALUE (signal ARG : BIT) return STRING is
 begin
  case ARG is
   when '0' => return("0");
   when '1' => return("1");
  end case;
 end VALUE;
end UTILS;

entity SR_LATCH is
 port
  (S, R     : in BIT;
   Q        : inout BIT := '0';
   Q_BAR : inout BIT := '1');
end SR_LATCH;

use WORK.UTILS.all;
architecture DATA_FLOW of SR_LATCH is
begin
 -- assertion statement defining S and R value limitations
 assert not (S='1' and R='1')
  report
   "Input Violation: S = " & VALUE(S) & " and R = " & VALUE(R);

 Q <= S or (not R and Q);
 Q_BAR <= not Q;
end DATA_FLOW;
```

17-14. Solution.

```
package UTILS is
 function HOW_BAD (constant SETUP : TIME;
                            signal   DATA : BIT) return SEVERITY_LEVEL;
end UTILS;
package body UTILS is
 function HOW_BAD (constant SETUP : TIME;
                            signal   DATA : BIT) return SEVERITY_LEVEL is
   variable MARGIN : NATURAL;
 begin
   MARGIN := (DATA'LAST_EVENT*100)/SETUP;

   if    (MARGIN <= 10)  then   return NOTE;
   elsif (MARGIN <= 30)  then   return WARNING;
   elsif (MARGIN <= 50)  then   return ERROR;
   else                          return FAILURE;
   end if;
 end HOW_BAD;
end UTILS;

entity LOGIC is
 generic (ONE_HOLD : TIME);
 port
   (CNTRL       : in    BIT;
    DATA_OUT    : out BIT);
end LOGIC;

use WORK.UTILS.all;
architecture DATA_FLOW of LOGIC is
begin
 -- assertion statement defining hold-1 timing requirement on CNTRL
 assert not (CNTRL='0' and not CNTRL'STABLE
             and not CNTRL'DELAYED'STABLE(ONE_HOLD))
  report "HOLD-1 Timing Violation"
  severity HOW_BAD(ONE_HOLD, CNTRL);

 DATA_OUT <= '1' when CNTRL='0' else
                     '0';
end DATA_FLOW;
```

17-15. Solution.

```
entity RISING_EDGE is
 port
  (X, CLK : in BIT;
   Z      : out BIT);
end RISING_EDGE;

architecture DATA_FLOW of RISING_EDGE is
  type FF_INDEX is (A, B, C);
  type FF_TYPE is array (FF_INDEX) of BIT;
  signal Q : FF_TYPE;

begin
 -- State : D flip-flops
 DFF: block (CLK = '1' and not CLK'STABLE)
 begin
  Q(A) <= guarded Q(B) and not X;

  Q(B) <= guarded Q(A) or Q(C);

  Q(C) <= guarded not Q(B) and not X;
 end block DFF;

 -- Output
 Z <= Q(A) and not Q(C);
end DATA_FLOW;
```

17-16. Solution.

```
entity SEQ_COMPARATOR is
 port
  (A, B : in BIT;
   CLK  : in BIT;
   Z    : out BIT_VECTOR(1 downto 0));
end SEQ_COMPARATOR;

architecture DATA_FLOW of SEQ_COMPARATOR is
  type FF_INDEX is (A, B);
  type FF_TYPE is array (FF_INDEX) of BIT;
  signal Q, D : FF_TYPE;

begin
 -- Next State Logic
 D(A) <= (not A and B and not Q(B)) or Q(A);
 D(B) <= (A and not B and not Q(A)) or Q(B);

 -- Output
 Z <= D(B) & D(A);

 -- State : D flip-flops
 DFF: block (CLK = '1' and not CLK'STABLE)
 begin
  Q(A) <= guarded D(A);
  Q(B) <= guarded D(B);
 end block DFF;
end DATA_FLOW;
```

17-17. Solution.

```
entity SEQ_COMPARATOR is
 port
  (A, B : in BIT;
   CLK  : in BIT;
   Z    : out BIT_VECTOR(1 downto 0));
end SEQ_COMPARATOR;
```

```
architecture DATA_FLOW of SEQ_COMPARATOR is
 type FF_INDEX is (A, B);
 type FF_TYPE is array (FF_INDEX) of BIT;
 signal Q, D : FF_TYPE;
begin
 -- Next State Logic
 process
 begin
  D(A) <= (not A and B and not Q(B)) or Q(A);
  wait on  A, B, Q(A), Q(B);
 end process;
process
 begin
  D(B) <= (A and not B and not Q(A)) or Q(B);
  wait on  A, B, Q(A), Q(B);
 end process;
 -- Output
 process
 begin
  Z <= D(B) & D(A);
  wait D(A), D(B);
 end process;
  -- State : D flip-flops
 DFF: block (CLK = '1' and not CLK'STABLE)
 begin
  process
  begin
   if GUARD then
     Q(A) <= D(A);
   end if;
   wait on GUARD, D(A);
  end process;
  process
  begin
   if GUARD then
     Q(B) <= D(B);
   end if;
   wait on GUARD, D(B);
  end process;
 end block DFF;
end DATA_FLOW;
```

17-18. Solution.

```
package COMPUTER_DEFS is
  type TR_TYPE is (ACTIVE, INACTIVE);
  type STATUS_TYPE is (IDLE, BUSY);
  subtype BIT_ARRAY8 is BIT_VECTOR(7 downto 0);
  type SOURCE_INPUTS is array (INTEGER range <>) of BIT_ARRAY8;

  function WIRED_OR8 (SOURCES : SOURCE_INPUTS) return BIT_ARRAY8;
end COMPUTER_DEFS;

package body COMPUTER_DEFS is
function WIRED_OR8 (SOURCES : SOURCE_INPUTS) return BIT_ARRAY8 is
  variable RESOLVE_VALUE : BIT_ARRAY8 := X"00";
begin
  for I in SOURCES'RANGE loop
   RESOLVE_VALUE := RESOLVE_VALUE or SOURCES(I);
  end loop;
  return RESOLVE_VALUE;
end WIRED_OR8;
end COMPUTER_DEFS;
```

```
use WORK.COMPUTER_DEFS.all;
entity COM_ARBITER is
  port
   (TR_1, TR_2          : in   TR_TYPE;
    DATA_1, DATA_2 : in    BIT_ARRAY8;
    STATUS                 : out  STATUS_TYPE;
    CHANNEL               : out WIRED_OR8 BIT_ARRAY8 bus);
end COM_ARBITER;

architecture DATA_FLOW of COM_ARBITER is
begin
  PORT_1: block (TR_1 = ACTIVE)
  begin
   CHANNEL <= guarded DATA_1;
  end block PORT_1;

  PORT_2: block (TR_2 = ACTIVE)
  begin
   CHANNEL <= guarded DATA_2;
  end block PORT_2;

  STATUS <= BUSY when ((TR_1 = ACTIVE) or (TR_2 = ACTIVE)) else
                    IDLE;
end DATA_FLOW;
```

17-19. Solution.

Clock Cycle 1:

Simulation Cycle:

Rising edge on CLK and RESET is 1. CLK'STABLE goes from TRUE to FALSE. The guard expression changes from FALSE to TRUE and all guarded concurrent signal assignment statements are executed. Z is scheduled to take on the value B"1000".

Simulation Cycle:

Z takes on the value B"1000". CLK'STABLE goes from FALSE to TRUE. The guard expression goes from TRUE to FALSE, disabling all guarded concurrent signal assignment statements.

Simulation Cycle:

CLK changes from 1 to 0.

Clock Cycle 2:

Simulation Cycle:

Rising edge on CLK and RESET is 0. CLK'STABLE goes from TRUE to FALSE. The guard expression changes from FALSE to TRUE and all guarded concurrent signal assignment statements are executed. Z is scheduled to take on the value B"0100".

Simulation Cycle:

Z takes on the value B"0100". CLK'STABLE goes from FALSE to TRUE. The guard expression goes from TRUE to FALSE, disabling all guarded concurrent signal assignment statements.

Simulation Cycle:

CLK changes from 1 to 0.

Clock Cycle 3:

Simulation Cycle:

Rising edge on CLK and RESET is 0. CLK'STABLE goes from TRUE to FALSE. The guard expression changes from FALSE to TRUE and all guarded concurrent signal assignment statements are executed. Z is scheduled to take on the value B"0010".

Simulation Cycle:

Z takes on the value B"0010". CLK'STABLE goes from FALSE to TRUE. The guard expression goes from TRUE to FALSE, disabling all guarded concurrent signal assignment statements.

Simulation Cycle:

CLK changes from 1 to 0.

Clock Cycle 4:

Simulation Cycle:

Rising edge on CLK and RESET is 0. CLK'STABLE goes from TRUE to FALSE. The guard expression changes from FALSE to TRUE and all guarded concurrent signal assignment statements are executed. Z is scheduled to take on the value B"0001".

Simulation Cycle:

Z takes on the value B"0001". CLK'STABLE goes from FALSE to TRUE. The guard expression goes from TRUE to FALSE, disabling all guarded concurrent signal assignment statements.

Simulation Cycle:

CLK changes from 1 to 0.

Clock Cycle 5:

Simulation Cycle:

Rising edge on CLK and RESET is 0. CLK'STABLE goes from TRUE to FALSE. The guard expression changes from FALSE to TRUE and all guarded concurrent signal assignment statements are executed. Z is scheduled to take on the value B"1000".

Simulation Cycle:

Z takes on the value B"1000". CLK'STABLE goes from FALSE to TRUE. The guard expression goes from TRUE to FALSE, disabling all guarded concurrent signal assignment statements.

Simulation Cycle:

CLK changes from 1 to 0.

17-20. Solution.

```
entity JK_FF is
 port
  (J, K, CLK  : in BIT;
   Q          : inout BIT := '0';
   Q_BAR      : inout BIT := '1');
end JK_FF;

architecture DATA_FLOW of JK_FF is
begin
 postponed assert (Q = not Q_BAR)
  report "Assertion Violation: Q and Q_BAR equal.";

 CLKED: block (CLK = '1' and not CLK'STABLE)
 begin
  -- guarded concurrent signal assignment
  Q <= guarded  Q  when (J='0' and K='0') else
             '0' when (J='0' and K='1') else
             '1' when (J='1' and K='0') else
             Q_BAR ;

  Q_BAR <= not Q;
 end block CLKED;
end DATA_FLOW;
```

17-21. Solution.

2's Complement	A	B	C	X	Y	Z
0	0	0	0	0	0	0
1	0	0	1	1	1	1
2	0	1	0	1	1	0
3	0	1	1	1	0	1
-4	1	0	0	-	-	-
-3	1	0	1	0	1	1
-2	1	1	0	0	1	0
-1	1	1	1	0	0	1

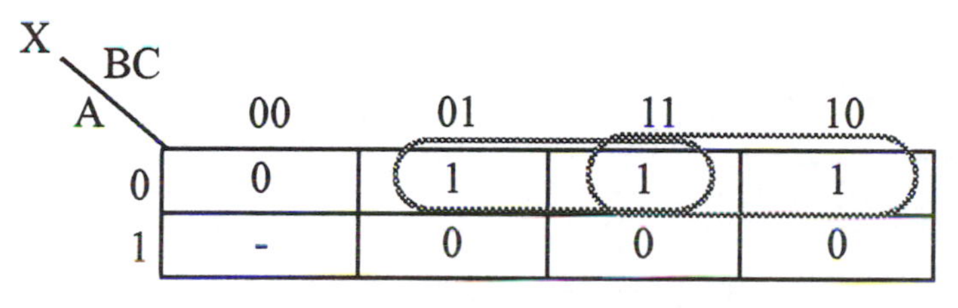

$X = (\bar{A} \bullet C) + (\bar{A} \bullet B)$

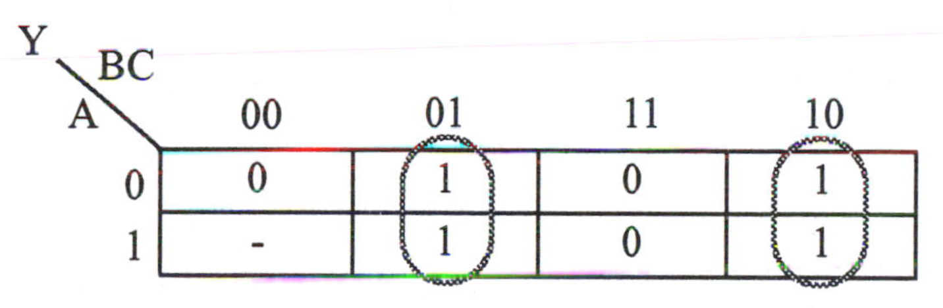

$Y = B \oplus C$

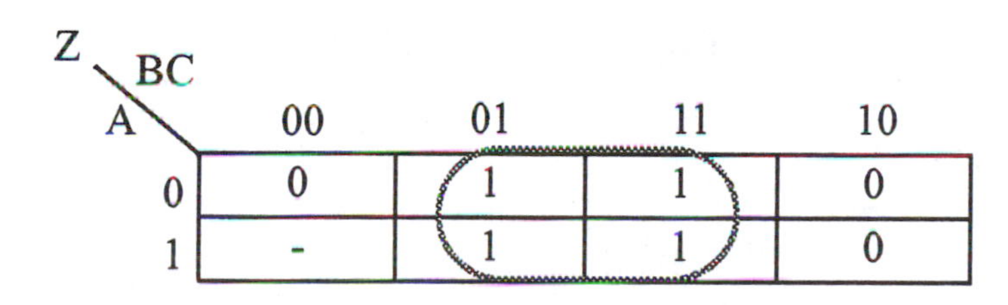

$Z = C$

a.
```
entity TWO_COMPL is
 port
  (A       : in   BIT_VECTOR(2 downto 0);
   A_BAR   : out BIT_VECTOR(2 downto 0));
end TWO_COMPL;

architecture DATA_FLOW of TWO_COMPL is
begin
  A_BAR(2) <= (not A(2) and A(0)) or (not A(2) and A(1));
  A_BAR(1) <= A(1) xor A(0);
  A_BAR(0) <= A(0)
end DATA_FLOW;
```

b.
```
entity TWO_COMPL is
 port
  (A       : in   BIT_VECTOR(2 downto 0);
   A_BAR   : inout BIT_VECTOR(2 downto 0));
end TWO_COMPL;

architecture DATA_FLOW of TWO_COMPL is
 function TO_INT (ARG : BIT_VECTOR) return INTEGER is
   variable INT_VALUE : INTEGER := 0;
 begin
   for I in ARG'RIGHT to ARG'LEFT-1 loop
    INT_VALUE := INT_VALUE + ARG(I)*(2**I);
   end loop;
   return (INT_VALUE - ARG(ARG'LEFT)*(2**ARG'LEFT));
 end TO_INT;
begin
 assert (A /= B"100")
  report "Assertion Violation: Illegal Input A = 100.";

 postponed assert (TO_INT(A) + TO_INT(A_BAR) = 0)
  report "Assertion Violation: Incorrect Two's Complement.";

  A_BAR(2) <= (not A(2) and A(0)) or (not A(2) and A(1));
  A_BAR(1) <= A(1) xor A(0);
  A_BAR(0) <= A(0)
end DATA_FLOW;
```

17-22. Solution.

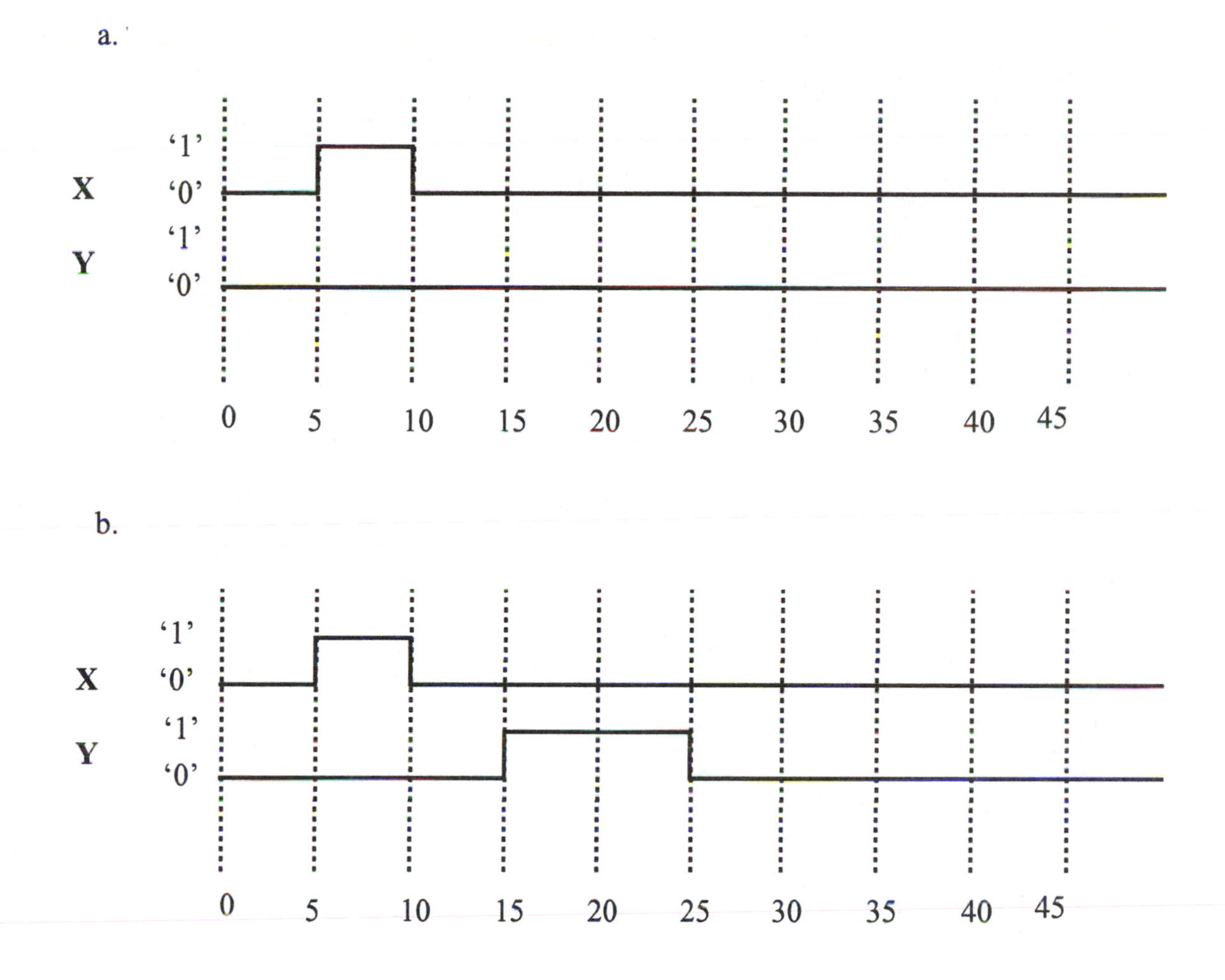

Chapter 18. Algorithmic Modeling in VHDL

18-1. Solution.

a.
```
entity COMB_LOGIC is
 port
  (IN_1, IN_2, IN_3, IN_4 : in BIT;
   OUT_1 : out BIT);
end COMB_LOGIC;
architecture DATA_FLOW of COMB_LOGIC is
begin
  OUT_1 <= not((IN_1 and IN_2) or (IN_3 and IN_4));
end DATA_FLOW;
```

b.
```
entity COMB_LOGIC is
 port
  (IN_1, IN_2, IN_3, IN_4 : in BIT;
   OUT_1 : out BIT);
end COMB_LOGIC;
architecture ALGORITHM of COMB_LOGIC is
begin
  process(IN_1, IN_2, IN_3, IN_4)
  begin
   OUT_1 <= not((IN_1 and IN_2) or (IN_3 and IN_4));
  end process;
end ALGORITHM;
```

18-2. Solution.

a.
```
entity COUNTER is
 port
  (CLK   : in BIT;
   COUNT : inout BIT_VECTOR(1 downto 0));
end COUNTER;
architecture DATA_FLOW of COUNTER is
begin
  -- Assume rising clock edge transition
  CLOCK:
  block (CLK='1' and not CLK'STABLE)
  begin
   COUNT <= guarded B"00" when (COUNT = B"11") else
                B"01" when (COUNT = B"00") else
                B"10" when (COUNT = B"01") else
                B"11"; --  (COUNT = B"10")
  end block CLOCK;
end DATA_FLOW;
```

b.
```
architecture STATE of COUNTER is
begin
  -- Assume rising clock edge transition
  process
  begin
    if (COUNT = B"00") then
      COUNT <= B"01"
    elsif (COUNT = B"01") then
      COUNT <= B"10"
    elsif (COUNT = B"10") then
      COUNT <= B"11"
    else                    -- (COUNT = B"11")
      COUNT <= B"00"
    end if;
    wait until CLK='1';
  end process;
end STATE;
```

c. architecture STATE of COUNTER is

```
begin
 -- Assume rising clock edge transition
 process
  function "+" (LARG : BIT_VECTOR; RARG : NATURAL) return BIT_VECTOR is
  begin
   if    (LARG = B"00") then   return(B"01");
   elsif (LARG = B"01") then  return(B"10");
   elsif (LARG = B"10") then  return(B"11");
   else                        return(B"00");  -- (LARG = B"11")
   end if;
  end "+";
 begin
   COUNT <= COUNT + 1;
   wait until CLK='1';
 end process;
end STATE;
```

18-3. Solution.

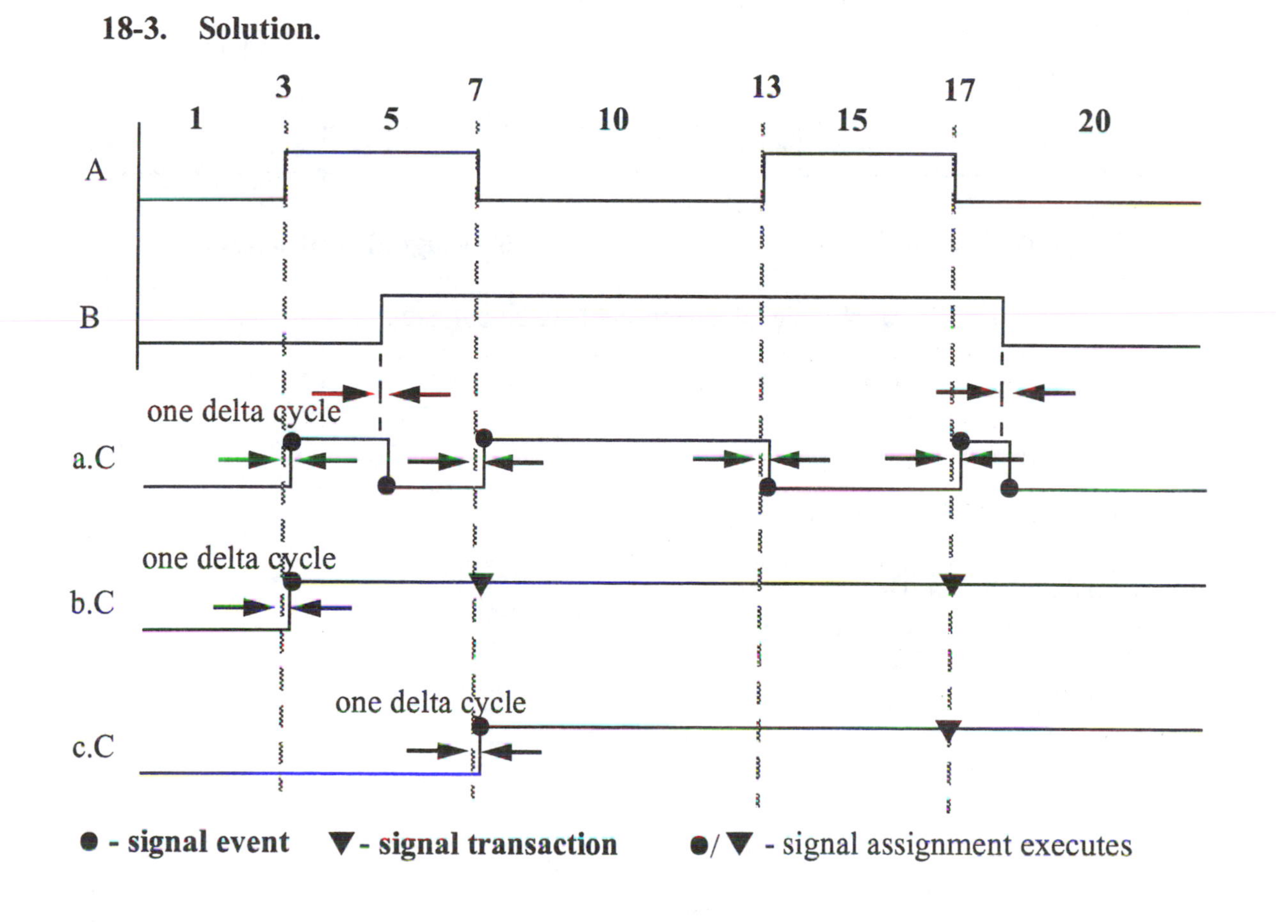

18-4. Solution.

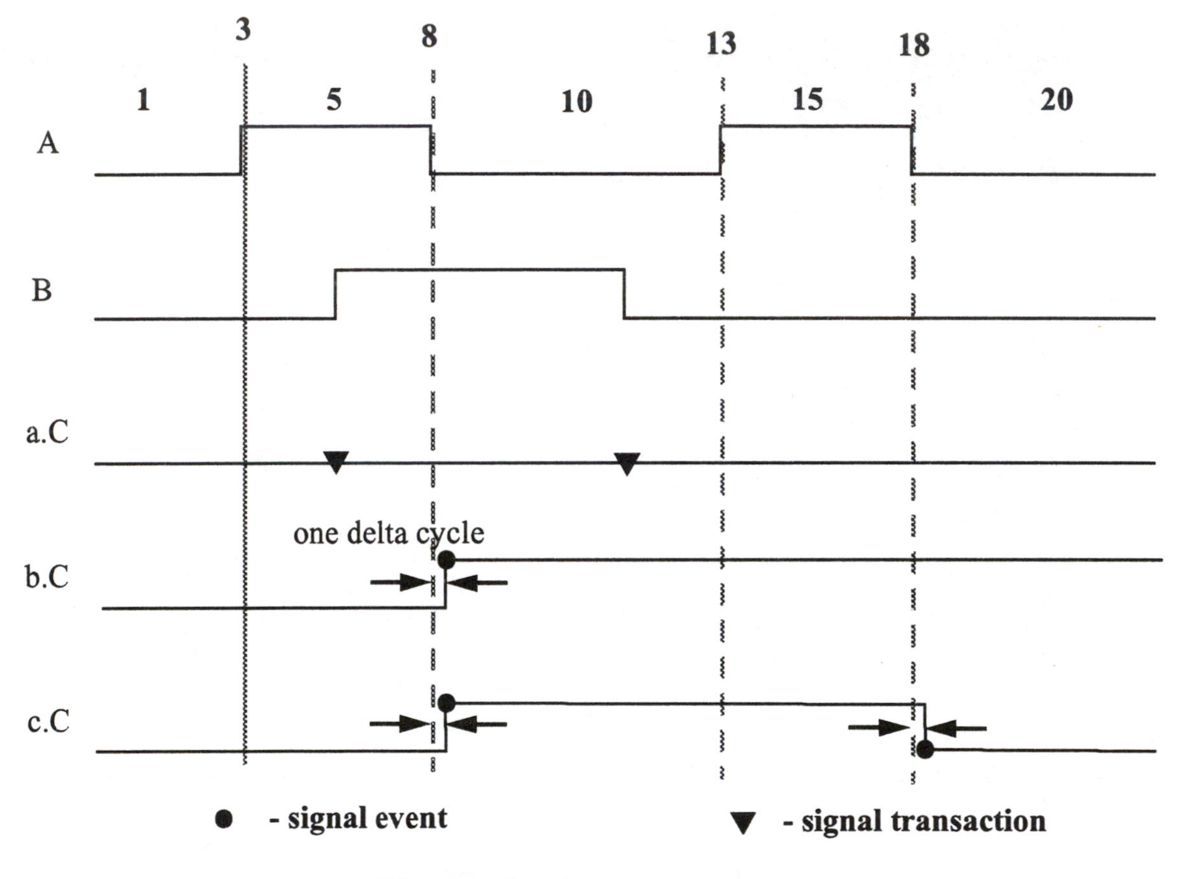

18-5. Solution.

a.
```
entity MEALY_MACHINE is
 port
  (DATA, CLK : in  BIT;
   Z         : out BIT);
end MEALY_MACHINE;
```

```
architecture PROCESS1 of MEALY_MACHINE is
begin
 -- Assume rising clock edge transition
 process (CLK, DATA)
  type STATE_TYPE is (A, B, C);
  variable STATE : STATE_TYPE := A;
 begin
  if (CLK'EVENT and CLK='1') then
   case STATE is
    when A =>
     if (DATA='0') then  STATE := A;
     else                STATE := B;
     end if;
    when B =>
     if (DATA='0') then  STATE := A;
     else                STATE := C;
     end if;
    when C =>
     if (DATA='0') then  STATE := A;
     else                STATE := C;
     end if;
   end case;
  end if;

   -- do not evaluate on the following edge of CLK
  if ((CLK'EVENT and CLK='1') or DATA'EVENT) then
   case STATE is
    when A =>
     if (DATA='0') then  Z <= '1';
     else                Z <= '0';
     end if;
    when B =>
     Z <= '0';
    when C =>
     if (DATA='0') then  Z <= '0';
     else                Z <= '1';
     end if;
   end case;
  end if;
 end process;
end PROCESS1;
```

b. architecture PROCESS2 of MEALY_MACHINE is

```
    type STATE_TYPE is (A, B, C);
    signal STATE : STATE_TYPE := A;
begin
  -- Assume rising clock edge transition
  process (CLK)
  begin
   if (CLK='1') then
    case STATE is
     when A =>
      if (DATA='0') then  STATE <= A;
      else                STATE <= B;
      end if;
     when B =>
      if (DATA='0') then  STATE <= A;
      else                STATE <= C;
      end if;
     when C =>
      if (DATA='0') then  STATE <= A;
      else                STATE <= C;
      end if;
    end case;
   end if;
  end process;

  process (DATA, STATE)
  begin
    case STATE is
     when A =>
      if (DATA='0') then  Z <= '1';
      else                Z <= '0';
      end if;
     when B =>
      Z <= '0';
     when C =>
      if (DATA='0') then  Z <= '0';
      else                Z <= '1';
      end if;
    end case;
  end process;
end PROCESS2;
```

c.
```
architecture PROCESS4 of MEALY_MACHINE is
    type STATE_TYPE is (A, B, C);
    type STATES_TYPE is array (NATURAL range <>) of STATE_TYPE;
    function MUX_RES (DRIVERS : STATES_TYPE) return STATE_TYPE is
    begin
     assert (DRIVERS'LENGTH=1)
      report "Error: Multiple States Active.";
     return (DRIVERS(DRIVERS'LEFT));
    end MUX_RES;
    signal STATE : MUX_RES STATE_TYPE := A;
begin
  -- Assume rising clock edge transition
  process (CLK)
  begin
   if (CLK='1' and STATE=A) then
     if (DATA='0') then  STATE <= A;
     else                STATE <= B;
     end if;
   else
     STATE <= null;
   end if;
  end process;

  process (CLK)
  begin
   if (CLK='1' and STATE=B) then
     if (DATA='0') then  STATE <= A;
     else                STATE <= C;
     end if;
   else
     STATE <= null;
   end if;
  end process;
```

```
process (CLK)
begin
 if (CLK='1' and STATE=C) then
   if (DATA='0') then  STATE <= A;
   else                STATE <= C;
   end if;
 else
   STATE <= null;
 end if;
end process;

process (DATA, STATE)
begin
  case STATE is
   when A =>
    if (DATA='0') then  Z <= '1';
    else                Z <= '0';
    end if;
   when B =>
    Z <= '0';
   when C =>
    if (DATA='0') then  Z <= '0';
    else                Z <= '1';
    end if;
  end case;
end process;
end PROCESS4;
```

18-6. Solution.

a.
```
entity MEALY_MACHINE is
 port
  (DATA, CLK : in  BIT;
   Z         : out BIT);
end MEALY_MACHINE;

architecture CON_SIG2 of MEALY_MACHINE is
   type STATE_TYPE is (A, B, C);
   signal STATE : STATE_TYPE := A;
begin
  -- Assume rising clock edge transition
  CLOCK:
  block (CLK='1' and not CLK'STABLE)
  begin
   STATE <= guarded A when (STATE=A and DATA='0') else
                    B when (STATE=A and DATA='1') else
                    A when (STATE=B and DATA='0') else
                    C when (STATE=B and DATA='1') else
                    A when (STATE=C and DATA='0') else
                    C;  -- (STATE=C and DATA='1')
  end block CLOCK;

  Z <= '1' when (STATE=A and DATA='0') else
       '0' when (STATE=A and DATA='1') else
       '0' when (STATE=B)              else
       '0' when (STATE=C and DATA='0') else
       '1';  -- (STATE=C and DATA='1')
end CON_SIG2;
```

b. architecture CON_SIG4 of MEALY_MACHINE is

```
    type STATE_TYPE is (A, B, C);
    type STATES_TYPE is array (NATURAL range <>) of STATE_TYPE;
    function MUX_RES (DRIVERS : STATES_TYPE) return STATE_TYPE is
    begin
     assert (DRIVERS'LENGTH=1)
      report "Error: Multiple States Active.";
     return (DRIVERS(DRIVERS'LEFT));
    end MUX_RES;
    signal STATE : MUX_RES STATE_TYPE register := A;
  begin
   -- Assume rising clock edge transition
   STATE_A:
   block (CLK='1' and not CLK'STABLE and STATE=A)
   begin
    STATE <= guarded A when (DATA='0') else
                B; -- (DATA='1')
   end block STATE_A;

   STATE_B:
   block (CLK='1' and not CLK'STABLE and STATE=B)
   begin
    STATE <= guarded A when (DATA='0') else
                C; -- (DATA='1')
   end block STATE_B;

   STATE_C:
   block (CLK='1' and not CLK'STABLE and STATE=C)
   begin
    STATE <= guarded A when (DATA='0') else
                C; -- (DATA='1')
   end block STATE_C;

   Z <= '1' when (STATE=A and DATA='0') else
        '0' when (STATE=A and DATA='1') else
        '0' when (STATE=B)              else
        '0' when (STATE=C and DATA='0') else
        '1'; -- (STATE=C and DATA='1')
  end CON_SIG4;
```

18-7. Solution.

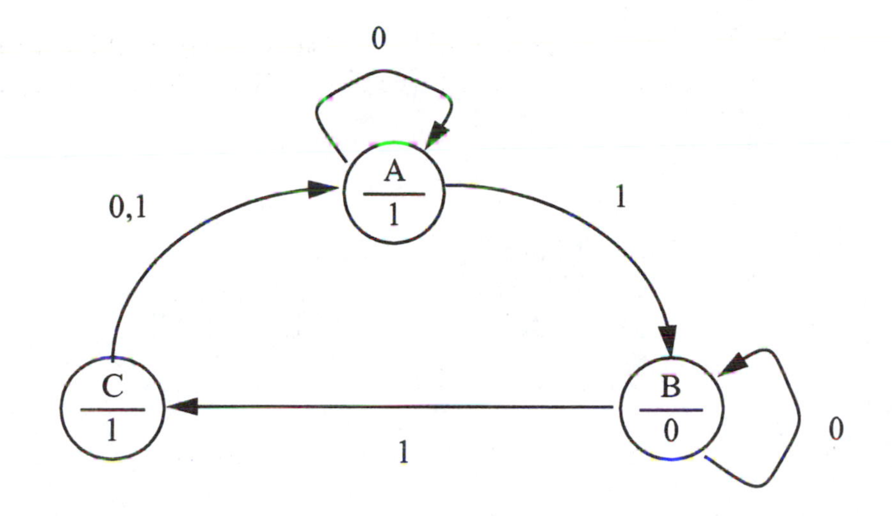

18-8. Solution.

```
entity SEQ_ADDER is
 port
  (X, Y, CLK : in  BIT;
   S         : out BIT );
end SEQ_ADDER;

architecture ALGORITHM of SEQ_ADDER is
  type STATE_TYPE is (S0, S1);
  signal PRESENT_STATE, NEXT_STATE : STATE_TYPE;

begin
 STATE: process
 begin
   wait until (CLK = '1');
   PRESENT_STATE <= NEXT_STATE;
 end process STATE;

 NS: process (PRESENT_STATE, X, Y)
 begin
    case PRESENT_STATE is
     when S0 => if (X&Y = B"11") then NEXT_STATE <= S1;
     when S1 => if (X&Y = B"00") then NEXT_STATE <= S0;
    end case;
 end process NS;

 OUTPUT: process (PRESENT_STATE, X, Y)
 begin
    case PRESENT_STATE is
     when S0 => if ((X&Y=B"00") or (X&Y=B"11")) then S <= '0';
                                  else S <= '1';
     when S1 => if ((X&Y=B"01") or (X&Y=B"10")) then S <= '0';
                                  else S <= '1';
    end case;
 end process OUTPUT;
end ALGORITHM;
```

18-9. Solution.

```
function MUX (DRIVERS : STATES_TYPE) return STATE_TYPE is
begin
  assert (DRIVERS'LENGTH=1)
    report "Assertion Violation: Multiple active states.";
  return (DRIVERS(DRIVERS'LEFT));
end MUX;
```

18-10. Solution.

```
entity BIN_COUNTER is
 port
   (CLK : in  BIT;
    CLR, SET : in BIT;
   Z   : out BIT_VECTOR(2 downto 0) );
end BIN_COUNTER;

architecture ALGORITHM of BIN_COUNTER is
begin
 process
  -- state information held in PRESENT_STATE
  variable PRESENT_STATE : BIT_VECTOR(2 downto 0) := B"111";

 begin
  if   (CLR='1') then   PRESENT_STATE := B"000";
  elsif (SET='1') then   PRESENT_STATE := B"111";
  else
   -- update present state
   case PRESENT_STATE is
    when B"000" =>   PRESENT_STATE := B"001";
    when B"001" =>   PRESENT_STATE := B"010";
    when B"010" =>   PRESENT_STATE := B"011";
    when B"011" =>   PRESENT_STATE := B"100";
    when B"100" =>   PRESENT_STATE := B"101";
    when B"101" =>   PRESENT_STATE := B"110";
    when B"110" =>   PRESENT_STATE := B"111";
    when B"111" =>   PRESENT_STATE := B"000";
   end case;
  end if;

  -- update output
  Z <= PRESENT_STATE after 10 ns;

  -- wait until next rising edge of clock
  wait until (CLK = '1');
 end process;
end ALGORITHM;
```

a.
```
entity BIN_COUNTER is
port
  (CLK : in  BIT;
   CLR, SET : in BIT;
   Z   : out BIT_VECTOR(2 downto 0) );
end BIN_COUNTER;

architecture ALGORITHM of BIN_COUNTER is
begin
 process
  -- state information held in PRESENT_STATE
  variable PRESENT_STATE : BIT_VECTOR(2 downto 0) := B"111";

 begin
  assert not (CLR = SET = '1')
   report "Assertion Violation: CLR and SET simultaneously asserted.";

  if     (CLR='1') then   PRESENT_STATE := B"000";
  elsif (SET='1') then   PRESENT_STATE := B"111";
  else
   -- update present state
   case PRESENT_STATE is
    when B"000" =>   PRESENT_STATE := B"001";
    when B"001" =>   PRESENT_STATE := B"010";
    when B"010" =>   PRESENT_STATE := B"011";
    when B"011" =>   PRESENT_STATE := B"100";
    when B"100" =>   PRESENT_STATE := B"101";
    when B"101" =>   PRESENT_STATE := B"110";
    when B"110" =>   PRESENT_STATE := B"111";
    when B"111" =>   PRESENT_STATE := B"000";
   end case;
  end if;

  -- update output
  Z <= PRESENT_STATE after 10 ns;

  -- wait until next rising edge of clock
  wait until (CLK = '1');
 end process;
end ALGORITHM;
```

b. entity BIN_COUNTER is

```
port
  (CLK : in  BIT;
   CLR, SET : in BIT;
  Z   : out BIT_VECTOR(2 downto 0) );
end BIN_COUNTER;

architecture ALGORITHM of BIN_COUNTER is
begin
 assert not (CLR = SET = '1')
  report "Assertion Violation: CLR and SET simultaneously asserted.";

 process
  -- state information held in PRESENT_STATE
  variable PRESENT_STATE : BIT_VECTOR(2 downto 0) := B"111";

 begin
  if     (CLR='1') then   PRESENT_STATE := B"000";
  elsif (SET='1') then   PRESENT_STATE := B"111";
  else
   -- update present state
   case PRESENT_STATE is
    when B"000" =>   PRESENT_STATE := B"001";
    when B"001" =>   PRESENT_STATE := B"010";
    when B"010" =>   PRESENT_STATE := B"011";
    when B"011" =>   PRESENT_STATE := B"100";
    when B"100" =>   PRESENT_STATE := B"101";
    when B"101" =>   PRESENT_STATE := B"110";
    when B"110" =>   PRESENT_STATE := B"111";
    when B"111" =>   PRESENT_STATE := B"000";
   end case;
  end if;

  -- update output
  Z <= PRESENT_STATE after 10 ns;

  -- wait until next rising edge of clock
  wait until (CLK = '1');
 end process;
end ALGORITHM;
```

18-11. Solution.

a.
```
function WIRED_AND (SOURCES : BIT_VECTOR) return BIT is
  variable RESULT : BIT := '1';
  variable I : INTEGER range SOURCES'RANGE := SOURCES'LOW;
begin
  loop
    RESULT := RESULT and SOURCES(I);
    exit when (I = SOURCES'HIGH);
    I := I + 1;
  end loop;
  return( RESULT );
end WIRED_AND;
```

b.
```
function WIRED_AND (SOURCES : BIT_VECTOR) return BIT is
  variable RESULT : BIT := '1';
  variable I : NATURAL := SOURCES'LOW;
begin
  while (I <= SOURCES'HIGH) loop
    RESULT := RESULT and SOURCES(I);
    I := I + 1;
  end loop;
  return( RESULT );
end WIRED_AND;
```

c. The for-loop seems most closely matched to looping through the finite range of the array index of WIRED_AND and thus, makes the code more readable.

18-12. Solution.

a.
```
package INT_ARITH_PKG is
  procedure ADDER (LARG, RARG : in  BIT_VECTOR;
                   SUM        : out BIT_VECTOR;
                   CARRY      : out BIT);
end INT_ARITH_PKG;

package body INT_ARITH_PKG is
  procedure ADDER (LARG, RARG : in  BIT_VECTOR;
                   SUM        : out BIT_VECTOR;
                   CARRY      : out BIT) is
   alias XLARG : BIT_VECTOR(LARG'LENGTH-1 downto 0) is LARG;
   alias XRARG : BIT_VECTOR(RARG'LENGTH-1 downto 0) is RARG;
   variable XSUM : BIT_VECTOR(XLARG'RANGE);  variable XCARRY : BIT := '0';
  begin
   for I in XLARG'LOW to XLARG'HIGH loop
     XSUM(I) := XCARRY xor XLARG(I) xor XRARG(I);
     XCARRY := (XCARRY and XLARG(I)) or (XCARRY and XRARG(I)) or
               (XLARG(I) and XRARG(I));
   end loop;
   SUM := XSUM;   CARRY := XCARRY;
  end ADDER;
end INT_ARITH_PKG;
```

b.
```
entity ADDRESS_CALC is
 port
  (ADDRESS, OFFSET : in  BIT_VECTOR(7 downto 0);
   VADDRESS        : out BIT_VECTOR(7 downto 0);
   OVERFLOW        : out BIT);
end ADDRESS_CALC;

use WORK.INT_ARITH_PKG.all;
architecture DATA_FLOW of ADDRESS_CALC is
begin
  -- concurrent procedure call
  ADDER(LARG=>ADDRESS, RARG=>OFFSET, SUM=>VADDRESS,
        CARRY=>OVERFLOW);
end DATA_FLOW;
```

18-13. Solution.

```
package INT_ARITH_PKG is
  procedure ADDER (LARG, RARG : in  BIT_VECTOR;
                   SUM        : out BIT_VECTOR;
                   CARRY      : out BIT);

  subtype BYTE is BIT_VECTOR(7 downto 0);
end INT_ARITH_PKG;

package body INT_ARITH_PKG is
  procedure ADDER (LARG, RARG : in  BIT_VECTOR;
                   SUM        : out BIT_VECTOR;
                   CARRY      : out BIT) is
   alias XLARG : BIT_VECTOR(LARG'LENGTH-1 downto 0) is LARG;
   alias XRARG : BIT_VECTOR(RARG'LENGTH-1 downto 0) is RARG;
   variable XSUM : BIT_VECTOR(XLARG'RANGE);  variable XCARRY : BIT := '0';
  begin
   for I in XLARG'LOW to XLARG'HIGH loop
     XSUM(I) := XCARRY xor XLARG(I) xor XRARG(I);
     XCARRY := (XCARRY and XLARG(I)) or (XCARRY and XRARG(I)) or
               (XLARG(I) and XRARG(I));
   end loop;
   SUM := XSUM;   CARRY := XCARRY;
  end ADDER;
end INT_ARITH_PKG;

use WORK.INT_ARITH_PKG.all;
entity ADDRESS_CALC is
 port
   (ADDRESS, OFFSET : in BYTE;
    VADDRESS        : out BYTE;
    OVERFLOW        : out BIT);
end ADDRESS_CALC;

architecture DATA_FLOW of ADDRESS_CALC is
begin
  -- concurrent procedure call
  ADDER(LARG=>ADDRESS, RARG=>OFFSET, SUM=>VADDRESS,
        CARRY=>OVERFLOW);
end DATA_FLOW;
```

18-14. Solution.

```
package INT_ARITH_PKG is
  procedure ADDER (LARG, RARG : in  BIT_VECTOR;
                   SUM        : out BIT_VECTOR;
                   CARRY      : out BIT);
end INT_ARITH_PKG;

package body INT_ARITH_PKG is
  procedure ADDER (LARG, RARG : in  BIT_VECTOR;
                   SUM        : out BIT_VECTOR;
                   CARRY      : out BIT) is
   alias XLARG : BIT_VECTOR(LARG'LENGTH-1 downto 0) is LARG;
   alias XRARG : BIT_VECTOR(RARG'LENGTH-1 downto 0) is RARG;
   variable XSUM : BIT_VECTOR(XLARG'RANGE);  variable XCARRY : BIT := '0';
  begin
   -- Only a scalar type may serve as a qualifier for IMAGE
   report "Entering procedure " & ADDER'SIMPLE_NAME;
   for I in XLARG'LOW to XLARG'HIGH loop
     report "LARG(" & INTEGER'IMAGE(I) & ") = " & BIT'IMAGE(LARG(I));
     report "RARG(" & INTEGER'IMAGE(I) & ") = " & BIT'IMAGE(RARG(I));
     XSUM(I) := XCARRY xor XLARG(I) xor XRARG(I);
     XCARRY := (XCARRY and XLARG(I)) or (XCARRY and XRARG(I)) or
               (XLARG(I) and XRARG(I));
   end loop;
   SUM := XSUM;   CARRY := XCARRY;
  end ADDER;
end INT_ARITH_PKG;
```

18-15. Solution.

a.
```
entity ROM_256x8 is
  port
   (ADDRESS : in  BIT_VECTOR(7 downto 0);
    CE_BAR  : in  BIT;
    NIB_MD  : in  BIT;
    DATA    : out BIT_VECTOR(7 downto 0));
end ROM_256x8;
```

```
architecture BINARY_FILE of ROM_256x8 is
begin
 process
  subtype BYTE is BIT_VECTOR(7 downto 0);
  type ROM_256x8_ARRAY_TYPE is array (0 to 255) of BYTE;
  variable ROM_256x8_ARRAY : ROM_256x8_ARRAY_TYPE;

  type ROM_256x8_FILE_TYPE is file of BYTE;
  file ROM_256x8_FILE : ROM_256x8_FILE_TYPE is in "ROM DATA A";

  variable XADDRESS : NATURAL;
 begin
  -- initialize ROM only at start of simulation
  if (NOW = 0 ns) then
    for I in 0 to 255 loop
      READ(ROM_256x8_FILE, ROM_256x8_ARRAY(I));
    end loop;
  end if;

  -- compute location address
  XADDRESS := 0;
  for I in ADDRESS'REVERSE_RANGE loop
   if (ADDRESS(I)='1') then XADDRESS := XADDRESS + 2**I;
  end loop;

  if (NIB_MD='0') then
    DATA <= ROM_256x8_ARRAY(XADDRESS);      -- read one byte
  else
    for I in 0 to 3 loop                    -- read nibble
     DATA <= ROM_256x8_ARRAY(XADDRESS+I);
     wait until (CE_BAR = '0');
    end loop;
  end if;

  wait until (CE_BAR = '0');
 end process;
end BINARY_FILE;
```

b. use STD.TEXTIO.all;

```
architecture TEXT_FILE of ROM_256x8 is
begin
 process
  subtype BYTE is BIT_VECTOR(7 downto 0);
  type ROM_256x8_ARRAY_TYPE is array (0 to 255) of BYTE;
  variable ROM_256x8_ARRAY : ROM_256x8_ARRAY_TYPE;

  file ROM_256x8_FILE : TEXT is in "ROM DATA A";
  variable BYTE_DATA : LINE;

  variable XADDRESS : NATURAL;
 begin
  -- initialize ROM only at start of simulation
  if (NOW = 0 ns) then
    for I in 0 to 255 loop
      READLINE(ROM_256x8_FILE, BYTE_DATA);
      READ(BYTE_DATA, ROM_256x8_ARRAY(I));
    end loop;
  end if;

  -- compute location address
  XADDRESS := 0;
  for I in ADDRESS'REVERSE_RANGE loop
   if (ADDRESS(I)='1') then XADDRESS := XADDRESS + 2**I;
  end loop;

  if (NIB_MD='0') then
    DATA <= ROM_256x8_ARRAY(XADDRESS);      -- read one byte
  else
    for I in 0 to 3 loop                    -- read nibble
     DATA <= ROM_256x8_ARRAY(XADDRESS+I);
     wait until (CE_BAR = '0');
    end loop;
  end if;

  wait until (CE_BAR = '0');
 end process;
end TEXT_FILE;
```

18-16. Solution.

a.
```
entity ROM_256x8 is
  port
   (ADDRESS : in  BIT_VECTOR(7 downto 0);
    CE_BAR  : in  BIT;
    NIB_MD  : in  BIT;
    DATA    : out BIT_VECTOR(7 downto 0));
end ROM_256x8;

use STD.TEXTIO.all;
architecture BINARY_FILE of ROM_256x8 is
begin
 process
  subtype BYTE is BIT_VECTOR(7 downto 0);
  type ROM_256x8_ARRAY_TYPE is array (0 to 255) of BYTE;
  variable ROM_256x8_ARRAY : ROM_256x8_ARRAY_TYPE;

  type ROM_256x8_FILE_TYPE is file of BYTE;
  file ROM_256x8_FILE : ROM_256x8_FILE_TYPE open READ_MODE is
       "ROM DATA A";

  variable XADDRESS : NATURAL;
 begin
  -- initialize ROM only at start of simulation
  if (NOW = 0 ns) then
    for I in 0 to 255 loop
      READ(ROM_256x8_FILE, ROM_256x8_ARRAY(I));
    end loop;
  end if;

  -- compute location address
  XADDRESS := 0;
  for I in ADDRESS'REVERSE_RANGE loop
   if (ADDRESS(I)='1') then XADDRESS := XADDRESS + 2**I;
  end loop;
```

```
    if (NIB_MD='0') then
      DATA <= ROM_256x8_ARRAY(XADDRESS);      -- read one byte
    else
      for I in 0 to 3 loop                              -- read nibble
       DATA <= ROM_256x8_ARRAY(XADDRESS+I);
       wait until (CE_BAR = '0');
      end loop;
    end if;

    wait until (CE_BAR = '0');
   end process;
  end BINARY_FILE;
```

b.
```
  use STD.TEXTIO.all;
  architecture TEXT_FILE of ROM_256x8 is
  begin
   process
    subtype BYTE is BIT_VECTOR(7 downto 0);
    type ROM_256x8_ARRAY_TYPE is array (0 to 255) of BYTE;
    variable ROM_256x8_ARRAY : ROM_256x8_ARRAY_TYPE;

    file ROM_256x8_FILE : TEXT open READ_MODE is "ROM DATA A";
    variable BYTE_DATA : LINE;

    variable XADDRESS : NATURAL;
   begin
    -- initialize ROM only at start of simulation
    if (NOW = 0 ns) then
      for I in 0 to 255 loop
        READLINE(ROM_256x8_FILE, BYTE_DATA);
        READ(BYTE_DATA, ROM_256x8_ARRAY(I));
      end loop;
    end if;

    -- compute location address
    XADDRESS := 0;
    for I in ADDRESS'REVERSE_RANGE loop
     if (ADDRESS(I)='1') then XADDRESS := XADDRESS + 2**I;
    end loop;
```

```
    if (NIB_MD='0') then
      DATA <= ROM_256x8_ARRAY(XADDRESS);        -- read one byte
    else
      for I in 0 to 3 loop                       -- read nibble
       DATA <= ROM_256x8_ARRAY(XADDRESS+I);
       wait until (CE_BAR = '0');
      end loop;
    end if;

    wait until (CE_BAR = '0');
  end process;
end TEXT_FILE;
```

18-17. Solution.

```
entity TEST_BENCH is
end TEST_BENCH;

use STD.TEXTIO.all;
architecture TEST_BENCH of TEST_BENCH is
 component ROM_256x8
  port
   (ADDRESS : in  BIT_VECTOR(7 downto 0);
    CE_BAR  : in  BIT;
    NIB_MD  : in  BIT;
    DATA    : out BIT_VECTOR(7 downto 0));
 end component;

 signal CLK, NIB_MD : BIT;  signal ADDRESS, DATA : BIT_VECTOR(7 downto 0);
begin
 CLK_GEN:
 process (CLK)
 begin
  CLK <= not CLK after 50 ns;  -- 100 ns clock cycle
 end process CLK_GEN;

 SIG_GEN:
 process
 begin
  NIB_MD <= '0';
  -- four byte reads
  wait until CLK='1';    ADDRESS <= X"00";    -- 0
  wait until CLK='1';    ADDRESS <= X"69";    -- 105
  wait until CLK='1';    ADDRESS <= X"C8";    -- 200

  NIB_MD <= '1';
  -- four nibble reads
  wait until CLK='1';    ADDRESS <= X"04";    -- 0
  wait until CLK='1';    wait until CLK='1';    wait until CLK='1';
  wait until CLK='1';    ADDRESS <= X"40";    -- 64
  wait until CLK='1';    wait until CLK='1';    wait until CLK='1';
  wait until CLK='1';    ADDRESS <= X"FC";    -- 252

  wait;
 end process SIG_GEN;
```

```
SIG_REC:
process
 subtype BYTE is BIT_VECTOR(7 downto 0);
 file TEST_FILE : TEXT is out "ROM TEST A";
 variable BYTE_DATA : LINE;
begin
 -- record input
 wait until CLK='0';
 WRITE(BYTE_DATA, NOW);       WRITE(BYTE_DATA, ADDRESS);

 -- record output
 wait on DATA'TRANSACTION;
 WRITE(BYTE_DATA, DATA);      WRITELINE(TEST_FILE, BYTE_DATA);
end process SIG_REC;

-- instantiate unit under test
UUT: ROM_256x8 port map(ADDRESS=>ADDRESS, CE_BAR=>CLK,
                        NIB_MD=>NIB_MD,   DATA=>DATA);
end TEST_BENCH;
```

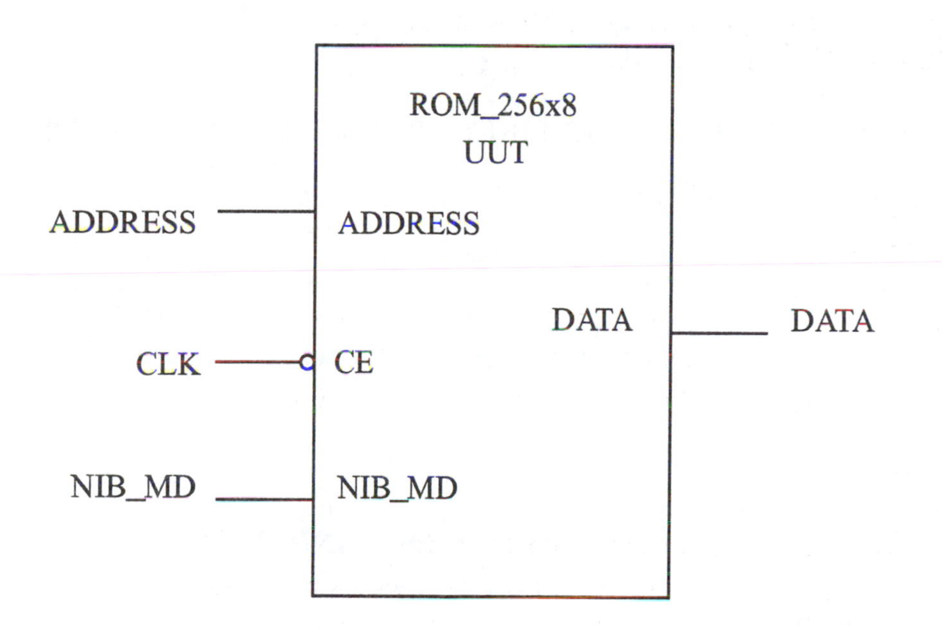

18-18. Solution.

```
entity TEST_BENCH is
end TEST_BENCH;
use STD.TEXTIO.all;
architecture TEST_BENCH of TEST_BENCH is
 component MEALY_MACHINE
  port (DATA, CLK : in  BIT;  Z        : out BIT);
 end component;

 signal CLK, DATA, Z : BIT;
begin
 CLK_GEN:
 process (CLK)
 begin
  CLK <= not CLK after 50 ns;  -- 100 ns clock cycle
 end process CLK_GEN;

 SIG_GEN:
 process
  file TEST_FILE     : TEXT is in "MEALY DATA A";
  variable TEST_LINE : LINE;
 begin
  READLINE(TEST_FILE, TEST_LINE);    READ(TEST_LINE, DATA);
  wait until CLK='0';
 end process SIG_GEN;

 SIG_REC:
 process
  file TEST_FILE     : TEXT is out "MEALY TEST A";
  variable TEST_LINE : LINE;
 begin
  -- record input
  wait until CLK='1';
  WRITE(TEST_LINE, NOW);    WRITE(TEST_LINE, DATA);

  -- record output
  wait on Z'TRANSACTION;
  WRITE(TEST_LINE, Z);      WRITELINE(TEST_FILE, TEST_LINE);
 end process SIG_REC;
 -- instantiate unit under test
 UUT: MEALY_MACHINE port map(DATA=>DATA, CLK=>CLK, Z=>Z);
end TEST_BENCH;
```

Chapter 19. VHDL: A Last Look

19-1. Solution.

a.

Present State $Q_A\ Q_B\ Q_C\ Q_D$	Next State $Q_A^+\ Q_B^+\ Q_C^+\ Q_D^+$	Output
0000	1001	1001
0001	0000	0000
0010	0001	0001
0011	0010	0010
0100	0011	0011
0101	0100	0100
0110	0101	0101
0111	0110	0110
1000	0111	0111
1001	1000	1000
1010	----	----
1011	----	----
1100	----	----
1101	----	----
1110	----	----
1111	----	----

$Q_A^+ = D_A$

Q_AQ_B \ Q_CQ_D	00	01	11	10
00	1	0	0	0
01	0	0	0	0
11	-	-	-	-
10	0	1	-	-

$Q_B^+ = D_B$

Q_AQ_B \ Q_CQ_D	00	01	11	10
00	0	0	0	0
01	0	1	1	1
11	-	-	-	-
10	1	0	-	-

$Q_C^+ = D_C$

Q_AQ_B \ Q_CQ_D	00	01	11	10
00	0	0	1	0
01	1	0	1	0
11	-	-	-	-
10	1	0	-	-

$Q_D^+ = D_D$

Q_AQ_B \ Q_CQ_D	00	01	11	10
00	1	0	0	1
01	1	0	0	1
11	-	-	-	-
10	1	0	-	-

$$D_A = (\overline{Q}_A \bullet \overline{Q}_B \bullet \overline{Q}_C \bullet \overline{Q}_D) + (Q_A \bullet Q_D)$$

$$D_B = (Q_A \bullet \overline{Q}_D) + (Q_B \bullet Q_C) + (Q_B \bullet Q_D)$$

$$D_C = (Q_C \bullet Q_D) + (Q_A \bullet \overline{Q}_D) + (Q_B \bullet \overline{Q}_C \bullet \overline{Q}_D)$$

$$D_D = \overline{Q}_D$$

b.
```
entity BCD_DWN is
   port
    (CLK : in BIT;
     CNT : out BIT_VECTOR(3 downto 0));
 end BCD_DWN;
 architecture MIXED of BCD_DWN is
  component DFF4
    port
     (D            : in BIT_VECTOR(3 downto 0);
      CLK          : in BIT;
      CLR_BAR : in BIT;
      Q            : out BIT_VECTOR(3 downto 0);
      Q_BAR      : out BIT_VECTOR(3 downto 0));
   end component;

   signal D, Q, Q_BAR : BIT_VECTOR(3 downto 0);
   signal TIE_ONE : BIT := '1';
 begin
  D(3) <= (Q_BAR(3) and Q_BAR(2) and Q_BAR(1) and Q_BAR(0)) or (Q(3) and Q(0));
  D(2) <= (Q(3) and Q_BAR(0)) or (Q(2) and Q(1)) or (Q(2) and Q(0));
  D(1) <= (Q(1) and Q(0)) or (Q(3) and Q_BAR(0)) or
              (Q(2) and Q_BAR(1) and Q_BAR(0));
  D(0) <= Q_BAR(0);


  DFFS : DFF4 port map(D=>D; CLK=>CLK; CLR_BAR=>TIE_ONE; Q=>Q;
                                       Q_BAR=>Q_BAR);
  CNT <= Q;
 end MIXED;
```

c.
```
entity DFF4
  port
   (D       : in  BIT_VECTOR(3 downto 0);
    CLK     : in  BIT;
    CLR_BAR : in  BIT;
    Q       : out BIT_VECTOR(3 downto 0);
    Q_BAR   : out BIT_VECTOR(3 downto 0));
end DFF4;

architecture BEHAVIOR of DFF4 is
begin
 process(CLK, CLR_BAR)
 begin
  if (CLR_BAR'EVENT and CLR_BAR='0') then
      Q <= B"0000";   Q_BAR <= B"1111";
  elsif (CLK'EVENT and CLK='1') then
      Q <= D;         Q_BAR <= not D;
  end if;
 end process;
end BEHAVIOR;
```

d.

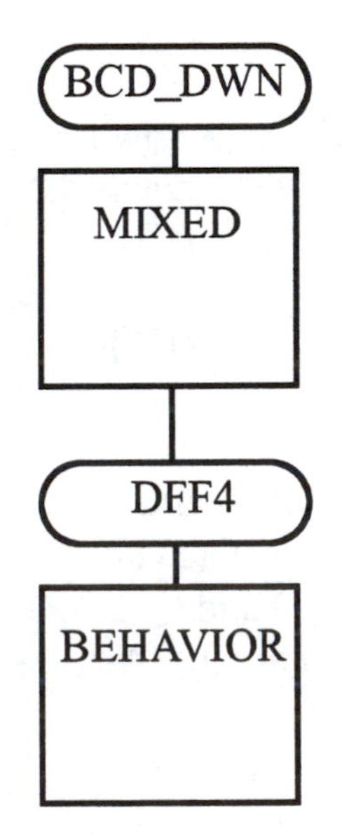

19-2. Solution.

a.

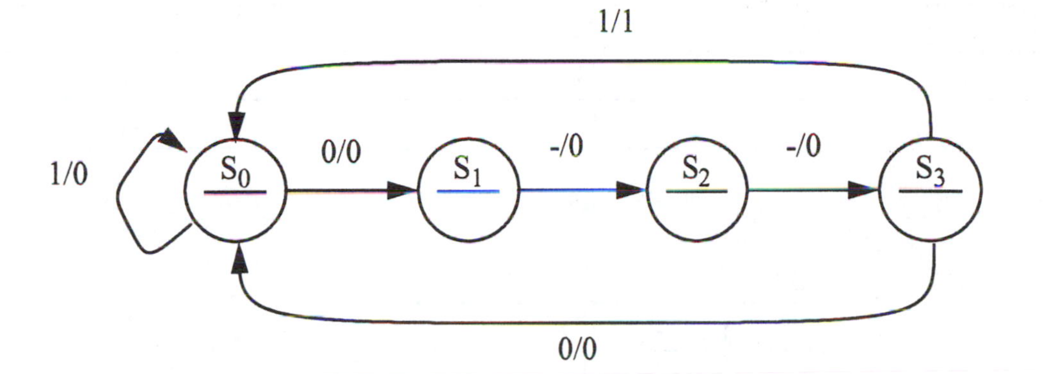

Present State	Next State		Output (Z)	
	X=0	X=1	X=0	X=1
S_0	S_1	S_0	0	0
S_1	S_2	S_2	0	0
S_2	S_3	S_3	0	0
S_3	S_0	S_0	0	1

Armstrong-Humphrey Rule #1: S_0,S_3

Armstrong-Humphrey Rule #2: S_0,S_1

Armstrong-Humphrey Rule #3: S_0,S_1,S_2,S_3 S_0,S_1,S_2

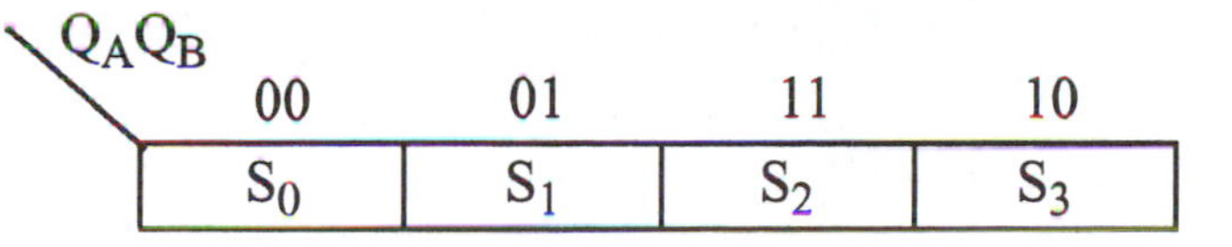

$Q_A Q_B$	$Q_A^+ Q_B^+$		Output (Z)	
	X=0	**X=1**	**X=0**	**X=1**
00	01	00	0	0
01	11	11	0	0
11	10	10	0	0
10	00	00	0	1

$Q_A^+ = D_A$

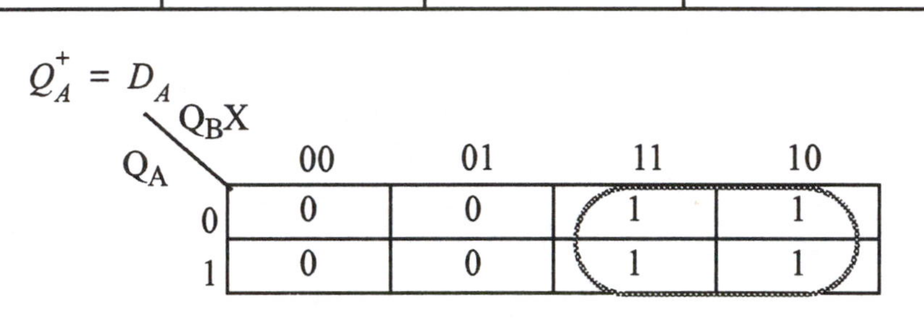

$Q_B^+ = D_B$

Q_A \ $Q_B X$	00	01	11	10
0	1	0	1	1
1	0	0	0	0

Z

Q_A \ $Q_B X$	00	01	11	10
0	0	0	0	0
1	0	1	0	0

$$D_A = Q_B$$

$$D_B = (\overline{Q}_A \bullet \overline{X}) + (\overline{Q}_A \bullet Q_B)$$

$$Z = (Q_A \bullet \overline{Q}_B \bullet X)$$

b.
```
entity BCD_PRIME is
 port
  (CLK : in BIT;
   X   : in BIT;
   Z   : out BIT);
end BCD_PRIME;
architecture MIXED of BCD_PRIME is
  component DFF4
   port
    (D3, D2, D1, D0 : in BIT;
     CLK : in BIT;
     CLR_BAR : in BIT;
     Q3, Q2, Q1, Q0 : out BIT;
     Q_BAR3, Q_BAR2, Q_BAR1, Q_BAR0 : out BIT);
  end component;

  signal D, Q, Q_BAR : BIT_VECTOR(3 downto 2);
  signal TIE_ONE : BIT := '1';
begin
 D(3) <= Q(2);
 D(2) <= (Q_BAR(3) and not X) or (Q_BAR(3) and Q(2));

 DFFS : DFF4 port map(D3=>D(3); D2=>D(2); D1=>open; D0=>open;
                CLK=>CLK; CLR_BAR=>TIE_ONE;
                Q3=>Q(3); Q2=>Q(2); Q1=>open; Q0=>open;
                Q_BAR3=>Q_BAR(3); Q_BAR2=>Q_BAR(2); Q_BAR1=>open;
                Q_BAR0=>open);

 Z <= Q(3) and Q_BAR(2) and X;
end MIXED;
```

c.

```
entity DFF4
  port
   (D3, D2, D1, D0 : in  BIT := '0';
    CLK     : in  BIT;
    CLR_BAR : in  BIT;
    Q3, Q2, Q1, Q0 : out BIT;
    Q_BAR3, Q_BAR2, Q_BAR1, Q_BAR0 : out BIT);
end DFF4;

architecture BEHAVIOR of DFF4 is
begin
 process(CLK, CLR_BAR)
 begin
  if (CLR_BAR'EVENT and CLR_BAR='0') then
   (Q3,Q2,Q1,Q0) <= B"0000";
   (Q_BAR3,Q_BAR2,Q_BAR1,Q_BAR0) <= B"1111";

  elsif (CLK'EVENT and CLK='1') then
   (Q3,Q2,Q1,Q0) <= (D3,D2,D1,D0);
   (Q_BAR3,Q_BAR2,Q_BAR1,Q_BAR0) <= (not D3, not D2, not D1, not D0);
  end if;
 end process;
end BEHAVIOR;
```

d.

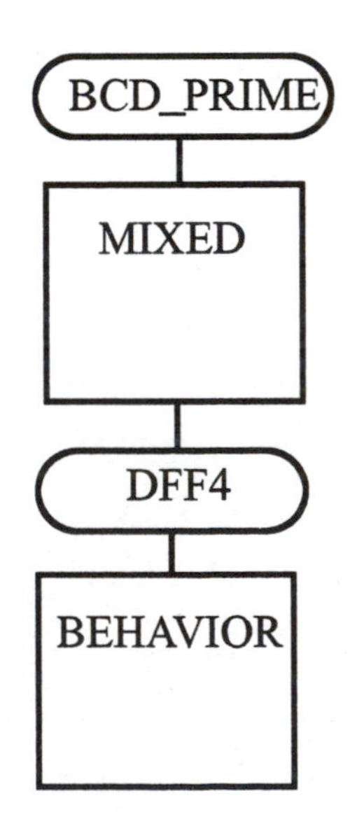

19-3. Solution.

a.

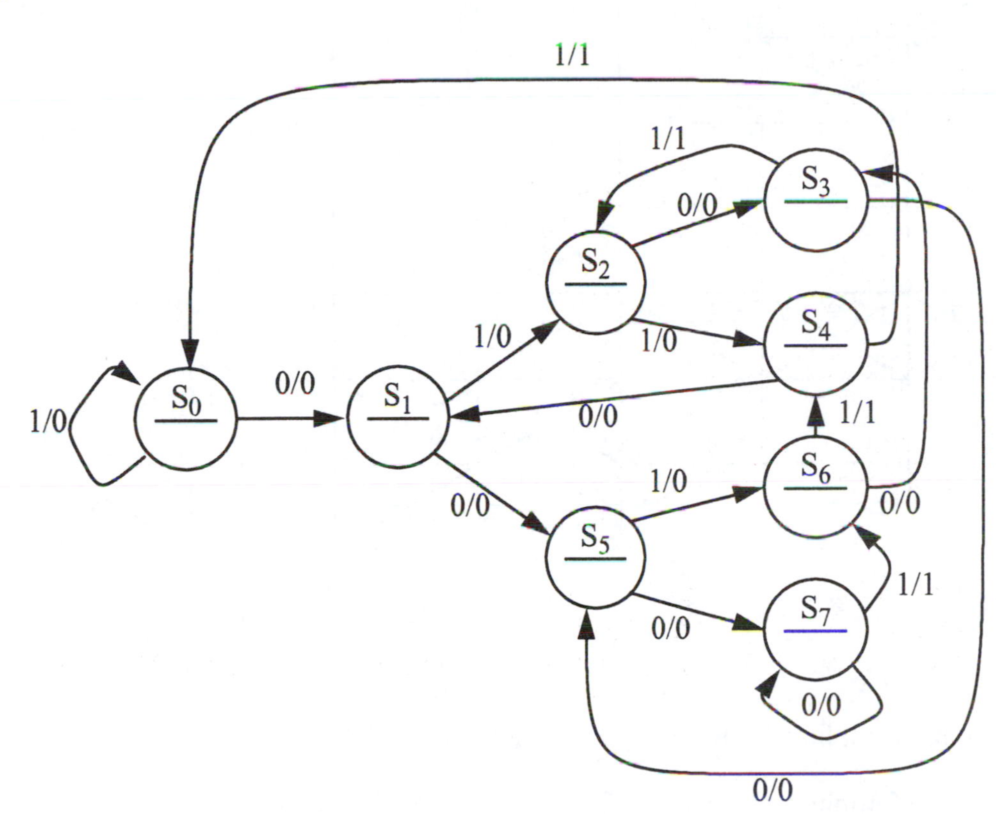

Present State	Next State		Output (Z)	
	X=0	X=1	X=0	X=1
S_0	S_1	S_0	0	0
S_1	S_5	S_2	0	0
S_2	S_3	S_4	0	0
S_3	S_5	S_2	0	1
S_4	S_1	S_0	0	1
S_5	S_7	S_6	0	0
S_6	S_3	S_4	0	1
S_7	S_7	S_6	0	1

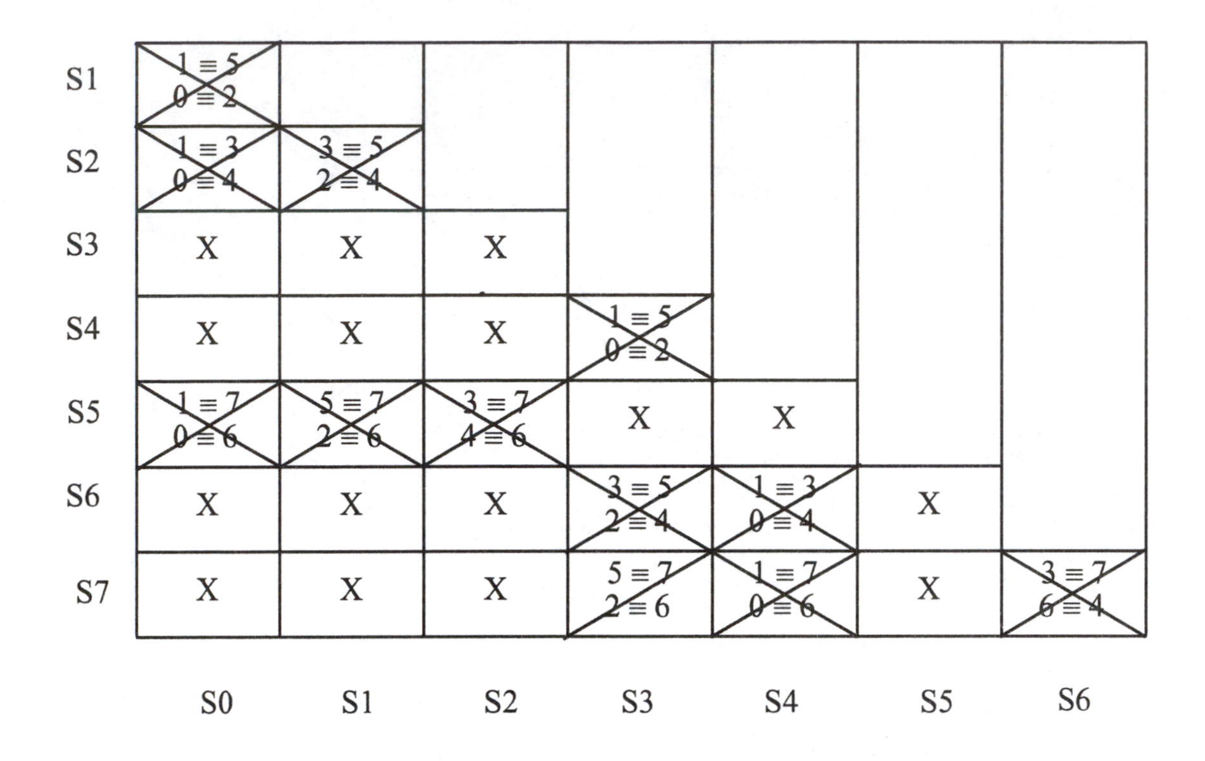

Armstrong-Humphrey Rule #1: S_1,S_3 S_0,S_4 S_5,S_7 S_2,S_6

Armstrong-Humphrey Rule #2: S_0,S_1 S_2,S_5 S_3,S_4 S_5,S_7 S_3,S_7

Armstrong-Humphrey Rule #3: $S_0,S_1,S_2,S_3,S_4,S_5,S_6,S_7$ S_0,S_1,S_2,S_5 S_3,S_4,S_6,S_7

$Q_A \backslash Q_BQ_C$	00	01	11	10
0	S_0	S_1	S_5	S_2
1	S_4	S_3	S_7	S_6

$Q_A\, Q_B\, Q_C$	$Q_A^+\, Q_B^+\, Q_C^+$		Output (Z)	
	X=0	**X=1**	**X=0**	**X=1**
000	001	000	0	0
001	011	010	0	0
010	101	100	0	0
101	011	010	0	1
100	001	000	0	1
011	111	110	0	0
110	101	100	0	1
111	111	110	0	1

$Q_A^+ = D_A$

Q_AQ_B \ Q_CX	00	01	11	10
00	0	0	0	0
01	1	1	1	1
11	1	1	1	1
10	0	0	0	0

$Q_B^+ = D_B$

Q_AQ_B \ Q_CX	00	01	11	10
00	0	0	1	1
01	0	0	1	1
11	0	0	1	1
10	0	0	1	1

$Q_C^+ = D_C$

Q_AQ_B \ Q_CX	00	01	11	10
00	1	0	0	1
01	1	0	0	1
11	1	0	0	1
10	1	0	0	1

Z

Q_AQ_B \ Q_CX	00	01	11	10
00	0	0	0	0
01	0	0	0	0
11	0	1	1	0
10	0	1	1	0

$$D_A = Q_B$$

$$D_B = Q_C$$

$$D_C = \overline{X}$$

$$Z = Q_A \bullet X$$

b. entity BCD_PRIME is

```
port
 (CLK : in BIT;
  X   : in BIT;
  Z   : out BIT);
end BCD_PRIME;
architecture MIXED of BCD_PRIME is
 component DFF4
  port
   (D3, D2, D1, D0 : in BIT;
    CLK : in BIT;
    CLR_BAR : in BIT;
    Q3, Q2, Q1, Q0 : out BIT;
    Q_BAR3, Q_BAR2, Q_BAR1, Q_BAR0 : out BIT);
 end component;

 signal D, Q, Q_BAR : BIT_VECTOR(3 downto 1);
 signal TIE_ONE : BIT := '1';
begin
 A: D(3) <= Q(2);

 B: D(2) <= Q(1);

 C: D(1) <= not X;

 DFFS : DFF4 port map(D3=>D(3); D2=>D(2); D1=>D(1); D0=>open;
                 CLK=>CLK; CLR_BAR=>TIE_ONE;
                 Q3=>Q(3); Q2=>Q(2); Q1=>Q(1); Q0=>open;
                 Q_BAR3=>Q_BAR(3); Q_BAR2=>Q_BAR(2); Q_BAR1=>Q_BAR(1);
                 Q_BAR0=>open);

 Z <= Q(3) and X;
end MIXED;
```

c. entity DFF4

```
  port
   (D3, D2, D1, D0 : in  BIT := '0';
    CLK     : in  BIT;
    CLR_BAR : in  BIT;
    Q3, Q2, Q1, Q0 : out BIT;
    Q_BAR3, Q_BAR2, Q_BAR1, Q_BAR0 : out BIT);
end DFF4;

architecture BEHAVIOR of DFF4 is
begin
 process(CLK, CLR_BAR)
 begin
  if (CLR_BAR'EVENT and CLR_BAR='0') then
   (Q3,Q2,Q1,Q0) <= B"0000";
   (Q_BAR3,Q_BAR2,Q_BAR1,Q_BAR0) <= B"1111";

  elsif (CLK'EVENT and CLK='1') then
   (Q3,Q2,Q1,Q0) <= (D3,D2,D1,D0);
   (Q_BAR3,Q_BAR2,Q_BAR1,Q_BAR0) <= (not D3, not D2, not D1, not D0);
  end if;
 end process;
end BEHAVIOR;
```

19-4. Solution.

```
package PLA_UTIL is
 type ARRAY_9X8 is array (1 to 9, 1 to 8) of BIT;
 type ARRAY_9X4 is array (1 to 9, 1 to 4) of BIT;
 constant AND_INIT : ARRAY_9X8 := (others=>(others=>'0'));
 constant OR_INT   : ARRAY_9X4 := (others=>(others=>'0'));
end PLA_UTIL;

use WORK.PLA_UTIL.all;
entity PLA is
 generic
  (APLN : ARRAY_9X8 := AND_INT; -- '1' means connection is programmed
   OPLN : ARRAY_9X4 := OR_INT);  -- '0' means connection is not programmed
 port
  (A, B, C, D : in BIT;
   W, X, Y, Z : out BIT);
end PLA;

architecture BEHAVIOR of PLA is
  signal A_BAR, B_BAR, C_BAR, D_BAR : BIT;
  signal AND_PLN : BIT_VECTOR(1 to 9);
  signal OR_PLN  : BIT_VECTOR(1 to 4);
begin
 A_BAR <= not A;   B_BAR <= not B;   C_BAR <= not C;   D_BAR <= not D;

 AND_PLANE: for I in 1 to 9 generate
   AND_PLN(I) <= (A or not APLN(I,1)) and (A_BAR or not APLN(I,2)) and
               (B or not APLN(I,3)) and (B_BAR or not APLN(I,4)) and
               (C or not APLN(I,5)) and (C_BAR or not APLN(I,6)) and
               (D or not APLN(I,7)) and (D_BAR or not APLN(I,8));
 end generate AND_PLANE;

 OR_PLANE: for I in 1 to 4 generate
   OR_PLN(I) <= (AND_PLN(1) and OPLN(1,I)) or (AND_PLN(2) and OPLN(2, I)) or
              (AND_PLN(3) and OPLN(3,I)) or (AND_PLN(4) and OPLN(4, I)) or
              (AND_PLN(5) and OPLN(5,I)) or (AND_PLN(6) and OPLN(8, I)) or
              (AND_PLN(7) and OPLN(7,I)) or (AND_PLN(8) and OPLN(8, I)) or
              (AND_PLN(9) and OPLN(9,I));
 end generate OR_PLANE;

 W <= OR_PLN(1);   X <= OR_PLN(2);   Y <= OR_PLN(3);   Z <= OR_PLN(4);
end BEHAVIOR;
```

19-5. Solution.

```
package PLA_UTIL is
 type ARRAY_9X8 is array (1 to 9, 1 to 8) of BIT;
 type ARRAT_9X4 is array (1 to 9, 1 to 4) of BIT;

 constant AND_INIT : ARRAY_9X8 := (others=>(others=>'0'));
 constant OR_INT   : ARRAY_9X4 := (others=>(others=>'0'));
end PLA_UTIL;

use WORK.PLA_UTIL.all;
entity BCD_DWN is
 port
  (CLK : in BIT;
   CNT : out BIT_VECTOR(3 downto 0));
end BCD_DWN;
```

```
architecture MIXED of BCD_DWN is
  component PLA
  generic
   (APLN : ARRAY_9X8; -- '1' means connection is programmed
    OPLN : ARRAY_9x4); -- '0' means connection is not programmed
  port
   (A, B, C, D : in BIT;
    W, X, Y, Z : out BIT);
  end component;
  signal Q, D : BIT_VECTOR(3 downto 0);
begin
 NS_LOGIC: PLA
        generic map (APLN=>(('0', '1', '0', '1', '0', '1', '0', '1'),
                            ('1', '0', '0', '0', '0', '0', '1', '0'),
                            ('1', '0', '0', '0', '0', '0', '0', '1'),
                            ('0', '0', '1', '0', '1', '0', '0', '0'),
                            ('0', '0', '1', '0', '0', '0', '1', '0'),
                            ('0', '0', '0', '0', '1', '0', '1', '0'),
                            ('1', '0', '0', '0', '0', '0', '0', '1'),
                            ('0', '0', '1', '0', '0', '1', '0', '1'),
                            ('0', '0', '0', '0', '0', '0', '0', '1')),
                     OPLN=>(('1', '0', '0', '0'),
                            ('1', '0', '0', '0'),
                            ('0', '1', '0', '0'),
                            ('0', '1', '0', '0'),
                            ('0', '1', '0', '0'),
                            ('0', '0', '1', '0'),
                            ('0', '0', '1', '0'),
                            ('0', '0', '1', '0'),
                            ('0', '0', '0', '1')))
        port map(A=>Q(3), B=>Q(2), C=>Q(1), D=>Q(0),
                 W=>D(3), X=>D(2), Y=>D(1), Z=>D(0));

DFF: block (CLK='1' and not CLK'STABLE)
begin
  Q <= guarded D;
end DFF;

CNT <= Q;
end MIXED;
```

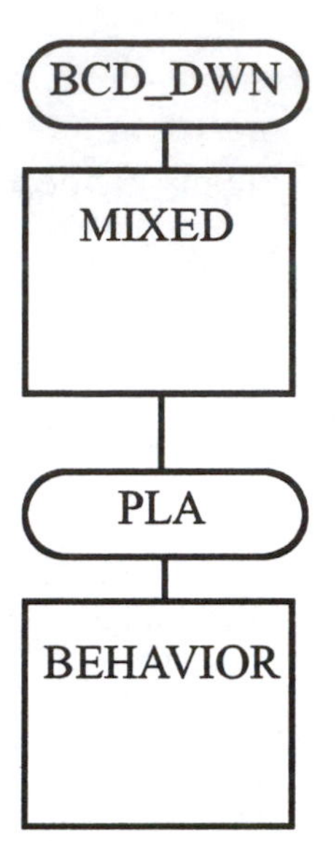

19-6. Solution.

```
package PLA_UTIL is
 type ARRAY_9X8 is array (1 to 9, 1 to 8) of BIT;
 type ARRAT_9X4 is array (1 to 9, 1 to 4) of BIT;

 constant AND_INIT : ARRAY_9X8 := (others=>(others=>'0'));
 constant OR_INT   : ARRAY_9X4 := (others=>(others=>'0'));
end PLA_UTIL;

use WORK.PLA_UTIL.all;
entity BCD_PRIME is
 port
  (CLK : in BIT;
   X   : in BIT;
   Z   : out BIT);
end BCD_PRIME;
```

```
architecture MIXED of BCD_PRIME is
  component PLA
  generic
   (APLN : ARRAY_9X8;  -- '1' means connection is programmed
    OPLN : ARRAY_9X4);  -- '0' means connection is not programmed
  port
   (A, B, C, D : in BIT;
    W, X, Y, Z : out BIT);
  end component;

  signal D, Q : BIT_VECTOR(3 downto 1);
begin

 NS_LOGIC: PLA
        generic map (APLN=>(('0', '0', '1', '0', '0', '0', '0', '0'),
                            ('0', '0', '0', '0', '1', '0', '0', '0'),
                            ('0', '0', '0', '0', '0', '0', '0', '1'),
                            ('1', '0', '0', '0', '0', '0', '1', '0'),
                            ('0', '0', '0', '0', '0', '0', '0', '0'),
                            ('0', '0', '0', '0', '0', '0', '0', '0'),
                            ('0', '0', '0', '0', '0', '0', '0', '0'),
                            ('0', '0', '0', '0', '0', '0', '0', '0'),
                            ('0', '0', '0', '0', '0', '0', '0', '0')),
                     OPLN=>(('1', '0', '0', '0'),
                            ('0', '1', '0', '0'),
                            ('0', '0', '1', '0'),
                            ('0', '0', '0', '1'),
                            ('0', '0', '0', '0'),
                            ('0', '0', '0', '0'),
                            ('0', '0', '0', '0'),
                            ('0', '0', '0', '0'),
                            ('0', '0', '0', '0')))
        port map(A=>Q(3), B=>Q(2), C=>Q(1), D=>X,
                 W=>D(3), X=>D(2), Y=>D(1), Z=>Z);

 DFF: block (CLK='1' and not CLK'STABLE)
 begin
  Q <= guarded D;
 end DFF;
end MIXED;
```

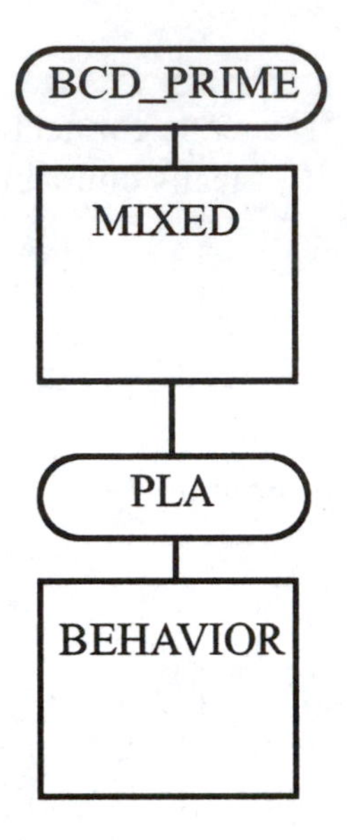
BCD_PRIME
MIXED
PLA
BEHAVIOR

19-7. Solution.

```
entity MACROCELL is
 generic
  (F1, F2 : BIT);
 port
  (OR_OUT, IO_CNTRL, CLK : in BIT;
   OR_IN : out BIT;
   IN_OUT : inout BIT);
end MACROCELL;
architecture BEHAVIOR of MACROCELL is
  signal Q, MUX : BIT;
begin
  DFF: block (CLK='1' and not CLK'STABLE)
  begin
   Q <= guarded OR_OUT;
  end DFF;

  with F1 select
   OR_IN <= not Q      when '0'
                  IN_OUT when '1';

  with (F1&F2) select
   MUX <= OR_OUT        when B"00",
               not OR_OUT when B"01",
               Q             when B"10",
               not Q         when B"11";

  IN_OUT <= MUX when (IO_CNTRL='1') else
                  IN_OUT;
end BEHAVIOR;

package PLA_UTIL is
 type ARRAY_13X16 is array (1 to 13, 1 to 16) of BIT;
 type ARRAY_9X4  is array (1 to 9, 1 to 4) of BIT;
 type ARRAY_4X2   is array (1 to 4, 1 to 2) of BIT;

 constant AND_INIT : ARRAY_13X16 := (others=>(others=>'0'));
 constant OR_INT     : ARRAY_9X4  := (others=>(others=>'0'));
 constant MC_INT     : ARRAY_4X2   := (others=>(others=>'0'));
end PLA_UTIL;
```

```
use WORK.PLA_UTIL.all;
entity SEQ_PLA is
 generic
  (APLN   : ARRAY_13X16 := AND_INT; -- '1' means connection is programmed
   OPLN   : ARRAY_9X4  := OR_INT -- '0' means connection is not programmed
   MCELL  : ARRAY_4X2   := MC_INT);
 port
  (A, B, C, D : in BIT := '0';
   CLK        : in BIT;
   W, X, Y, Z : inout BIT := '0');
end SEQ_PLA;

architecture BEHAVIOR of SEQ_PLA is
 component MACROCELL
  generic
   (F1, F2 : BIT);
  port
   (OR_OUT, IO_CNTRL, CLK : in BIT;
    OR_IN : out BIT;
    IN_OUT : inout BIT);
 end component;

  signal A_BAR, B_BAR, C_BAR, D_BAR : BIT;
  signal AND_PLN : BIT_VECTOR(1 to 13);
  signal OR_PLN  : BIT_VECTOR(1 to 4);
  signal IW, IW_BAR, IX, IX_BAR, IY, IY_BAR, IZ, IZ_BAR : BIT;
begin
 A_BAR <= not A;   B_BAR <= not B;   C_BAR <= not C;   D_BAR <= not D;
 IW <= not IW_BAR; IX <= not IX_BAR; IY <= not IY_BAR; IZ <= not IZ_BAR;

 AND_PLANE: for I in 1 to 13 generate
   AND_PLN(I) <= (A or not APLN(I,1))  and (A_BAR or not APLN(I,2)) and
               (B or not APLN(I,3))  and (B_BAR or not APLN(I,4)) and
               (C or not APLN(I,5))  and (C_BAR or not APLN(I,6)) and
               (D or not APLN(I,7))  and (D_BAR or not APLN(I,8)) and
               (IW or not APLN(I,9)  and (IW_BAR or not APLN(I,10)) and
               (IX or not APLN(I,11) and (IX_BAR or not APLN(I,12)) and
               (IY or not APLN(I,13) and (IY_BAR or not APLN(I,14)) and
               (IZ or not APLN(I,15) and (IZ_BAR or not APLN(I,16));
 end AND_PLANE;
```

```
OR_PLANE: for I in 1 to 4 generate
  OR_PLN(I) <= (AND_PLN(1) and OPLN(1,I)) or (AND_PLN(2) and OPLN(2, I)) or
          (AND_PLN(3) and OPLN(3,I)) or (AND_PLN(4) and OPLN(4, I)) or
          (AND_PLN(5) and OPLN(5,I)) or (AND_PLN(6) and OPLN(6, I)) or
          (AND_PLN(7) and OPLN(7,I)) or (AND_PLN(8) and OPLN(8, I)) or
          (AND_PLN(9) and OPLN(9,I));
end OR_PLANE;

M1 : MACROCELL generic map (F1=>MCELL(1, 1), F2=>MCELL(1, 2))
          port map(OR_OUT=>OR_PLN(1), IO_CNTRL=>AND_PLN(10),
                    CLK=>CLK, OR_IN=>IW_BAR, IN_OUT=>W);

M2 : MACROCELL generic map (F1=>MCELL(2, 1), F2=>MCELL(2, 2))
          port map(OR_OUT=>OR_PLN(2), IO_CNTRL=>AND_PLN(11),
                    CLK=>CLK, OR_IN=>IX_BAR, IN_OUT=>X);

M3 : MACROCELL generic map (F1=>MCELL(3, 1), F2=>MCELL(3, 2))
          port map(OR_OUT=>OR_PLN(3), IO_CNTRL=>AND_PLN(12),
                    CLK=>CLK, OR_IN=>IY_BAR, IN_OUT=>Y);

M4 : MACROCELL generic map (F1=>MCELL(4, 1), F2=>MCELL(4, 2))
          port map(OR_OUT=>OR_PLN(4), IO_CNTRL=>AND_PLN(13),
                    CLK=>CLK, OR_IN=>IZ_BAR, IN_OUT=>Z);
end BEHAVIOR;
```

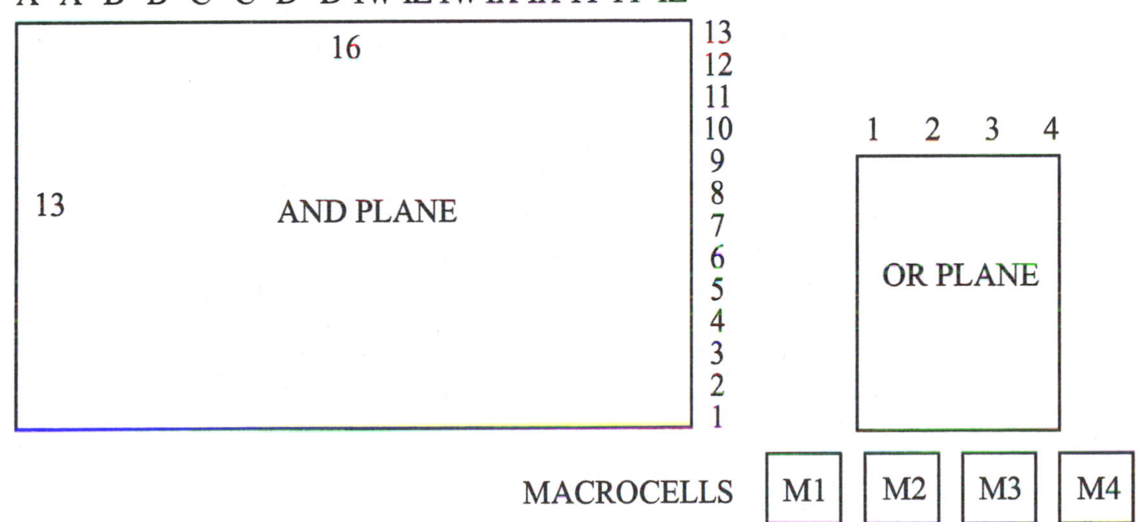

19-8. Solution.

a.

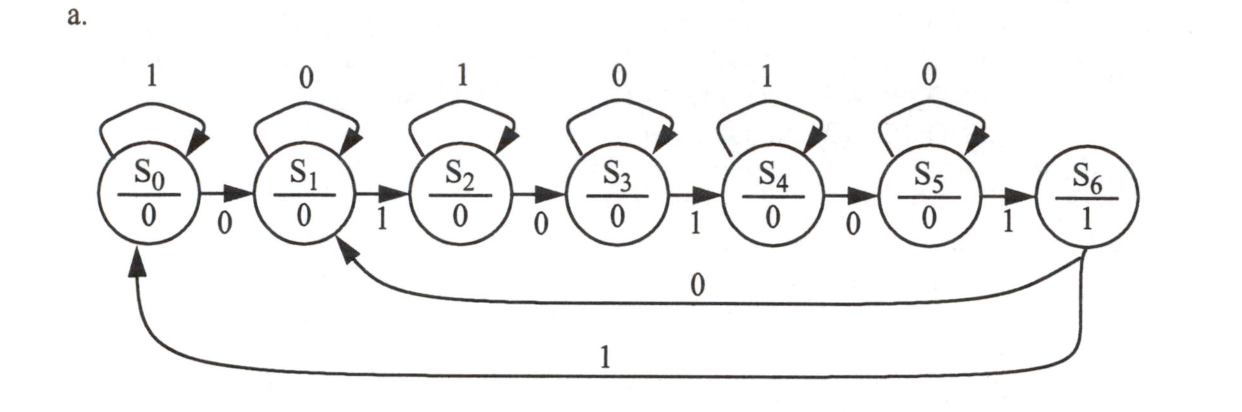

Present State	Next State		Output (Z)
	X=0	X=1	
S_0	S_1	S_0	0
S_1	S_1	S_2	0
S_2	S_3	S_2	0
S_3	S_3	S_4	0
S_4	S_5	S_4	0
S_5	S_5	S_6	0
S_6	S_1	S_0	1

Armstrong-Humphrey Rule #1: S_0,S_1,S_6 S_2,S_3 S_4,S_5 S_1,S_2 S_3,S_4 S_0,S_6

Armstrong-Humphrey Rule #2: S_0,S_1 S_1,S_2 S_2,S_3 S_3,S_4 S_4,S_5 S_5,S_6

Armstrong-Humphrey Rule #3: S_0,S_1,S_2,S_3,S_4,S_5

Q_A \ Q_BQ_C	00	01	11	10
0	S_0	S_1	S_2	S_3
1	S_5	S_6		S_4

$Q_A\, Q_B\, Q_C$	$Q_A^+\, Q_B^+\, Q_C^+$		Output (Z)
	X=0	**X=1**	
000	001	000	0
001	001	011	0
011	010	011	0
010	010	110	0
110	100	110	0
100	100	101	0
101	001	000	1

$Q_A^+ = D_A$

Q_AQ_B \ Q_CX	00	01	11	10
00	0	0	0	0
01	0	1	0	0
11	1	1	-	-
10	1	1	0	0

$Q_B^+ = D_B$

Q_AQ_B \ Q_CX	00	01	11	10
00	0	0	1	0
01	1	1	1	1
11	0	1	-	-
10	0	0	0	0

$Q_C^+ = D_C$

Q_AQ_B \ Q_CX	00	01	11	10
00	1	0	1	1
01	0	0	1	0
11	0	0	-	-
10	0	1	0	1

Z

Q_A \ Q_BQ_C	00	01	11	10
0	0	0	0	0
1	0	1	-	0

$$D_A = (Q_A \bullet \overline{Q}_C) + (Q_B \bullet \overline{Q}_C \bullet X)$$

$$D_B = (\overline{Q}_A \bullet Q_B) + (Q_B \bullet X) + (\overline{Q}_A \bullet Q_C \bullet X)$$

$$D_C = (Q_A \bullet \overline{Q}_B \bullet \overline{Q}_C \bullet X) + (\overline{Q}_A \bullet \overline{Q}_B \bullet \overline{X}) + (\overline{Q}_A \bullet Q_C \bullet X) + (Q_A \bullet Q_C \bullet \overline{X})$$

$$Z = Q_A \bullet Q_C$$

b.
```
entity RISE3 is
 port
  (X   : in BIT;
   CLK : in BIT;
   Z   : out BIT);
end RISE3;
architecture ALGORITHM of RISE3 is
begin
 process
   variable RISE_CNT : INTEGER range 0 to 3 := 0;
   variable XOLD : BIT := '0';
 begin
  if (X='1' and XOLD='0') then
    RISE_CNT := RISE_CNT + 1;
  end if;
  if (RISE_CNT=3) then
    Z <= '1';        RISE_CNT := 0;
  else
    Z <= '0';
  end if;

  XOLD := X;
  wait until CLK='1';
 end process;
end ALGORITHM;
```

```
c.  entity RISE3 is
     port
      (X : in BIT;
       CLK : in BIT;
       Z   : out BIT);
    end RISE3;

    architecture STRUCTURE of RISE3 is
     component SEQ_PLA
     generic
      (APLN   : ARRAY_13X16;  -- ‘1’ means connection is programmed
       OPLN   : ARRAY_9X4;     -- ‘0’ means connection is not programmed
       MCELL  : ARRAY_4X2);
     port
      (A, B, C, D : in BIT;
       CLK        : in BIT;
       W, X, Y, Z : inout BIT);
     end component;

    signal TIE_ONE : BIT := ‘1’;

    begin
      C1 : SEQ_PLA
         generic map (APLN=>(
         --  A       B       C       D       W       X       Y       Z
            (‘0’,’0’,’0’,’0’,’0’,’0’,’0’,’0’,’1’,’0’,’0’,’0’,’0’,’1’,’0’,’0’),
            (‘1’,’0’,’0’,’0’,’0’,’0’,’0’,’0’,’0’,’0’,’1’,’0’,’0’,’1’,’0’,’0’),
            (‘0’,’0’,’0’,’0’,’0’,’0’,’0’,’0’,’0’,’1’,’1’,’0’,’0’,’0’,’0’,’0’),
            (‘1’,’0’,’0’,’0’,’0’,’0’,’0’,’0’,’0’,’0’,’1’,’0’,’0’,’0’,’0’,’0’),
            (‘1’,’0’,’0’,’0’,’0’,’0’,’0’,’0’,’0’,’1’,’0’,’0’,’1’,’0’,’0’,’0’),
            (‘1’,’0’,’0’,’0’,’0’,’0’,’0’,’0’,’1’,’0’,’0’,’1’,’0’,’1’,’0’,’0’),
            (‘0’,’1’,’0’,’0’,’0’,’0’,’0’,’0’,’0’,’1’,’0’,’1’,’0’,’0’,’0’,’0’),
            (‘0’,’1’,’0’,’0’,’0’,’0’,’0’,’0’,’1’,’0’,’0’,’0’,’1’,’0’,’0’,’0’),
            (‘0’,’0’,’0’,’0’,’0’,’0’,’0’,’0’,’1’,’0’,’0’,’0’,’1’,’0’,’0’,’0’),
            (‘0’,’0’,’0’,’0’,’0’,’0’,’0’,’0’,’0’,’0’,’0’,’0’,’0’,’0’,’0’,’0’),
            (‘0’,’0’,’0’,’0’,’0’,’0’,’0’,’0’,’0’,’0’,’0’,’0’,’0’,’0’,’0’,’0’),
            (‘0’,’0’,’0’,’0’,’0’,’0’,’0’,’0’,’0’,’0’,’0’,’0’,’0’,’0’,’0’,’0’),
            (‘0’,’0’,’1’,’0’,’0’,’0’,’0’,’0’,’0’,’0’,’0’,’0’,’0’,’0’,’0’,’0’)),
```

```
                OPLN=>(('1','0','0','0'),
                       ('1','0','0','0'),
                       ('0','1','0','0'),
                       ('0','1','0','0'),
                       ('0','1','1','0'),
                       ('0','0','1','0'),
                       ('0','0','1','0'),
                       ('0','0','1','0'),
                       ('0','0','0','1')),

                MCELL=>(('0','0'),
                        ('0','0'),
                        ('0','0'),
                        ('0','0')))

        port map    (A=>X, B=>TIE_ONE, C=>open, D=>open,
                     CLK=>CLK,
                     W=>open, X=>open, Y=>open, Z=>Z);
    end STRUCTURE;
```

d. Simulate.

e.

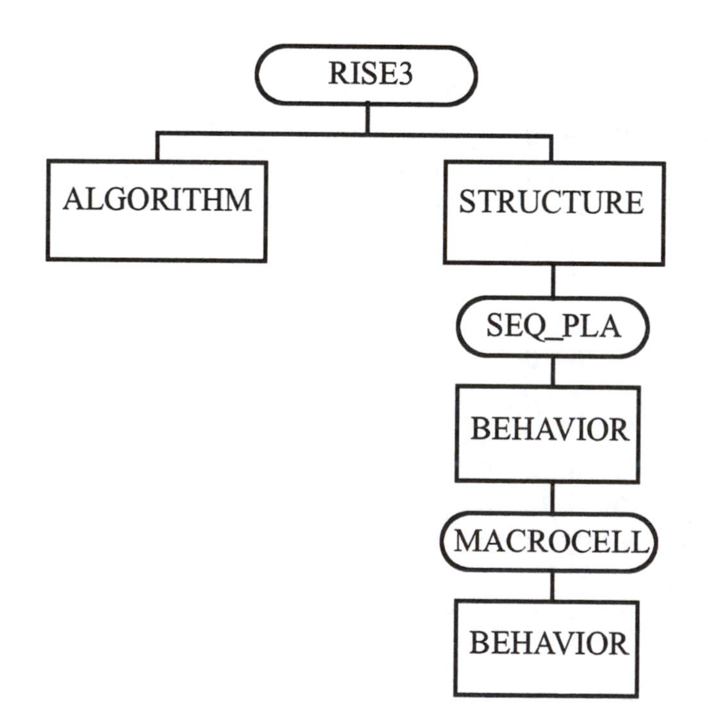

19-9. Solution.

```
configuration RISE3_ALGO of RISE3 is
 for ALGORITHM
 end for;
end RISE3;
```

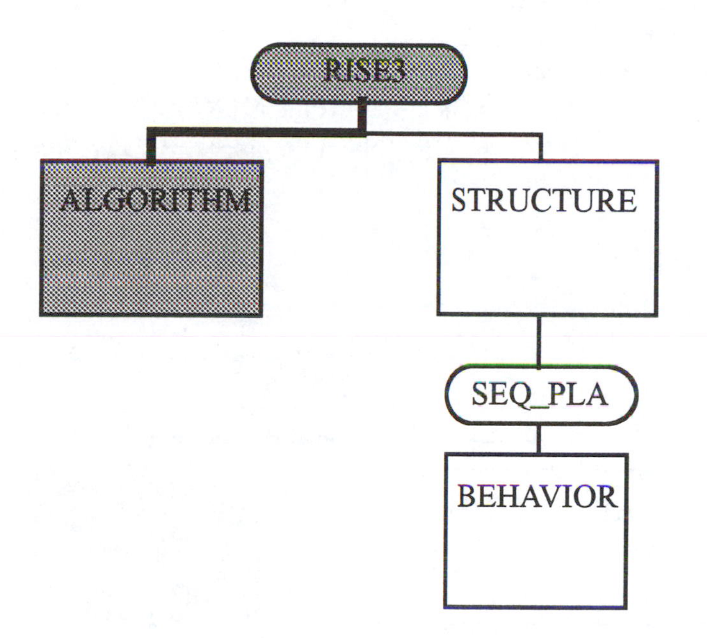

```
configuration RISE3_STRUC of RISE3 is
 for STRUCTURE
  for C1 : SEQ_PLA
   use entity SEQ_PLA
      generic map(APLN=>APLN, OPLN=>OPLN, MCELL=>MCELL)
      port map(A=>A, B=>B, C=>C, D=>D, CLK=>CLK, W=>W, X=>X, Y=>Y, Z=>Z);
      for all: MACROCELL
        use entity MACROCELL
          generic map (F1=>F1, F2=>F2)
          port map (OR_OUT=>OR_OUT, IO_CNTRL=>IO_CNTRL,
                    CLK=>CLK, OR_IN=>OR_IN, IN_OUT=>IN_OUT);
      end for;
  end for;
end RISE3_STRUC;
```

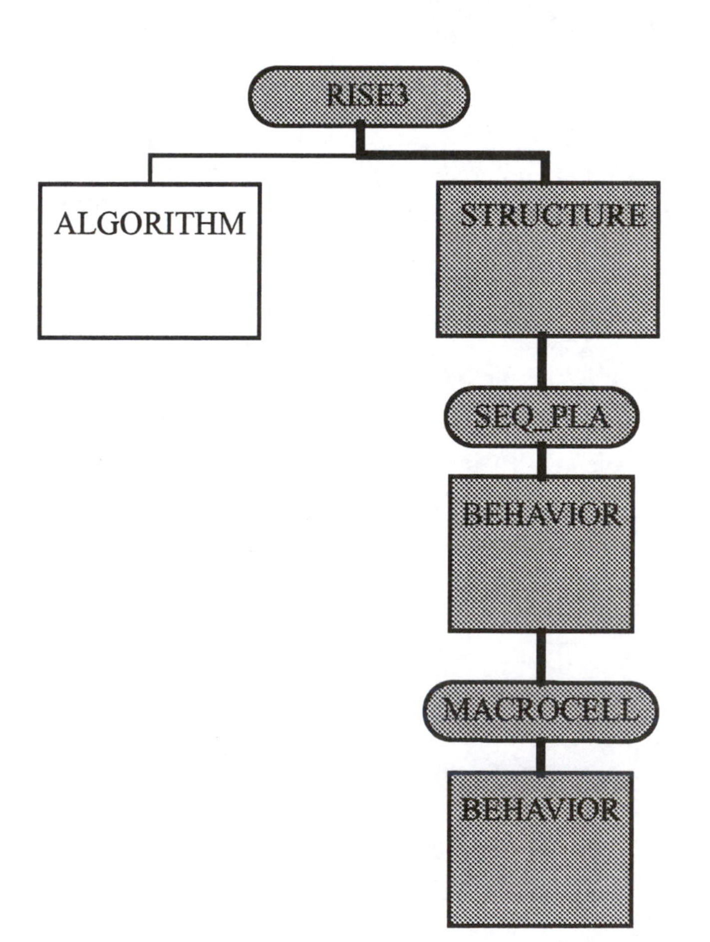

19-10. Solution.

a.
```
entity SINV is
 port
  (DI : in BIT;
   DO : out BIT);
end SINV;
architecture STRUCTURE of SINV is
 component BUF_INV
  generic (BUF_OP : BOOLEAN);
  port (A in BIT; B : out BIT);
 end component;

 signal INT_SIG : BIT;
begin
 BUFFER: BUF_INV
  generic map (BUF_OP=>TRUE)
  port map (A=>DI, B=>INT_SIG);

 INVERT: BUF_INV
  generic map (BUF_OP=>FALSE)
  port map (A=>INT_SIG, B=>DO);
end STRUCTURE;
```

b.
```
configuration SINV_CONFIG of SINV is
 for STRUCTURE
  for all : BUF_INV
   use entity WORK.EQ_NOT(BEHAVIOR)
      generic map(EQ_OP=>BUF_OP, DELAY=>10 ns);
      port map(A=>A, Z=>B);
  end for;
 end for;
end SINV_CONFIG;
```

c. Simulate

19-11. Solution.

```
entity SINV is
 port
  (DI : in BIT;
   DO : out BIT);
end SINV;
architecture STRUCTURE of SINV is
 component BUF_INV
  generic (BUF_OP : BOOLEAN);
  port (A in BIT; B : out BIT);
 end component;

 for all : BUF_INV
  use entity WORK.EQ_NOT(BEHAVIOR)
      generic map(EQ_OP=>BUF_OP, DELAY=>10 ns);
      port map(A=>A, Z=>B);

 signal INT_SIG : BIT;
begin
 BUFFER: BUF_INV
  generic map (BUF_OP=>TRUE)
  port map (A=>DI, B=>INT_SIG);

 INVERT: BUF_INV
  generic map (BUF_OP=>FALSE)
  port map (A=>INT_SIG, B=>DO);
end STRUCTURE;
```

19-12. Solution.

```
configuration EQ_NOT_CONFIG of EQ_NOT is
 for BEHAVIOR
 end for;
end EQ_NOT_CONFIG;

entity SINV is
 port
  (DI : in BIT;
   DO : out BIT);
end SINV;
architecture STRUCTURE of SINV is
  signal INT_SIG : BIT;
begin
 BUFFER: configuration WORK.EQ_NOT_CONFIG
  generic map (EQ_OP=>TRUE, DELAY=>10 ns)
  port map (A=>DI, Z=>INT_SIG);

 INVERT: configuration WORK.EQ_NOT_CONFIG
  generic map (EQ_OP=>FALSE, DELAY=>10 ns)
  port map (A=>INT_SIG, Z=>DO);
end STRUCTURE;
```

19-13. Solution.

a.
```
entity SBUF is
 port (D1 : in BIT; O1 : out BIT);
end SBUF;
architecture STRUCTURE of SBUF is
 component SINV
  port (DI : in BIT; DO : out BIT);
 end component;

 signal INT_SIG : BIT;
begin
 I1: SINV port map(DI=>D1,        DO=>INT_SIG);
 I2: SINV port map(DI=>INT_SIG, DO=>O1);
end STRUCTURE;

configuration SBUF_CONFIG of SBUF is
 for STRUCTURE
  for all : SINV use WORK.SINV_CONFIG;
 end for;
end SBUF_CONFIG;
```

19-14. Solution.

```
package UTILITIES_PKG is
 -- record type declaration
 type CONTROL_TYPE is
  record
   EN_BAR : BIT;
   ADDR   : BIT_VECTOR(3 downto 0);
  end record;

 -- subtype declaration
 subtype BYTE is BIT_VECTOR(7 downto 0);

 -- function declaration
 function TO_INT ( ARG : BIT_VECTOR ) return INTEGER;
end UTILITIES_PKG;

package body UTILITIES_PKG is
 function TO_INT ( ARG : BIT_VECTOR ) return INTEGER is
  variable INT : NATURAL := 0;
  alias   NEW_ARG : BIT_VECTOR(ARG'LENGTH-1 downto 0) is ARG;
 begin
  -- assume MSB .... LSB
  for I in NEW_ARG'REVERSE_RANGE loop
   if (NEW_ARG(I)='1') then
    INT := INT + 2**I;
   end if;
  end loop;
  return INT;
 end TO_INT;
end UTILITIES_PKG;
```

19-15. Solution.

```
configuration CONFIG_DFF of D_FF is
 for STRUCTURE
  for G3 : NAND3_GATE
   use entity WORK.NAND_GATE(DATA_FLOW);
   port map ( I1=>A, I2=>B, I3=>C, O1=>D );
  end for;
  for all : NAND2_GATE
   use entity WORK.NAND_GATE(DATA_FLOW)
   port map ( I1=>A, I2=>B, I3=>open, O1=>C );
  end for;
 end for;
end CONFIG_DFF;


configuration CONFIG of DUAL_PORT_REG is
 for DATA_FLOW_STRUCTURE -- block configuration

  for BIT_LOOP(A)   -- block configuration
   for D_FFX : D_FF  -- component configuration
    use entity WORK.D_FF(DATA_FLOW);
   end for;
  end for;

  for BIT_LOOP(B to C) -- block configuration
   for D_FFX : D_FF      -- component configuration
    use configuration WORK.CONFIG_DFF;
   end for;
  end for;

  for BIT_LOOP(D)   -- block configuration
   for D_FFX : D_FF  -- component configuration
    use entity WORK.D_FF(DATA_FLOW);
   end for;
 end for;
end CONFIG;
```

19-16. Solution.

```
package UTILITIES_PKG is
 -- record type declaration
 type MESSAGE_TYPE is
  record
   TR_ADDR  : BIT_VECTOR(3 downto 0);
   RCV_ADDR : BIT_VECTOR(3 downto 0);
   TEXT     : STRING(1 to 20);
   PRIORITY : NATURAL;
  end record;
end UTILITIES_PKG;

use WORK.UTILITIES_PKG.all;
entity FIFO is
 port
  (W_BAR, R_BAR : in BIT;
   CLK   : in BIT;
   EMPTY : out BIT;
   DATA  : inout MESSAGE_TYPE);
end FIFO;
```

```
architecture INF_LENGTH of FIFO is
begin
 process
  type ENTRY;
  type ENTRY_PTR is access ENTRY;
  type ENTRY is record
    VALUE : MESSAGE_TYPE;
    NXT   : ENTRY_PTR;
  end record;

  variable FIRST_ENTRY, LAST_ENTRY, OLD_ENTRY : ENTRY_PTR;
 begin
  wait until (CLK = '1');

  -- if W_BAR is asserted active-0, write FIFO
  if ( W_BAR='0' ) then
    if ( FIRST_ENTRY=null ) then
      FIRST_ENTRY := new ENTRY'(DATA, null);
      LAST_ENTRY := FIRST_ENTRY;
    else
      LAST_ENTRY.NXT := new ENTRY'(DATA, null);
      LAST_ENTRY := LAST_ENTRY.NXT;
    end if;
    EMPTY <= '0' after 20 ns;

  -- if R_BAR is asserted active-0 and FIFO is not empty, read FIFO
  elsif ( R_BAR='0' and FIRST_ENTRY/=null ) then
    DATA <= FIRST_ENTRY.VALUE after 35 ns;
    OLD_ENTRY := FIRST_ENTRY;
    FIRST_ENTRY := FIRST_ENTRY.NXT;
    DEALLOCATE(OLD_ENTRY);
    if ( FIRST_ENTRY=null ) then
      LAST_ENTRY := FIRST_ENTRY;
      EMPTY <= '1' after 20 ns;
    end if;
  end if;
 end process;
end INF_LENGTH;
```

19-17. Solution.

```
package UTILITIES_PKG is
 -- record type declaration
 type MESSAGE_TYPE is
  record
   TR_ADDR  : BIT_VECTOR(3 downto 0);
   RCV_ADDR : BIT_VECTOR(3 downto 0);
   TEXT     : STRING(1 to 20);
   PRIORITY : NATURAL;
  end record;
end UTILITIES_PKG;

use WORK.UTILITIES_PKG.all;
entity PRIORITY_FIFO is
 port
  (W_BAR, R_BAR : in BIT;
   CLK   : in BIT;
   EMPTY : out BIT;
   DATA  : inout MESSAGE_TYPE);
end PRIORITY_FIFO;

architecture INF_LENGTH of PRIORITY_FIFO is
begin
 process
  type ENTRY;
  type ENTRY_PTR is access ENTRY;
  type ENTRY is record
    VALUE : MESSAGE_TYPE;
    NXT   : ENTRY_PTR;
  end record;

  -- dummy head element
  variable FIRST_ENTRY : ENTRY_PTR := new ENTRY'((B"0000",
                                       B"0000",
                                       (others=>'Z'),
                                       0),
                                       null);
  variable PTR : ENTRY_PTR;
```

```
begin
 wait until (CLK = '1');

 -- if W_BAR is asserted active-0, write FIFO
 if ( W_BAR='0' ) then
   PTR := FIRST_ENTRY;
   while ( PTR.NXT /= null ) loop
    if (DATA.PRIORITY > PTR.NXT.VALUE.PRIORITY) then
      PTR.NXT := new ENTRY'(DATA, PTR.NXT);
      exit;
    elseif (DATA.PRIORITY = PTR.NXT.VALUE.PRIORITY) then
      while ((PTR.NXT/=null) and
              (DATA.PRIORITY=PTR.NXT.VALUE.PRIORITY)) loop
       PTR := PTR.NXT;
      end loop;
      PTR.NXT := new ENTRY'(DATA, PTR.NXT);
      exit;
    end if;
    PTR := PTR.NXT;
   end loop;
   if (PTR.NXT = null) then
     PTR.NXT := new ENTRY'(DATA, PTR.NXT);
   end if;
  end if;
  EMPTY <= '0' after 20 ns;

 -- if R_BAR is asserted active-0 and FIFO is not empty, read FIFO
 elsif ( R_BAR='0' and FIRST_ENTRY.NXT/=null ) then
  DATA <= FIRST_ENTRY.NXT.VALUE after 35 ns;
  PTR := FIRST_ENTRY.NXT;
  FIRST_ENTRY.NXT := FIRST_ENTRY.NXT.NXT;
  DEALLOCATE(PTR);
  if ( FIRST_ENTRY.NXT=null ) then
    EMPTY <= '1' after 20 ns;
  end if;
 end if;
end process;
end INF_LENGTH;
```

19-18. Solution.

a.
```
entity LIFO is
 port
  (W_BAR, R_BAR : in BIT;
   CLK   : in BIT;
   EMPTY : out BIT;
   DATA  : inout BIT_VECTOR(7 downto 0));
end LIFO;
architecture LINKED_LIST of LIFO is
begin
 process
  type ENTRY;
  type ENTRY_PTR is access ENTRY;
  type ENTRY is record
   VALUE : BIT_VECTOR(0 to 7);
   NEXT : ENTRY_PTR;
  end record;
  variable LIFO, OLD_ENTRY : ENTRY_PTR;
 begin
  wait until (CLK = '1');

  -- if W_BAR is asserted active-0, write LIFO
  if (W_BAR = '0') then
   LIFO := new ENTRY'(DATA, LIFO);
   EMPTY <= '0' after 20 ns;

  -- if R_BAR is asserted active-0 and LIFO is not empty, read LIFO
  elsif (R_BAR='0' and LIFO/=null) then
   OLD_ENTRY := LIFO;
   LIFO := OLD_ENTRY.NEXT;
   DATA <= OLD_ENTRY.VALUE after 50 ns;
   DEALLOCATE(OLD_ENTRY);
   if ( LIFO=null ) then
     EMPTY <= '1' after 20 ns;
   end if;
  end if;
 end process;
end LINKED_LIST;
```

b. entity LIFO is

```
port
 (W_BAR, R_BAR : in BIT;
  CLK    : in BIT;
  EMPTY  : out BIT;
  DATA   : inout BIT_VECTOR(7 downto 0));
end LIFO;
architecture ARRAY_LIST of LIFO is
begin
 process
  subtype BYTE is BIT_VECTOR(7 downto 0);
  constant SIZE : POSITIVE := 100;
  type ARRAY_LIST_TYPE is array (0 to SIZE-1) of BYTE;
  variable LIFO : ARRAY_LIST_TYPE;
  variable INDEX, START : INTEGER range 0 to SIZE-1 := 0;
  variable WRAP : BOOLEAN := FALSE;
 begin
  wait until (CLK = '1');

  -- if W_BAR is asserted active-0, write LIFO
  if (W_BAR = '0') then
   LIFO(INDEX) := DATA;
   if (INDEX=START and WRAP=TRUE) then
     START := (START + 1) mod SIZE;
   end if;
   if (INDEX=SIZE) then
     WRAP := TRUE;
   end if;
   INDEX := (INDEX + 1) mod SIZE;
   EMPTY <= '0' after 20 ns;

  -- if R_BAR is asserted active-0 and LIFO is not empty, read LIFO
  elsif (R_BAR='0' and EMPTY/='1') then
   INDEX := (INDEX - 1) mod SIZE;
   DATA <= LIFO(INDEX) after 50 ns;
   if ( INDEX=START ) then
     EMPTY <= '1' after 20 ns;
     WRAP := FALSE;
   end if;
  end if;
 end process;
end ARRAY_LIST;
```